Stochastic Electromagnetic Image Propagation

Optical and Electro-Optical Engineering Series
Robert E. Fischer and Warren J. Smith, *Series Editors*

Published

ALLARD • *Fiber Optics Handbook*
HECHT • *The Laser Guidebook*
NISHIHARA, HARUNA, SUHARA • *Optical Integrated Circuits*
RANCOURT • *Optical Thin Films Users' Handbook*
SIBLEY • *Optical Communications*
SMITH • *Modern Optical Engineering*
SMITH/GENESEE • *Modern Lens Design*
STOVER • *Optical Scattering*
WYATT • *Electro-Optical System Design*

Other Published Books of Interest

CSELT • *Fiber Optic Communications Handbook*
KAO • *Optical Fiber Systems*
KEISER • *Optical Fiber Communications*
MACLEOD • *Thin Film Optical Filters*
OPTICAL SOCIETY OF AMERICA • *Handbook of Optics*

Stochastic Electromagnetic Image Propagation

And Adaptive Compensation

Robert M. Manning

McGraw-Hill, Inc.
New York St. Louis San Francisco Auckland Bogotá
Caracas Lisbon London Madrid Mexico Milan
Montreal New Delhi Paris San Juan São Paulo
Singapore Sydney Tokyo Toronto

To my parents
Robert J. and Lillian A. Manning
and my wife Susan

Library of Congress Cataloging in Publication Data

Manning, Robert Michael.
Stochastic electromagnetic image propagation : and adaptive compensation / Robert M. Manning.
p. cm.—(Optical and electro-optical engineering series)
Includes index.
ISBN 0-07-039929-8
1. Laser beams—Atmospheric effects. 2. Electromagnetic waves—Transmission. 3. Laser communication systems. I. Title. II. Series.
QC976.L36M36 1993
621.36′6—dc20 92-32903
CIP

1 2 3 4 5 6 7 8 9 0 DOC/DOC 9 8 7 6 5 4 3 2

ISBN 0-07-039929-8

The sponsoring editor for this book was Daniel A. Gonneau, the editing supervisor was David E. Fogarty, and the production supervisor was Donald Schmidt. This book was set in Century Schoolbook by Techna Type.

Printed and bound by R.R. Donnelley & Sons Company.

Contents

Preface vii

Chapter 1. The Fundamental Equations for the Analysis of Electromagnetic Wave Propagation in the Turbulent and Turbid Atmosphere 1

1.1 Introduction 1
1.2 The Application of the Maxwell Equations to the Random Atmospheric Propagation Environment—The Vector Stochastic Wave Equation 2
1.3 The Scalar Stochastic Wave Equation 8
D1.1 Random Functions, Ensemble Averages, and Statistical Moments 10
1.4 The Stochastic Parabolic Wave Equation and the Spatial Moments of the Propagating Wave Field 21
D1.2 Characteristic Functionals 36
1.5 The Green Function Solution of the Stochastic Parabolic Wave Equation 39
1.6 The Rytov Transformation and the Theory of Weak Fluctuations 42
D1.3 The Spectral Representation of Random Functions 48
D1.4 Locally Homogeneous Random Fields 53
1.7 The Extended Huygens-Fresnel Principle and Its Phase Approximation 59
References 66

Chapter 2. The Statistical Modeling of the Atmospheric Permittivity Field 69

2.1 Introduction 69
2.2 The Description of the Refractive Index Field at Visible Wavelengths 70
2.3 Fluctuations and Statistics of the Gaseous Refractivity Field 77
2.4 The Effective Refractive Index of Atmospheric Aerosols at Optical Wavelengths 94

2.5 A Unified Statistical Model of the Atmospheric Electromagnetic Propagation Environment 99
References 108

Chapter 3. The Formation and Characterization of Images Viewed through Random Media 111

3.1 Introduction 111
3.2 The Analysis of Diffraction Image Formation by a Lens of an Object in a Random Medium 112
3.3 The Statistical Analysis of the Random Intensity Distribution in the Imaging Plane—The First-Order Moment 119
3.4 Short-Term versus Long-Term Imaging through a Random Medium—Statistical Zernike Polynomial Expansions 126
3.5 The Statistical Analysis of the Random Intensity Distribution in the Imaging Plane—Higher-Order Moments 137
3.6 Image Resolution and Its Assessment 139
References 147

Chapter 4. The Analysis of Image Propagation in the Atmosphere 149

4.1 Introduction 149
4.2 Application of the Rytov Method 150
4.3 Application of the Parabolic Equation Method 174
4.4 Analysis of the Temporal Spectrum of Wavefront Tilt and Astigmatism 187
References 192

Chapter 5. Methods of Compensation for Image Degradation Due to the Atmosphere 195

5.1 Introduction 195
5.2 Two Forms of the Instantaneous Image Transfer Function 196
5.3 Object Intensity Distribution Estimation by Phase Compensation within the Imaging Optics 199
5.4 Object Intensity Distribution Reconstruction via Speckle Imaging and Interferometry 202
5.5 Application of the Maximum Likelihood Method to the Analysis and Subsequent Reconstruction of Speckle Images 209
References 213

Appendix A. Comparison of the Depolarized and Scattered Fields Due to Random Atmospheric Permittivity Fluctuations 215

Appendix B. Analysis of the Applicability of the Stochastic Parabolic Wave Equation 221

Index 227

Preface

This book is intended for physicists and engineers, whether they be graduate students or professionals working in the area of, or on applications related to, statistical electromagnetic wave propagation through the atmosphere. Although the example application of the subject matter is by way of statistical image propagation and analysis, the basic theory is developed in a complete and rigorous manner so as to provide a basis for the analysis of any of the numerous applications that rely on transmission of short-wavelength electromagnetic fields through the atmospheric environment; such an environment is characterized by random variations of the refractive index due to the presence of temperature and humidity turbulence and aerosol or hydrometeor turbidity.

The subject of the book is timely since there is a proliferation of applications related to such wave transmission through the atmosphere, particularly optical image propagation. The state-of-the-art imaging and target acquisition equipment in use today is such that the effect of the atmosphere is, in many instances, the determining factor in the design and implementation of such equipment. The purpose of this book is threefold: (1) to present the derivation from first principles of various theoretical formalisms for analyzing electromagnetic wave propagation in random media; (2) to apply such analysis to the description of image degradation due to transmission through the atmosphere; and (3) to provide a rigorous basis for the development of various image correction techniques to compensate for such degradation.

The intent of the work is to treat the individual theoretical derivations in a step-by-step fashion that can be easily followed by the beginner in the subject and yet provide support to the reader interested in just the applications of the methodology. Through its presentation, the work will foster an appreciation of the prevailing constraints and limitations of the underlying theoretical structure and determine exactly where they arise, thus aiding the reader in the proper and correct application of the results.

The emphasis of the relevant theory employed in the analysis of wave propagation through a turbulent and turbid atmosphere has established the order of presentation within the book. Thus, Chap. 1 covers the development of the various forms of stochastic wave equations that are the starting point of investigations of wave propagation in random media. Beginning with the Maxwell equations, where the electric permittivity is a random function of space and time, conditions and palatable assumptions are noted that allow one to pass through a rigorous development of the following stochastic wave equations and the methods for working with them: the vector stochastic wave equation; the scalar stochastic wave equation; the parabolic stochastic wave equation and the Markov approximation; the Rytov transformation and the method of weak fluctuations; and the phase approximation of the extended Huygens-Fresnel principle. Although the description of image transmission through a random medium requires the use of the second-order moment and, to a lesser extent, the fourth-order moment of the electromagnetic wave field, the general case of arbitrary moments is considered at this stage for completeness. The well-known Markov δ-function approximation of refractive index fluctuations is established at the outset and is employed with the method of weak fluctuations just as naturally as with the parabolic stochastic wave equation for which it was historically developed. So as not to severely deviate from the flow of the material but, at the same time, to aid in the completeness of the technical presentation, mathematical digressions are employed where needed to provide necessary background material.

In Chap. 2, the major components that determine the atmospheric permittivity field are discussed in terms of the associated refractive index field. This allows one to identify the mechanisms, i.e., temperature, humidity, and aerosol concentration that cause the refractive index to be a random function, and to analyze the statistical structure of the atmospheric refractive index in terms of the statistical behavior of these contributing mechanisms. Thus, after a brief overview of the fluid mechanics governing the situation, it is shown how the temperature and humidity fields are governed by the Kolmogorov spectrum and the aerosol concentration is treated as a Poisson field. Methods are then introduced that allow one to rigorously treat the refractive index field in terms of a general characteristic functional of these perturbing mechanisms. The results of this chapter allow one to connect the descriptions of the random electromagnetic field established in Chap. 1 with those of the basic components of the atmosphere.

As a prelude to the use of the foregoing in the investigation of image propagation, the basic foundations are established in Chap. 3 for image formation in the presence of random fluctuations of the propagating wave field. The general case is initially considered: obtaining an image of an object, characterized by various degrees of surface roughness (and thus diffusivity), which is illuminated by light characterized by various degrees of coherence. Expressions are then developed which apply to the specialized yet common case of incoherent imaging, and it is directly shown how and where such a statistical image formation problem necessarily employs the results of Chap. 1. The concepts of long-term versus short-term imaging are established along with the idea of decomposing the randomly perturbed phase front of an image entering an optical system, thus motivating the development of statistical Zernike polynomial expansions. Parameters are derived and discussed which are used to describe image quality, e.g., resolution and optical transfer functions.

Finally, the results of the first three chapters are synthesized in Chap. 4 for the analysis of several atmospheric imaging scenarios. Following the historical development of the subject, the results which follow from the application of the method of weak fluctuations are derived for both the turbulence and hydrometeor cases. In addition to the classical long-exposure mutual coherence function for atmospheric turbulence, expressions are also developed for the short-term imaging case. The application of the parabolic equation method is then considered, along with its attendant techniques of analysis. As is well known, the final results are identical to those of the weak fluctuation method at the level of the second-order moment of the electric field of the propagating wave, with the exception, of course, of the range of applicability in strong fluctuations of the random medium. The small scattering angle approximation of the radiative transfer equation is then derived from the parabolic equation for the second-order moment. It is then shown how contact is made between the Zernike polynomial expansion of a random phase front entering an optical system and the relevant descriptive equations of Chap. 1, thus yielding a quantitative line of demarcation between the qualitative descriptors of "short-term" versus "long-term."

Although the short exposure of a random image is an implicit form of an image compensation technique whereby the deleterious effects of the atmosphere can partially be removed, Chap. 5 covers the derivation of the fundamental basis for three types of image compensation techniques that form the rudiments of all those that have found application in the literature. These are the mitigation techniques based on phase compensation, target intensity reconstruction based on speckle phenomena, and image estimation based on maximum likelihood methods.

No other person is as aware of the shortcomings of this book as the author. It may be effectively argued that the order of presentation of material in Chaps. 1 and 3 should be reversed, i.e., motivating the need for particular statistical moments of a random electromagnetic field through the requirements of image formation in a basic optical system. Although such an approach is more logically satisfying, it falls short of helping produce the desired emphasis of Chap. 1 on the various theoretical techniques and results that have been developed for the general analysis of wave propagation in the atmosphere. It is considered (at least by the author) to be eminently more rigorous, as well as instructive, to motivate the entire statistical treatment given to wave propagation by commencing with the problems encountered with the solution to the Dyson-type equation which results from the straightforward statistical treatment of the Maxwell equations, and then derive, as well as interrelate, all of the well-known approximate methods in use today that were necessarily developed to make tractable the analysis of random electromagnetic wave propagation.

Some readers may find objectionable the time afforded to the development of the Kolmogorov-Obukhov turbulence spectrum directly from fluid mechanical considerations. However, those other readers who are not as well grounded in this interdisciplinary connection will find the treatment very instructive. The same can be said of Chap. 4, where the results which are obtained from the weak fluctuation method are formally identical to those derived from the parabolic equation method. This is deemed to be warranted in exchange for the wealth of differing mathematical techniques that are employed; just as in Chap. 1, the emphasis is on enabling the reader to master, not just become vaguely familiar with, the mathematical methods required of a competent researcher in the subject area of statistical wave propagation.

The author would like to thank Dr. Val Zavorotnyi for reading and critically commenting on portions of the manuscript. Gratitude is also extended to my McGraw-Hill editor, Daniel Gonneau, for his patience and understanding during the production of the manuscript, with its attendant problems and delays. Finally, and most importantly, I would like to thank my wife Susan and our daughter Morgan for their unfailing support even during the many hours I had to devote to this book rather than to them.

Robert M. Manning

Chapter 1

The Fundamental Equations for the Analysis of Electromagnetic Wave Propagation in the Turbulent and Turbid Atmosphere

1.1 Introduction

This initial chapter will be devoted to the systematic development of the various approaches and modeling techniques which have proven useful in the analysis of stochastic wave propagation in the atmosphere. After the results of application of the Maxwell equations are simplified to within the approximations that are easily met in realistic atmospheric propagation scenarios, a vector stochastic wave equation is obtained in Sec. 1.2 which is capable of describing electromagnetic wave propagation over a wide range of wavelengths. This equation is then specialized in Sec. 1.3 to a scalar stochastic wave equation that holds for optical propagation conditions within the atmosphere. The parametric nature of this equation is discussed as well as the attendant problems encountered in its statistical analysis. This provides the setting and motivation for development of the four major approaches to such a solution that are detailed in the remaining sections. In particular, Sec. 1.4 deals with the formation of the well-known parabolic wave equation that, with use of the Markov approximation, yields a closed-form differential equation for the arbitrary statistical moments of the random wave field. Section 1.5 details the useful two-dimensional

Green function form of the parabolic wave equation, from which the extended Huygens-Fresnel principle is obtained. Section 1.6 treats the Rytov transformation technique and the well-known Rytov approximation solution for the random wave field. Finally, Sec. 1.7 discusses the "phase" approximation version of the extended Huygens-Fresnel principle, which was the original version of this approach, and its inherent limitations. The order of presentation is in terms of level of approximation, the more restrictive forms following from the less restrictive ones.

1.2 The Application of the Maxwell Equations to the Random Atmospheric Propagation Environment—The Vector Stochastic Wave Equation

The analysis of electromagnetic wave propagation through the atmosphere at optical wavelengths necessarily has its basis in the application of the Maxwell equations with prevailing assumptions that are realized in the troposphere of the earth, the most important of these being the stochastic character of the electric permittivity. In their full form, these well-known equations are

$$\nabla \cdot \mathcal{D} = 4\pi\rho \tag{1.1a}$$

$$\nabla \times \mathcal{E} + \frac{1}{c}\frac{\partial \mathcal{B}}{\partial t} = 0 \tag{1.1b}$$

$$\nabla \cdot \mathcal{B} = 0 \tag{1.1c}$$

$$\nabla \times \mathcal{H} - \frac{1}{c}\frac{\partial \mathcal{D}}{\partial t} = \frac{4\pi}{c}\mathcal{J} \tag{1.1d}$$

where $\mathcal{D} = \mathcal{D}(\mathbf{r}, t)$ and $\mathcal{E} = \mathcal{E}(\mathbf{r}, t)$ are, respectively, the electric induction (or displacement) and electric field; $\mathcal{B} = \mathcal{B}(\mathbf{r}, t)$ and $\mathcal{H} = \mathcal{H}(\mathbf{r}, t)$ are, respectively, the magnetic induction and magnetic field; and $\mathcal{J} = \mathcal{J}(\mathbf{r}, t)$ and $\rho = \rho(\mathbf{r}, t)$ are, respectively, the current density and charge density, all of which are functions of position $\mathbf{r}$ and time t. As usual, c is the velocity of light. The first five of these quantities are also coupled by the constitutive relationships

$$\mathcal{D} = \varepsilon\mathcal{E} \tag{1.2a}$$

$$\mathcal{B} = \mu\mathcal{H} \tag{1.2b}$$

$$\mathcal{J} = \sigma\mathcal{E} \tag{1.2c}$$

where $\varepsilon(\mathbf{r})$, $\mu(\mathbf{r})$, and $\sigma(\mathbf{r})$ are the electric permittivity, the magnetic permeability, and the conductivity that are characteristics of the position $\mathbf{r}$ in the medium in which propagation is considered. Note that in conditions prevalent in the troposphere, one has, at least at the field intensities dealt with in image propagation, that the atmosphere is nonconducting ($\sigma = 0$), that it does not contain free charges ($\rho = 0$), and that $\mu = 1$. Substituting Eqs. (1.2*a*) to (1.2*c*) into Eqs. (1.1*a*) to (1.1*d*) and applying these conditions yields the set of relations

$$\nabla \cdot (\varepsilon \mathcal{E}) = 0 \tag{1.3a}$$

$$\nabla \times \mathcal{E} + \frac{1}{c}\frac{\partial \mathcal{H}}{\partial t} = 0 \tag{1.3b}$$

$$\nabla \cdot \mathcal{H} = 0 \tag{1.3c}$$

$$\nabla \times \mathcal{H} - \frac{1}{c}\frac{\partial (\varepsilon \mathcal{E})}{\partial t} = 0 \tag{1.3d}$$

These equations have been written with the assumption that the electric permittivity ε can, in general, be a function of position $\mathbf{r}$ and time t, i.e., $\varepsilon = \varepsilon(\mathbf{r}, t)$. Such variations can be caused by the gaseous and aerosol content of the atmosphere; random temperature fluctuations due to local heating or cooling and the subsequent diffusion give rise to corresponding fluctuations in the absorption bands that characterize the molecular structure of the gaseous components, thus causing random dispersion and refraction. In the same manner, aerosols will give rise to diffraction and adsorption of a propagating electromagnetic wave. Suffice it to say that at this point of the development, whatever the sources of the random nature of $\varepsilon(\mathbf{r}, t)$ in space and time, the characteristic spatial and temporal frequencies of the fluctuations are negligible as compared to those of the electric or magnetic fields. In this circumstance, one can decompose the $\mathcal{E}(\mathbf{r}, t)$ and $\mathcal{H}(\mathbf{r}, t)$ fields into the frequency ω of vibrations of the field due to the oscillatory nature of the electromagnetic radiation (assuming, of course, monochromatic or at least very narrow-band radiation), simply given by exp $(-i\omega t)$, and those due to the much slower variation of $\varepsilon(\mathbf{r}, t)$, simply represented by the complex amplitudes $\mathbf{E}(\mathbf{r}, t)$ and $\mathbf{H}(\mathbf{r}, t)$, by the following relationships:

$$\mathcal{E}(\mathbf{r}, t) = \mathbf{E}(\mathbf{r}, t) \exp(-i\omega t) \tag{1.4a}$$

$$\mathcal{H}(\mathbf{r}, t) = \mathbf{H}(\mathbf{r}, t) \exp(-i\omega t) \tag{1.4b}$$

Substituting these representations into Eqs. (1.3*a*), (1.3*b*), and (1.3*d*) yields

$$\nabla \cdot (\varepsilon \mathbf{E}) = 0 \tag{1.5a}$$

$$\nabla \times \mathbf{E} - ik_0 \mathbf{H} + \frac{1}{c}\frac{\partial \mathbf{H}}{\partial t} = 0 \tag{1.5b}$$

$$\nabla \times \mathbf{H} + ik_0 \varepsilon \mathbf{E} - \frac{1}{c}\frac{\partial(\varepsilon \mathbf{E})}{\partial t} = 0 \tag{1.5c}$$

where $k_0 = \omega/c$ is the free space wave number. Taking the "curl" of Eq. (1.5*b*) and using Eq. (1.5*c*) solved for $\nabla \times \mathbf{H}$ gives, after simplification,

$$\nabla \times \nabla \times \mathbf{E} = k_0^2 \varepsilon \mathbf{E} + \frac{2ik_0}{c}\frac{\partial(\varepsilon \mathbf{E})}{\partial t} - \frac{1}{c^2}\frac{\partial^2(\varepsilon \mathbf{E})}{\partial t^2} \tag{1.6}$$

Finally, using the vector identity $\nabla \times \nabla \times \mathbf{E} = \nabla(\nabla \cdot \mathbf{E}) - \nabla^2 \mathbf{E}$ and noting that from Eq. (1.5*a*) one has

$$\nabla \cdot \mathbf{E} = -\frac{\mathbf{E} \cdot \nabla \varepsilon}{\varepsilon} = -\mathbf{E} \cdot \nabla(\ln \varepsilon)$$

Eq. (1.6) can be written

$$\nabla^2 \mathbf{E} + k_0^2 \varepsilon \mathbf{E} = -\nabla[\mathbf{E} \cdot \nabla(\ln \varepsilon)] - \frac{2ik_0}{c}\frac{\partial(\varepsilon \mathbf{E})}{\partial t} + \frac{1}{c^2}\frac{\partial^2(\varepsilon \mathbf{E})}{\partial t^2} \tag{1.7}$$

Since the temporal dependence of the $\mathbf{E}(\mathbf{r}, t)$ field due to the natural oscillations of the propagating wave has been accounted for through Eq. (1.4), the only time dependence remaining in $\mathbf{E}(\mathbf{r}, t)$, the evolution of which is described by Eq. (1.7), is that due solely to the random fluctuations of $\varepsilon(\mathbf{r}, t)$.

Equation (1.7) is the most general form that an equation can take that purports to describe the spatial and temporal structure of an electromagnetic field [which satisfied Eq. (1.4)] that propagates in a medium characterized by spatial and temporal variations in the permittivity field $\varepsilon(\mathbf{r}, t)$. In the case that $\varepsilon(\mathbf{r}, t)$ is a random function of position and time, Eq. (1.7) can be considered the *vector stochastic wave equation in its most general form for atmospheric propagation scenarios,* i.e., when the assumptions stated at the outset of this section are satisfied, viz., $\sigma = 0$, $\rho = 0$, and $\mu = 1$. Equation (1.7) will now be simplified by applying conditions that are prevalent in the earth's troposphere at the low power levels entailed in imaging situations.

The fluctuations in the permittivity field can be represented in the form

$$\varepsilon(\mathbf{r}, t) = \langle\varepsilon\rangle[1 + \tilde{\varepsilon}(\mathbf{r}, t)] \tag{1.8}$$

where $\langle\varepsilon\rangle$ is the average value of $\varepsilon(\mathbf{r}, t)$, which is taken to be a constant, and $\tilde{\varepsilon}(\mathbf{r}, t)$ is the fluctuating (random) part, taken to be such that the condition $|\tilde{\varepsilon}(r, t)| \ll 1$ prevails. Substituting Eq. (1.8) into Eq. (1.7) and using the expansion $\ln(1 + \tilde{\varepsilon}) \approx \tilde{\varepsilon}$, one obtains

$$\begin{aligned}\nabla^2\mathbf{E} + k^2\mathbf{E} = {} & -k^2\tilde{\varepsilon}\mathbf{E} - \nabla(\mathbf{E}\cdot\nabla\tilde{\varepsilon}) \\ & - \frac{2ik}{c}\frac{\partial}{\partial t}[(1 + \tilde{\varepsilon})\mathbf{E}] + \frac{\langle\varepsilon\rangle}{c^2}\frac{\partial^2}{\partial t^2}[(1 + \tilde{\varepsilon})\mathbf{E}]\end{aligned} \tag{1.9}$$

where $k^2 = k_0^2\langle\varepsilon\rangle$ is the wave number in the propagation medium. The two terms on the left side of this expression make up the well-known Helmholtz operator. On the right side of the equation, the first term is the source term of the copolarized scattered radiation. The second term on the right is the source term of the depolarized radiation resulting from the cross-coupled scattering of the field components. The third term describes the phase delay of the wave field that results from temporal variations of $\tilde{\varepsilon}(\mathbf{r}, t)$. Finally, the fourth term gives the temporal variations of the wave field amplitude. Equation (1.9) can be simplified even further if one takes into account the magnitude of the temporal variations of $\tilde{\varepsilon}$ relative to the speed of light c. To this end, consider the third term on the right in Eq. (1.9):

$$\begin{aligned}\frac{\partial}{\partial t}[(1 + \tilde{\varepsilon})\mathbf{E}] &= \frac{\partial\mathbf{E}}{\partial t}(1 + \tilde{\varepsilon}) + \mathbf{E}\frac{\partial\tilde{\varepsilon}}{\partial t} \\ &\approx \frac{\partial\mathbf{E}}{\partial t} + \mathbf{E}\frac{\partial\tilde{\varepsilon}}{\partial t}\end{aligned} \tag{1.10}$$

As will be discussed in Chap. 2, the random variations of the electric permittivity of the atmosphere have associated with them a characteristic spatial scale l and a characteristic velocity v. Now, noting that the time derivative of the electric field in the first term on the right of Eq. (1.10) contains a spatial "convective" part due to its spatial variation along a wavelength in the presence of this velocity, one has

$$\frac{\partial\mathbf{E}}{\partial t} \longrightarrow \frac{\partial\mathbf{E}}{\partial t} + \mathbf{v}\cdot\nabla\mathbf{E} \sim \frac{E}{\tau} + v\frac{E}{\lambda} \tag{1.11}$$

The derivative in the second member of Eq. (1.10), on the other hand, is simply represented by spatial variations over l with velocity v, viz.,

$$\left| \mathbf{E} \frac{\partial \tilde{\varepsilon}}{\partial t} \right| \sim E \frac{\tilde{\varepsilon}}{\tau} = Ev \frac{\tilde{\varepsilon}}{l} \tag{1.12}$$

where $\tau = l/v$ is the associated characteristic time. Taking the ratio of the first term to the second term of Eq. (1.10), i.e., the ratio of Eq. (1.11) to (1.12), one obtains

$$\frac{\partial E/\partial t}{E\ \partial\tilde{\varepsilon}/\partial t} = \begin{cases} \dfrac{1}{|\tilde{\varepsilon}|} \gg 1 & l \ll \lambda \\ \dfrac{l}{\lambda|\tilde{\varepsilon}|} \gg 1 & l \gg \lambda \end{cases}$$

since $|\tilde{\varepsilon}| \ll 1$. Thus, the second term in Eq. (1.10) is negligible only because of the level of variations considered for $\tilde{\varepsilon}$. Hence, the third term of Eq. (1.9) becomes

$$\frac{2ik}{c}\frac{\partial}{\partial t}[(1 + \tilde{\varepsilon})E] \approx \frac{2ik}{c}\frac{\partial E}{\partial t} \sim \begin{cases} \dfrac{Ev}{l\lambda c} & l \ll \lambda \\ \dfrac{Ev}{\lambda^2 c} & l \gg \lambda \end{cases} \tag{1.13}$$

The magnitude of this term should now be compared to those of the first two terms on the right of Eq. (1.9). The first one is simply

$$k^2\tilde{\varepsilon}|\tilde{\mathbf{E}}| \sim \frac{\tilde{\varepsilon}E}{\lambda^2}$$

Similarly, the magnitude of the second term is

$$\nabla(\mathbf{E} \cdot \nabla\tilde{\varepsilon}) \sim \begin{cases} \dfrac{E\tilde{\varepsilon}}{l^2} & l \ll \lambda \\ \dfrac{E\tilde{\varepsilon}}{l\lambda} & l \gg \lambda \end{cases}$$

where, in the case of $l \ll \lambda$, changes on the scale of l are taken to dominate over those of λ. Taking the ratio of the sum of these two terms with Eq. (1.13) yields

$$\frac{|k^2\tilde{\varepsilon}\mathbf{E} + \nabla(\mathbf{E} \cdot \nabla\tilde{\varepsilon})|}{\dfrac{2ik}{c}\dfrac{\partial \mathbf{E}}{\partial t}} \sim \begin{cases} \tilde{\varepsilon}\dfrac{\lambda}{l}\left(\dfrac{c}{v}\right) & l \ll \lambda \\ \tilde{\varepsilon}\left(\dfrac{c}{v}\right) & l \gg \lambda \end{cases} \tag{1.14}$$

Since $l \ll \lambda$ in the first case, $\tilde{\varepsilon}(\lambda/l)(c/v) \gg \tilde{\varepsilon}(c/v)$, and so if $\tilde{\varepsilon} \gg v/c$, the first two terms of Eq. (1.9) will dominate over the third. In the case of the atmosphere, $v/c \leq 10^{-8}$ and, as will be shown in Chap. 2, $\tilde{\varepsilon} \sim 10^{-6}$. Therefore, this condition is always satisfied.

A similar examination of the last term of Eq. (1.9) reveals that it is of the next order of smallness from that of Eq. (1.13) by another factor of v/c and so too may be neglected. Thus Eq. (1.9) can be written

$$\nabla^2 \mathbf{E} + k^2 \mathbf{E} = -k^2 \tilde{\varepsilon} \mathbf{E} - \nabla(\mathbf{E} \cdot \nabla \tilde{\varepsilon}) \tag{1.15}$$

Equation (1.15), called the *vector stochastic wave equation*, contains as its source terms just the copolarized and depolarized scattered fields which, because of the presence of the random function $\tilde{\varepsilon}$, make Eq. (1.15) a stochastic differential equation for the wave field $\mathbf{E}(\mathbf{r}, t)$. It is, of course, the depolarization term that gives Eq. (1.15) its vector character as it describes the cross-coupling of the various $\mathbf{E}(\mathbf{r}, t)$ field components as a result of the spatially random variations of $\tilde{\varepsilon}(\mathbf{r}, t)$.

Note that Eq. (1.15) is devoid of time derivatives and only contains time as a parameter, which occurs in the random function $\tilde{\varepsilon}(\mathbf{r}, t)$. This is simply due to the fact that, as shown earlier for both $l \ll \lambda$ and $l \gg \lambda$, one has

$$\frac{\partial \tilde{\varepsilon}}{\partial t} \ll \left(\frac{1}{E}\right)\left(\frac{\partial E}{\partial t}\right)$$

i.e., the level of the temporal variation or evolution of the inhomogeneities in the permittivity field is much smaller than that which is characteristic of the electric field, a fact which is due to the assumed condition $|\tilde{\varepsilon}| \ll 1$. Also, the assumption that allowed Eq. (1.4) to be written, i.e., that the characteristic temporal variation of $\tilde{\varepsilon}$ is negligible relative to that of the propagating field, helps determine this form. Both of these circumstances give rise to the *quasi-static approximation*, i.e., the approximation that allows Eq. (1.9) to assume the form of Eq. (1.15), which is parametric in time.

The vector stochastic wave equation, Eq. (1.15), forms the basis of analysis of wave propagation in random media, in which the assumptions that lead to its derivations prevail, and where the wavelength is such that depolarization is appreciable, i.e., in cases where $\lambda > l_0$, l_0 being the smallest size of the inhomogeneities $\tilde{\varepsilon}$.[1–4] Thus, this equation finds application in long-range (i.e., over the horizon) tropospheric propagation at VHF wavelengths, propagation of millimeter waves (~300 GHz) through atmospheric dust and hydrometeors, etc.

However, for situations where $\lambda \leq l_0$, which encompasses the case of image propagation in the atmosphere, the power scattered into the depolarized component is negligible compared to that of the copolarized component. This circumstance allows a considerable simplification of Eq. (1.15) into what is known as the *scalar stochastic wave equation.* A rigorous analysis of this situation will be given in Appendix A (the analysis is deferred pending the development of required concepts and analytical results). For now, however, one can develop the following argument.

1.3 The Scalar Stochastic Wave Equation

When the condition $\lambda \leq l_0$ prevails, one has, from the type of considerations used earlier,

$$k^2\tilde{\varepsilon}\mathbf{E} \gg \nabla(\mathbf{E} \cdot \nabla\tilde{\varepsilon})$$

thus allowing the depolarization term to be neglected relative to that of the copolarization and thus decoupling the vector relationship of Eq. (1.15) into three scalar equations of the form

$$\nabla^2 E + k^2 E = -k^2\tilde{\varepsilon}E \tag{1.16}$$

for each component $E(\mathbf{r}, t)$ of the electric field. This scalar stochastic wave equation, although simpler than the vector version above, still possesses a major obstacle that is common to Eq. (1.15), i.e., that in the source terms of Eqs. (1.15) and (1.16), the random function $\tilde{\varepsilon}$ is a multiplicative factor of the unknown (and, necessarily, random) wave field $E(\mathbf{r}, t)$ which one wishes to determine. The effect that this parametric relationship has on obtaining closed solutions for the statistics of the wave field will now be shown.

1.3.1 A perturbation-theoretical series solution

The solution to Eq. (1.16) takes the well-known form

$$E(\mathbf{r}) = \int G_0(\mathbf{r}, \mathbf{r}')f(\mathbf{r}')\, d^3r' \tag{1.17}$$

where the kernel is composed of the Green function satisfying the relationship

$$\nabla^2 G_0(\mathbf{r}, \mathbf{r}') + k^2 G_0(\mathbf{r}, \mathbf{r}') = \delta(\mathbf{r} - \mathbf{r}') \tag{1.18}$$

which is just the form that Eq. (1.16) takes in the case of a homogeneous propagation medium, i.e., when $\tilde{\varepsilon} = 0$, and where the scattering term is given by

$$f(\mathbf{r}) = -k^2\tilde{\varepsilon}(\mathbf{r})E(\mathbf{r}) \tag{1.19}$$

The solution of Eq. (1.18) that satisfies the appropriate radiation conditions is

$$G_0(\mathbf{r}, \mathbf{r}') = G_0(\mathbf{r} - \mathbf{r}') = -\frac{\exp(ik|\mathbf{r} - \mathbf{r}'|)}{4\pi|\mathbf{r} - \mathbf{r}'|} \tag{1.20}$$

An iterative solution will now be developed for Eq. (1.16) with an arbitrary source distribution at positions **R** which is easily incorporated into the stochastic differential equation, i.e.,

$$\nabla^2 E + k^2 E = -k^2\tilde{\varepsilon}E + \delta(\mathbf{r} - \mathbf{R}) \tag{1.21}$$

which, via Eqs. (1.17) to (1.20), has the solution

$$E(\mathbf{r}) = G_0(\mathbf{r} - \mathbf{R}) - k^2 \int G_0(\mathbf{r} - \mathbf{r}')\tilde{\varepsilon}(\mathbf{r}')E(\mathbf{r}')\, d^3r' \tag{1.22}$$

Iteratively solving Eq. (1.22) yields, for the first iterate,

$$E(\mathbf{r}) = G_0(\mathbf{r} - \mathbf{R}) - k^2 \int G_0(\mathbf{r} - \mathbf{r}')\tilde{\varepsilon}(\mathbf{r}'E(\mathbf{r}')\, d^3r'$$
$$+ k^4 \int\int G_0(\mathbf{r} - \mathbf{r}')\tilde{\varepsilon}(\mathbf{r}')G_0(\mathbf{r}' - \mathbf{r}'')\tilde{\varepsilon}(\mathbf{r}'')E(\mathbf{r}'')\, d^3r'\, d^3r'' \tag{1.23}$$

Continuing with this process, one obtains an infinite series, viz.,

$$E(\mathbf{r}) = G_0(\mathbf{r} - \mathbf{R}) + \sum_{n=1}^{\infty} (-k^2)^n \int \cdots (n) \cdots \int G_0(\mathbf{r} - \mathbf{r}_1)$$
$$\cdots G_0(\mathbf{r}_{n-1} - \mathbf{r}_n)G_0(\mathbf{r}_n - \mathbf{R})\tilde{\varepsilon}(\mathbf{r}_1)\tilde{\varepsilon}(\mathbf{r}_2) \cdots \tilde{\varepsilon}(\mathbf{r}_n)\, d^3r_1 \cdots d^3r_n \tag{1.24}$$

Even though this is an exact solution of the scalar stochastic wave equation, Eq. (1.16) (a similar solution can be written for the vector stochastic wave equation using dyadic Green functions to account for the depolarization[5]), it too is a random function because of the products of the random functions $\tilde{\varepsilon}(\mathbf{r})$ that enter into the individual terms; the nth term within the summation of Eq. (1.24) describes nth-order scattering of the initial, deterministic wave field $G_0(\mathbf{r} - \mathbf{R})$.

Since $E(\mathbf{r})$ is a random function, it can only be characterized by the various statistical quantities that can be formed from it. For example, in the case of direct field quantities, it is useful to obtain the statistical moments, i.e., the first-order moment $\langle E(\mathbf{r})\rangle$ (the average field at the point $\mathbf{r}$), the second-order moment $\langle E(\mathbf{r})E^*(\mathbf{r}')\rangle$ (the correlation of the field and its complex conjugate at two points $\mathbf{r}$ and $\mathbf{r}'$), etc.

Digression 1.1: Random Functions, Ensemble Averages, and Statistical Moments This book will invariably deal with random functions and the mathematical methodology needed to describe them. Thorough but concise discussions of this subject can be found in Refs. 6 and 7. The mathematically rigorous treatment of random functions, however, is not applicably practical. The experimental determination of the multidimensional probability densities needed to rigorously characterize the random field would be extremely complex if not impossible. For this reason, simpler statistical concepts, such as ergodicity, and statistical parameters, such as moments, that can, within the required bounds, sufficiently characterize the random field are used in practice.

Consider the special case of two random fields $u(\mathbf{r}, t)$ and $v(\mathbf{r}, t)$ that may indeed be functions of one another (e.g., via a relationship from electrodynamics or fluid mechanics) but are otherwise random due to the randomness of one of the functions. Adopting the notation $u_i = u(\mathbf{r}_i, t_i)$ for the field u at the ith space-time point, and similarly for the field v, the first- and second-order moments of the field u are defined, respectively, as

$$\langle u(\mathbf{r}, t)\rangle \equiv \int_{-\infty}^{\infty} u(\mathbf{r}, t)p[u(\mathbf{r}, t)]\, du \tag{D1.1.1}$$

and

$$\langle u(\mathbf{r}_1, t_1)u(\mathbf{r}_2, t_2)\rangle \equiv B_u[u(\mathbf{r}_1, t_1), u(\mathbf{r}_2, t_2)]$$

$$\equiv \int_{-\infty}^{\infty}\int_{-\infty}^{\infty} u(\mathbf{r}_1, t_1)u(\mathbf{r}_2, t_2)p[u(\mathbf{r}_1, t_1); u(\mathbf{r}_2, t_2)]\, du_1\, du_2 \tag{D1.1.2}$$

where the probability density $p(..)$ that appears in Eq. (D1.1.1) is such that, for *every* space-time point $\mathbf{r}$, t,

$$P[u < u(\mathbf{r}, t) < u + du] = p[u(\mathbf{r}, t)]\, du \tag{D1.1.3}$$

where $P[u < u(\mathbf{r}, t) < u + du]$ is the probability that the field $u(\mathbf{r}, t)$ has a value in the range bracketed by u and $u + du$, and similarly, for the joint density that appears in Eq. (D1.1.2),

$$P[u_1 < u(\mathbf{r}_1, t_1) < u_1 + du_1, u_2 < u(\mathbf{r}_2, t_2) < u_2 + du_2]$$
$$= p[u_1(\mathbf{r}_1, t_1); u_2(\mathbf{r}_2, t_2)]\, du_1\, du_2 \tag{D1.1.4}$$

where $P[u_1 < u(\mathbf{r}_1, t_1) < u_1 + du_1, u_2 < u(\mathbf{r}_2, t_2) < u_2 + du_2]$ is the probability that the field u has a value in the range u_1 to $u_1 + du_1$ at the point $\mathbf{r}_1$, t_1, and a value in the range u_2 to $u_2 + du_2$ at the point $\mathbf{r}_2$, t_2. The quantities on the left sides of Eqs. (D1.1.1) and (D1.1.2) are called the *probability or ensemble averages* of the quantities indicated. One can proceed along analogous lines and define third-, fourth-, etc., order moments at three, four, etc., space-time points. In general, the nth space-time moment of the field $u(\mathbf{r}, t)$ is given by

$$\langle u(\mathbf{r}_1, t_1)u(\mathbf{r}_2, t_2) \cdots u(\mathbf{r}_n, t_n)\rangle \equiv B_u[u(\mathbf{r}_1, t_1), u(\mathbf{r}_2, t_2), \ldots, u(\mathbf{r}_n, t_n)]$$

$$\equiv \int_{-\infty}^{\infty} \cdots (n) \cdots \int_{-\infty}^{\infty} u(r_1, t_1)\, u(r_2, t_2) \cdots u(r_n, t_n)$$

$$\times\, p[u(r_1, t_1); u(r_2, t_2); \ldots ; u(r_n, t_n)]\, du_1\, du_2 \ldots du_n$$

In fact, the requirement of having such a description for all orders of moments is sufficient for defining the entire random field $u(\mathbf{r}, t)$. As is immediately obvious, it is implicit that one knows the functional form of the corresponding joint probability densities.

One can form not only the first-, second-, and higher-order moments for the field $v(\mathbf{r}, t)$ in the same way, but also moments involving *both* fields, e.g.,

$$\langle u(\mathbf{r}_1, t_1)v(\mathbf{r}_2, t_2)\rangle \equiv B_{uv}[u(\mathbf{r}_1, t_1), v(\mathbf{r}_2, t_2)]$$

$$\equiv \int_{-\infty}^{\infty}\int_{-\infty}^{\infty} u(r_1, t_1)v(r_2, t_2)p[u(r_1, t_1); v(r_2, t_2)]\, du_1\, dv_2 \tag{D1.1.5}$$

and so on for the higher-order moments; the nth-order moment is more complicated than that for the single fields u or v, viz.,

$$\langle w_1(\mathbf{r}_1, t_1)w_2(\mathbf{r}_2, t_2) \ldots w_n(\mathbf{r}_n, t_n)\rangle$$

$$\equiv B_{w_1,w_2\ldots w_n}[w_1(\mathbf{r}_1, t_1), w_2(\mathbf{r}_2, t_2), \ldots, w_n(\mathbf{r}_n, t_n)]$$

$$\equiv \int_{-\infty}^{\infty} \ldots (n) \ldots \int_{-\infty}^{\infty} w_1(r_1, t_1)w_2(r_2, t_2) \cdots w_n(r_n, t_n)$$

$$\times\, p[w_1(r_1, t_1); w_2(r_2, t_2); \ldots ; w_n(r_n, t_n)]\, dw_1\, dw_2 \ldots dw_n$$

where the field designation w_i denotes either the u or v field at the space-time location $\mathbf{r}_i$, t_i. One then has rigorously defined the combined random fields $u(\mathbf{r}, t)$ and $v(\mathbf{r}, t)$ if, in addition to all the probability densities governing the individual fields, one also has all the joint probability densities governing both fields. For example, the probability density employed in Eq. (D1.1.5) is defined by

$$P[u_1 < u(\mathbf{r}_1, t_1) < u_1 + du_1, v_2 < v(\mathbf{r}_2, t_2) < v_2 + dv_2]$$

$$= p[u_1(\mathbf{r}_1, t_1); v_2(\mathbf{r}_2, t_2)]\, du_1\, dv_2 \tag{D1.1.6}$$

giving the probability $P[u_1 < u(\mathbf{r}_1, t_1) < u_1 + du_1, v_2 < v(\mathbf{r}_2, t_2) < v_2 + dv_2]$ that the field u has a value in the range u_1 to $u_1 + du_1$ at the point $\mathbf{r}_1$, t_1, and that the field v has a value in the range v_2 to $v_2 + dv_2$ at the point $\mathbf{r}_2$, t_2. (In the case of fields that have deterministic functional relationships between them, one can, in principle, obtain any of the joint statistics involving one field if one knows the probability densities that govern the other individual field.)

Although, as indicated above, one can proceed to define higher-order moments, much emphasis, as will be discussed below, is placed on the first-order moment of a field at one point (the mean) and the second-order moment of a field at two points. Such second-order moments at two points are also known as *correlation functions.*

The application of the formal definitions given above is straightforward when one is dealing with theoretical treatments. However, the unrealistic requirement to obtain all of the joint probability distributions introduced above so as to obtain the moments that characterize the particular random processes necessitates that one consider other approaches, created with complete regard to experimental determination, that will allow one to secure such statistical moments, particularly in the case where the theoretical predictions are to be compared with observational data. In particular, it is of interest to be able to determine the required averages of quantities over their probability densities (as introduced above) simply by averaging the same such quantities observed in an experimental scenario over time. This is the situation that is encountered in statistical mechanics whereby one replaces the averages over all states of a thermodynamic system (i.e., the ensemble) with an observed time or space average, where the length of the averaging interval is much larger than the largest time period or space interval of the process being observed. In some cases, the replacement of the ensemble average with a corresponding time or space average can be rigorously proved, and one is then dealing with the well-known *ergodic theorem*. In other situations, such as those encountered in statistical electromagnetic wave propagation through the atmosphere, where a strict proof cannot be given, one usually adopts the *ergodic hypothesis*. Thus, the consideration of the ensemble average in the problems to be encountered in this book will always employ the ergodic hypothesis; the corresponding time intervals over which the averages are formed will, by the nature of the problems addressed, be taken to be much greater than a corresponding characteristic time determined by the velocity of the atmosphere across the imaging path and the largest scale of the atmospheric permittivity fluctuations (i.e., the outer scale of turbulence).

The use of the ergodic hypothesis, however, places certain restrictions on the statistics of the random fields. Consider the particular case of obtaining the temporal mean and correlation function, i.e., that obtained over two points separated in time but not in space. By the ergodic theorem, one must have

$$\langle u(\mathbf{r}, t)\rangle = \overline{u(\mathbf{r}, t)} \equiv \lim_{T\to\infty} \frac{1}{T} \int_{-T/2}^{T/2} u(\mathbf{r}, t + \tau)\, d\tau \tag{D1.1.7}$$

where $\overline{u(\mathbf{r}, t)}$ denotes the time average of $u(\mathbf{r}, t)$. Now one of the many requirements placed on such an identification is that the function $\overline{u(\mathbf{r}, t)}$ must be a bounded function, i.e.,

$$\overline{u(\mathbf{r}, t)} - \overline{u(\mathbf{r}, t_1)} \equiv \lim_{T\to\infty} \frac{1}{T} \left[\int_{-T/2}^{T/2} u(\mathbf{r}, t + \tau)\, d\tau - \int_{-T/2}^{T/2} u(\mathbf{r}, t_1 + \tau)\, d\tau\right] \longrightarrow 0$$

Thus, $\overline{u(\mathbf{r}, t)} = \overline{u(\mathbf{r}, t_1)}$, thereby requiring that $\overline{u(\mathbf{r}, t)}$ be independent of t. But, in the general case, the ensemble average $\langle u(\mathbf{r}, t)\rangle$ is a function of t. Therefore, if ergodicity is to prevail, the first-order time average of the field u must be independent of time, i.e.,

$$\langle u(\mathbf{r}, t)\rangle = U_r(\mathbf{r}) \tag{D1.1.8}$$

where $U_r(\mathbf{r})$ is some function of the spatial position. Applying a similar argument to the second-order temporal moment, one finds that the temporal correlation function must depend only on the difference between the two times t and t_1 and not on t or t_1 individually. Thus,

$$B_u[u(\mathbf{r}, t_1), u(\mathbf{r}, t_2)] \equiv B_u(\mathbf{r}, t_1; \mathbf{r}, t_2) = B_u(\mathbf{r}, t_1 - t_2) \tag{D1.1.9}$$

with a similar relation holding for the random field $v(\mathbf{r}, t)$, and

$$B_{uv}[u(\mathbf{r}, t_1), v(\mathbf{r}, t_2)] \equiv B_{uv}(\mathbf{r}, t_1; \mathbf{r}, t_2) = B_{uv}(\mathbf{r}, t_1 - t_2) \tag{D1.1.10}$$

Hence, the fields $u(\mathbf{r}, t)$ and $v(\mathbf{r}, t)$ are required to possess *stationary* statistics. In Eq. (D1.1.9), the notation is such that if one field is designated as a subscript, it is obvious that this is the second moment with itself, contrasted with the second moment with respect to the two fields in Eq. (D1.1.10).

In terms of the rigorous definitions of Eqs. (D1.1.1) and (D1.1.2), if one wants to obtain the first- and second-order moments via simple time averaging, it is necessary that the requisite probability densities that define the statistics of the fields satisfy

$$\begin{aligned} p[u(\mathbf{r}, t)] &= p[u(\mathbf{r}, t + \tau)] \\ p[u(\mathbf{r}, t_1); u(\mathbf{r}, t_2)] &= p[u(\mathbf{r}, t_1 + \tau); u(\mathbf{r}, t_2 + \tau)] \\ p[u(\mathbf{r}, t_1); v(\mathbf{r}, t_2)] &= p[u(\mathbf{r}, t_1 + \tau); v(\mathbf{r}, t_2 + \tau)] \end{aligned} \tag{D1.1.11}$$

A similar situation exists for obtaining the spatial mean and correlation function, i.e., that obtained over two points separated in space but not in time. Here, one is required to have a spatial mean which is independent of position in the fields being considered, and a correlation function that is only a function of the spatial separation of the two points, i.e.,

$$\langle u(\mathbf{r}, t)\rangle = U_t(t) \tag{D1.1.12}$$

where $U_t(t)$ is some function of the time at which the spatial average is performed, and

$$B_u[u(\mathbf{r}_1, t), u(\mathbf{r}_2, t)] \equiv B_u[\mathbf{r}_1, t; \mathbf{r}_2, t] = B_u(\mathbf{r}_1 - \mathbf{r}_2, t) \tag{D1.1.13}$$

with a similar relation holding for the random field $v(\mathbf{r}, t)$, and

$$B_{uv}[u(\mathbf{r}_1, t), v(\mathbf{r}_2, t)] \equiv B_{uv}(\mathbf{r}_1, t; \mathbf{r}_2, t) = B_{uv}(\mathbf{r}_1 - \mathbf{r}_2, t) \tag{D1.1.14}$$

Hence, the random field is required to possess *homogeneous* statistics. The rigorous definition of this situation is again obtained from Eqs. (D1.1.1) and (D1.1.2):

$$\begin{aligned} p[u(\mathbf{r}, t)] &= p[u(\mathbf{r} + \mathbf{q}, t)] \\ p[u(\mathbf{r}_1, t); u(\mathbf{r}_2, t)] &= p[u(\mathbf{r}_1 + \mathbf{q}, t); u(\mathbf{r}_2 + \mathbf{q}, t)] \\ p[u(\mathbf{r}_1, t); v(\mathbf{r}_2, t)] &= p[u(\mathbf{r}_1 + \mathbf{q}, t); v(\mathbf{r}_2 + \mathbf{q}, t)] \\ &\;\;\cdots \end{aligned} \tag{D1.1.15}$$

where $\mathbf{q}$ is an arbitrary vector displacement within the field.

Combining both the requirements as specified in Eqs. (D1.1.11) and (D1.1.15) defines random fields that are both statistically stationary and homogeneous, a sufficient, general requirement for one to apply ergodicity to evaluate the required moments. In this case, the functions $U_t(t)$ and $U_r(\mathbf{r})$ become constants. Random fields that are taken to satisfy these requirements *only* for the first and second moments but *not* for the remaining moments are said to be *stationary and homogeneous in the wide sense*. Most of the random fields encountered in the applications are assumed to possess this property. Hence, the relative importance that is placed on the first and second moments rather than the remaining ones. Although one can (and, because of the problem constructions that are encountered, must) define moments of higher order, they are usually reduced to expressions that incorporate the first and second moments via the hypotheses that define the particular problem.

One can further specialize the description of random fields in space. Stationarity and homogeneity of the field statistics allow one to write for the second-order space-time moment

$$\langle u(\mathbf{r}_1, t_1), u(\mathbf{r}_2, t_2)\rangle = B_u(\mathbf{r}_1 - \mathbf{r}_2, t_1 - t_2) = B_u(\mathbf{q}, \tau) \quad \text{(D1.1.16)}$$

The spatial portion of the second moment is now only a function of the vector separation $\mathbf{q}$. In the most general case of a homogeneous field, this spatial correlation is a function of both the magnitude and direction of the displacement within the field. However, if one can assume a correlation function that is *only* a function of the magnitude and *independent* of the direction, one is dealing with an *isotropic* random field. Thus, when the statistics of the field are not only homogeneous but also isotropic, one has

$$B_u(\mathbf{q}, \tau) = B_u(|\mathbf{q}|, \tau) = B_u(q, \tau) \quad \text{(D1.1.17)}$$

Finally, there exists a class of space-time random fields where all the temporal changes that are observed to occur are associated with the translation of the spatial portion of the random field with a characteristic velocity v. This concept of a frozen field, where during the spatial translation, the field does not evolve temporally (e.g., no mixing occurs), is defined by considering the value of the field $f(\mathbf{r}, t + t')$ at a particular spatial location $\mathbf{r}$ at a time $t + t'$ and relating it to the field value $f(\mathbf{r}, t)$ at an earlier time t via the relation

$$f(\mathbf{r}, t + t') = f(\mathbf{r} - \mathbf{v}t', t)$$

In particular, but without loss of generality, one can let $t = 0$ and obtain

$$f(\mathbf{r}, t') = f(\mathbf{r} - \mathbf{v}t', 0) \quad \text{(D1.1.18)}$$

But since, using the left side of this equation,

$$\langle f(\mathbf{r}_1, t_1) f(\mathbf{r}_2, t_2)\rangle = B_f(\mathbf{q}, \tau) \qquad \mathbf{q} \equiv \mathbf{r}_1 - \mathbf{r}_2,\ \tau \equiv t_1 - t_2$$

by stationarity and homogeneity, one has for the space-time correlation using the right side of Eq. (D1.1.18)

$$\langle f(\mathbf{r}_1 - \mathbf{v}t_1) f(\mathbf{r}_2 - \mathbf{v}t_2)\rangle = B_f(\mathbf{r}_1 - \mathbf{v}t_1 - \mathbf{r}_2 + \mathbf{v}t_2, 0) = B_f(\mathbf{q} - \mathbf{v}\tau, 0)$$

Hence,

$$B_f(\mathbf{q}, \tau) = B_f(\mathbf{q} - \mathbf{v}\tau, 0) \quad \text{(D1.1.19)}$$

which is a property of frozen space-time random fields that will prove to be useful later. Among other simplifications, it allows one to obtain spatial correlation functions simply by time-averaging observations at two points spatially separated in the field.

In general, one has for the m,nth moment $\Gamma_{nm}(\mathbf{r}_1, \mathbf{r}_2, \ldots, \mathbf{r}_n; \mathbf{r}'_1, \mathbf{r}'_2, \ldots, \mathbf{r}'_m)$ of the random electric field

$$\Gamma_{nm}(\mathbf{r}_1, \mathbf{r}_2, \ldots, \mathbf{r}_n; \mathbf{r}'_1, \mathbf{r}'_2, \ldots, \mathbf{r}'_m) = \left\langle \prod_{i=1}^{n} E(\mathbf{r}_i) \prod_{j=n+1}^{n+m} E^*(\mathbf{r}_j) \right\rangle \quad (1.25)$$

with the proviso that in order to accommodate the above definition, the opposite case is just the complex conjugate of Eq. (1.25). Consider the first-order moment using Eq. (1.24); averaging this equation and noting that the functions $G_0(\mathbf{r}_{n-1} - \mathbf{r}_n)$ in the integrand are deterministic, one obtains

$$\langle E(\mathbf{r})\rangle = G_0(\mathbf{r} - \mathbf{R}) + \sum_{n=1}^{\infty} (-k^2)^n \int \dots (n) \dots \int G_0(\mathbf{r} - \mathbf{r}_1)$$

$$\cdots G_0(\mathbf{r}_{n-1} - \mathbf{r}_n) G_0(\mathbf{r}_n - \mathbf{R}) \langle \tilde{\varepsilon}(\mathbf{r}_1)\tilde{\varepsilon}(\mathbf{r}_2) \cdots \tilde{\varepsilon}(\mathbf{r}_n)\rangle \, d^3r_1 \cdots d^3r_n \quad (1.26)$$

thus indicating that one must know all the nth-order moments of the permittivity field fluctuations just to be able to determine the first moment $\langle E(\mathbf{r})\rangle$. Hence, one cannot, from this approach, derive a closed-form solution for even the first-order moment of the propagating wave field. This circumstance is directly attributable to the parametric form of Eq. (1.16). This does not mean, however, that Eq. (1.26) is not useful. Before going on to the further developments that will lead to closed-form solutions for the statistics of the wave field, methods with which Eq. (1.26) can be dealt with will be briefly presented.

1.3.2 Applications of the diagrammatic methods of quantum field theory

The problem defined by Eq. (1.26) is analogous to that of quantum field theory where one is required to determine the average of a field within which there are quantum fluctuations of arbitrary sources. The techniques that have been developed in quantum field theory to analytically study such situations can therefore be adapted to the study of Eq. (1.26), where the factors

$$G_0(\mathbf{r} - \mathbf{r}_1) \cdots G_0(\mathbf{r}_{n-1} - \mathbf{r}_n) G_0(\mathbf{r}_n - \mathbf{R}) \langle \tilde{\varepsilon}(\mathbf{r}_1)\tilde{\varepsilon}(\mathbf{r}_2) \cdots \tilde{\varepsilon}(\mathbf{r}_n)\rangle$$

represent the fluctuating sources of the field. Although more general situations can be considered,[8] one usually assumes that the fluctuations of the permittivity field are governed by gaussian (i.e., "normal") statistics. Such a fluctuation field has the property that all odd-order moments such as $\langle\tilde{\varepsilon}(\mathbf{r})\rangle$ or $\langle\tilde{\varepsilon}(\mathbf{r}_1)\tilde{\varepsilon}(\mathbf{r}_2)\tilde{\varepsilon}(\mathbf{r}_3)\rangle$ are equal to zero and all even-order moments are given in terms of second-order moments, e.g.,

$$\langle\tilde{\varepsilon}(\mathbf{r}_1)\tilde{\varepsilon}(\mathbf{r}_2)\tilde{\varepsilon}(\mathbf{r}_3)\tilde{\varepsilon}(\mathbf{r}_4)\rangle = \langle\tilde{\varepsilon}(\mathbf{r}_1)\tilde{\varepsilon}(\mathbf{r}_2)\rangle\langle\tilde{\varepsilon}(\mathbf{r}_3)\tilde{\varepsilon}(\mathbf{r}_4)\rangle$$

$$+ \langle\tilde{\varepsilon}(\mathbf{r}_1)\tilde{\varepsilon}(\mathbf{r}_3)\rangle\langle\tilde{\varepsilon}(\mathbf{r}_2)\tilde{\varepsilon}(\mathbf{r}_4)\rangle + \langle\tilde{\varepsilon}(\mathbf{r}_1)\tilde{\varepsilon}(\mathbf{r}_4)\rangle\langle\tilde{\varepsilon}(\mathbf{r}_2)\tilde{\varepsilon}(\mathbf{r}_3)\rangle \quad (1.27)$$

In general, a $2n$th-order moment of a gaussian random variable can be given as a sum of $(2n - 1)!!$ terms of n products of second-order (i.e., 2-point) correlation functions over all possible permutations of the $2n$ positions. Using this fact and evoking the operator notation

$$M_0(\mathbf{r}, \mathbf{r}')f(\mathbf{r}') \equiv \int G_0(\mathbf{r} - \mathbf{r}')f(\mathbf{r}')\, d^3r' \tag{1.28}$$

Eq. (1.26) can be written as

$$\begin{aligned}\langle E(\mathbf{r})\rangle &= G_0(\mathbf{r} - \mathbf{R}) + k^4 M_0(\mathbf{r}, \mathbf{r}_1)M_0(\mathbf{r}_1, \mathbf{r}_2)B_\varepsilon(\mathbf{r}_1, \mathbf{r}_2)\\ &\quad \times G_0(\mathbf{r}_2 - \mathbf{R}) + k^8 M_0(\mathbf{r}, \mathbf{r}_1)M_0(\mathbf{r}_1, \mathbf{r}_2)M_0(\mathbf{r}_2, \mathbf{r}_3)M_0(\mathbf{r}_3, \mathbf{r}_4)\\ &\quad \times M_0(\mathbf{r}_3, \mathbf{r}_4)[B_\varepsilon(\mathbf{r}_1, \mathbf{r}_2)B_\varepsilon(\mathbf{r}_3, \mathbf{r}_4) + B_\varepsilon(\mathbf{r}_1, \mathbf{r}_3)B_\varepsilon(\mathbf{r}_2, \mathbf{r}_4)\\ &\quad + B_\varepsilon(\mathbf{r}_1, \mathbf{r}_4)B_\varepsilon(\mathbf{r}_2, \mathbf{r}_3)]G_0(\mathbf{r}_4 - \mathbf{R}) + \cdots\end{aligned} \tag{1.29}$$

where $B_\varepsilon(\mathbf{r}_j, \mathbf{r}_k) \equiv \langle\tilde{\varepsilon}(\mathbf{r}_j)\tilde{\varepsilon}(\mathbf{r}_k)\rangle$ is the 2-point correlation function of the gaussian permittivity field.

At this point, it proves fruitful to employ the well-known (at least in quantum field theory) diagrammatic representation of the terms and factors of Eq. (1.29) to elucidate the structural order of the equation and facilitate a regrouping of terms so as to make a method of solution obvious. To this end, the typical Green function within the operator expression defined in Eq. (1.28) is represented as shown in Fig. 1.1*a*. It is important to note that this solid line represents only the factor $G_0(\mathbf{r}_{i-1} - \mathbf{r}_i')$ and not the operator of Eq. (1.28); as will be seen shortly,

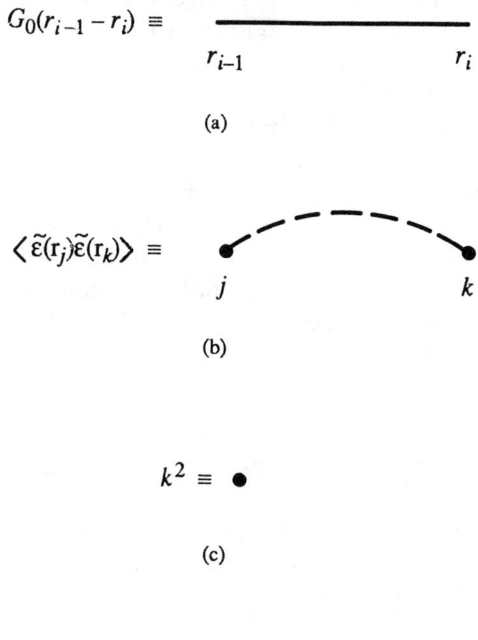

Figure 1.1 Diagrammatic equivalents of operators and factors.

$k^4 \langle \tilde{\varepsilon}(r_j)\tilde{\varepsilon}(r_k) \rangle k^4 B_\varepsilon(r_1, r_2) \equiv$

(a)

$k^4 G_0(r, r_j) G_0(r_1, r_2) B_\varepsilon(r_1, r_2) G_0(r_2 - R)$

$\equiv$

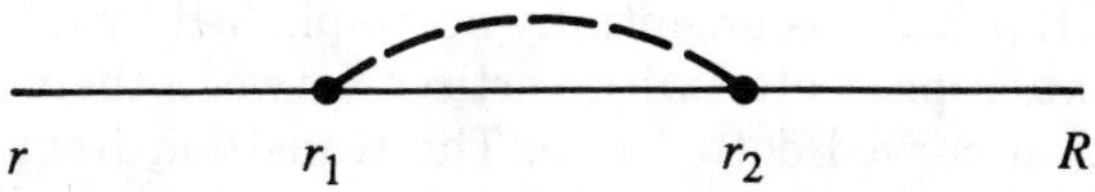

(b)

Figure 1.2 Examples of (*a*) second-order and (*b*) fourth-order correlation terms.

the operator relation of Eq. (1.28) will essentially be "induced" by the juxtaposition of these lines for various indices i. The correlation function $B_\varepsilon(\mathbf{r}_j, \mathbf{r}_k)$ is represented by the curved dotted line between the points $\mathbf{r}_j$ and $\mathbf{r}_k$ as shown in Fig. 1.1*b*. Finally, with each factor $\tilde{\varepsilon}(\mathbf{r}_j)$ there is necessarily associated, by the nature of the problem, a factor k^2 which is simply represented by a black dot (a vertex) as shown in Fig. 1.1*c*. Thus, the combination $k^2 B_\varepsilon(\mathbf{r}_j, \mathbf{r}_k)$ of factors is as presented in Fig. 1.2*a* and combination $k^4 G_0(\mathbf{r} - \mathbf{r}_1) G_0(\mathbf{r}_1 - \mathbf{r}_2) \times B_\varepsilon(\mathbf{r}_1, \mathbf{r}_2) G_0(\mathbf{r}_2 - \mathbf{R})$ is depicted in Fig. 1.2*b*. If one now adopts the rule that *at all points at which the vertices appear (i.e., the black dots representing k^2) one is to perform a spatial integration,* then the diagram of Fig. 1.2*b* becomes the second member of the right side of Eq. (1.29). With this rule, one essentially induces the application of the operator relationship defined by Eq. (1.28). The application of this rule is reiterated in Fig. 1.3. Continuing in the same manner, the third member of the right side of Eq. (1.29) is as depicted in Fig. 1.4. Because of the nature of how a $2n$-order gaussian correlation function decomposes into a product of n 2-

$k^4 M_0(r, r_1) M_0(r, r_1) B_\varepsilon(r_1, r_2) G_0(r_2 - R)$

$\equiv$

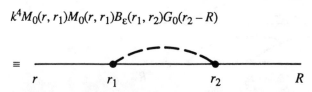

Figure 1.3 Application of operator definition of Eq. (1.28) for fourth-order correlation term.

point correlation functions, of which there are $(2n - 1)!!$ terms, one has for this case of $n = 2$ a total of $(4 - 1)!! = 3 \times 1 = 3$ separate diagrams representing this term. Also, since the $(2n - 1)!!$ terms represent all the permutations of 2-point correlation functions that can be selected from the total $2n$, each diagram in Fig. 1.4 is formed by selecting all combinations of pairs of points out of the total of $2n$ and placing the dotted lines between them. Finally, note that each graph of Fig. 1.4 has the observation point $\mathbf{r}$ at the left and the source point $\mathbf{R}$ at the right, just as the diagram for the first member of Eq. (1.29) would have. Hence, the diagrammatic transition from the first term to the second term of Eq. (1.29) is essentially accomplished by adding two intermediate points represented with vertices between them and connecting them with a curved dotted line. The transition from the second to the third term is carried out by adding two more intermediate points and connecting pairs of them at the internal vertices and adding other such diagrams obtained by permuting over all combinations.

Thus, one can immediately and rapidly construct as many terms in the series of Eq. (1.29) as is desired simply by employing this graphical method. Figure 1.5 is the diagrammatic representation of Eq. (1.29); the heavy line on the left of the equal sign in this figure is $\langle E(\mathbf{r})\rangle$, i.e., the representation for the sum total of all the terms generated by the procedure outlined above.

Much more important than providing an intuitive way of constructing higher-order terms in Eq. (1.29), the diagrammatic method allows one to regroup classes of terms in Eq. (1.29) to obtain a form for it that is more palatable. In particular, one notes from Fig. 1.5 that two classes of terms can be identified: the "strongly" connected graphs, characterized by the topology that does not permit one to break that graph between two vertices without breaking a dotted line between the ver-

$$k^8 M_0(r, r_j) M_0(r_1, r_2) M_0(r_2, r_3)$$
$$\times\ M_0(r_3, r_4) \langle \tilde{\varepsilon}(r_1)\tilde{\varepsilon}(r_2)\tilde{\varepsilon}(r_3)\tilde{\varepsilon}(r_4)\rangle G_0(r_2 - R)$$

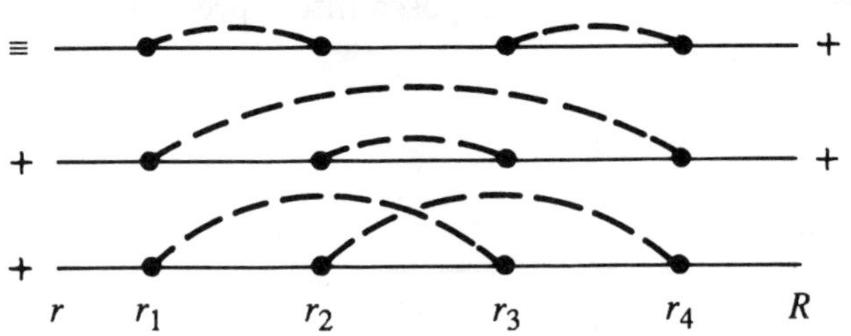

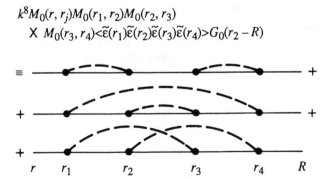

Figure 1.4 Diagrammatic representation of all fourth-order correlation terms for gaussian statistics.

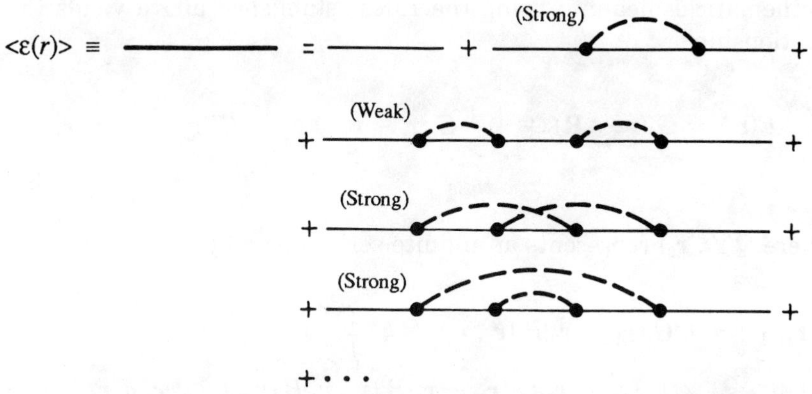

Figure 1.5 Diagram of Eq. (1.29) distinguishing strongly and weakly connected members.

tices, and the "weakly" connected graphs, where one can break a graph without simultaneously breaking a dotted line. The second, fourth, and fifth graphs of the diagram in Fig. 1.5 are strongly connected graphs, and the third is a weakly connected graph. All other higher-order terms, of course, also possess strongly and weakly connected graphs.

In fact, these higher-order terms can have graphs that are weak between more than one pair of vertices. For example, consider two sixth-order terms of Eq. (1.29) [there are $(2 \times 3 - 1)!! = 5 \times 3 \times 1 = 15$ in all] as shown in Fig. 1.6. The first graph depicted is weak between two pairs of vertices, and the second is weak between just one pair. Higher-order graphs can possess higher-order weaknesses.

Using this property of weak connectedness, one can diagrammatically group together all those graphs that are not weak between any pair of vertices, group together those that are weak between just one pair of vertices, group together those that are weak between two pairs of vertices, etc. Figure 1.7 displays such a graphical regrouping of the diagram of Fig. 1.5. At this point, one can make an immediate transition by graphically "inverse iterating" Fig. 1.5 to the diagram shown in Fig. 1.6, where the heavy line is a slightly more general definition as before; it denotes the quantity $\langle E(\mathbf{r}_i)\rangle$ at the point $\mathbf{r}_i$ due to a source at $\mathbf{R}$; of course, one has $\mathbf{r}_0 \equiv \mathbf{r}$. Converting this diagram back into its

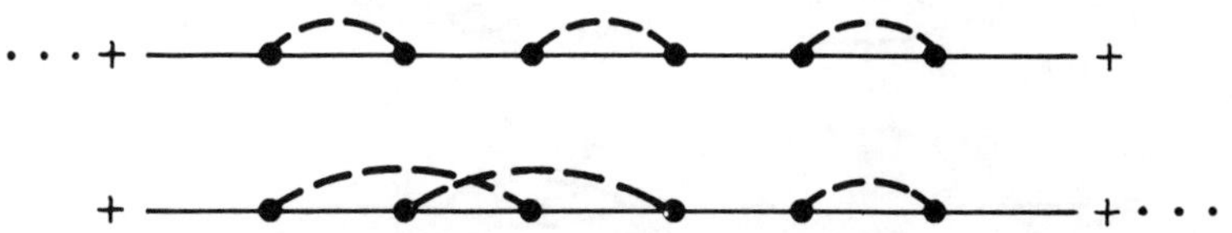

Figure 1.6 Examples of strongly connected and weakly connected sixth-order terms.

mathematical meaning using the rules established above yields the relationship

$$\langle E(\mathbf{r})\rangle = G_0(\mathbf{r} - \mathbf{R}) + \iint G_0(\mathbf{r} - \mathbf{r}_1)Q(\mathbf{r}_1, \mathbf{r}_2)\langle E(\mathbf{r}_2)\rangle \, d^3r_1 \, d^3r_2 \tag{1.30}$$

where $Q(\mathbf{r}_1, \mathbf{r}_2)$ represents an infinite series given by

$$Q(\mathbf{r}_1, \mathbf{r}_2) = k^4 G_0(\mathbf{r}_1 - \mathbf{r}_2)B_\varepsilon(\mathbf{r}_1, \mathbf{r}_2) + k^8 \iint G_0(\mathbf{r}_1 - \mathbf{r}_3) \times G_0(\mathbf{r}_3 - \mathbf{r}_4)G_0(\mathbf{r}_4 - \mathbf{r}_2)B_\varepsilon(\mathbf{r}_1, \mathbf{r}_4)B_\varepsilon(\mathbf{r}_3, \mathbf{r}_2) \, d^3r_3 \, d^3r_4 + \cdots \tag{1.31}$$

Equation (1.30) is the analog of Dyson's equation in quantum field theory, and $Q(\mathbf{r}_1, \mathbf{r}_2)$ is analogous to its mass operator. The integral equation Eq. (1.30) is indeed a closed relation in the quantity $\langle E(\mathbf{r}_i)\rangle$. In fact, in the event that the statistics of the permittivity field are homogeneous, i.e., $B_\varepsilon(\mathbf{r}_j, \mathbf{r}_k) = B_\varepsilon(\mathbf{r}_j - \mathbf{r}_k)$, one has through Eq. (1.31) that $Q(\mathbf{r}_1, \mathbf{r}_2) = Q(\mathbf{r}_1 - \mathbf{r}_2)$, thus allowing Eq. (1.30) to be easily solved via a double Fourier convolution. However, the fact still remains that the quantity $Q(\mathbf{r}_1, \mathbf{r}_2)$ is an infinite series that cannot be evaluated

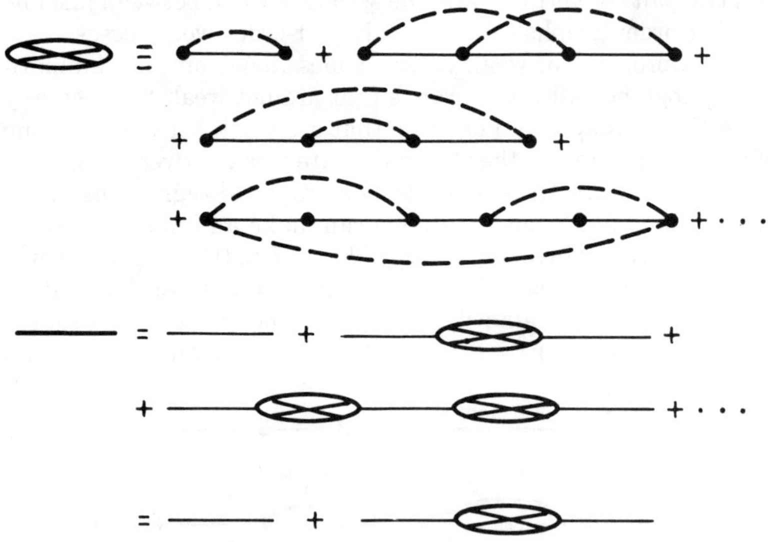

Figure 1.7 Formation of subseries of strongly coupled diagrams, graphical refactoring and inverse iteration yielding the Dyson equation.

explicitly. One must resort to representing $Q(\mathbf{r}_1, \mathbf{r}_2)$ with tractable approximate expressions.

A similar and more complicated situation exists in the case of the spatial correlation function of the wave field. A diagrammatic analysis leads to a relationship that is the analog of the Bethe-Salpeter equation.[2,5] It too, however, is fraught with difficulties similar to those associated with Eq. (1.30).

Even though the use of diagrammatic techniques gives exact relations (within, of course, the confines of the assumptions employed, e.g., gaussian statistics of the permittivity field) that can only be dealt with on an approximate basis, the relations possess the ability to give a full multiplicative scattering treatment in media not limited to relatively small fluctuations $\tilde{\varepsilon}(\mathbf{r}_i)$ [so long as $\tilde{\varepsilon}(\mathbf{r}_i) \ll 1$] where backscatter phenomena are prevalent.[9] In fact, they can be used as a basis for the derivation of the radiative transfer equation, which is usually constructed on a phenomenological foundation.

1.4 The Stochastic Parabolic Wave Equation and the Spatial Moments of the Propagating Wave Field

As seen from the considerations above, the parametric structure of the stochastic wave equation poses fundamental problems in its solution. In this and the following two sections, treatments of Eq. (1.16) will be presented that allow one to obtain closed-form, albeit approximate, solutions for the statistics of the random wave field. In what will be presented in this section, the parametric form of the equation will be retained, but its form will be simplified so that it can be used to obtain a differential equation that directly describes the behavior of the spatial moments of the propagating wave field. Section 1.5 will concern itself with a different form of the solution obtained here, and finally, Sec. 1.6 will concern itself with the transformation of Eq. (1.16) into a relationship that is not fraught with the troublesome parametric term but, instead, is a nonlinear differential equation that must be solved via perturbation expansions.

1.4.1 Derivation of the parabolic wave equation and its regions of applicability

Consider the decomposition of the random electric field $E(\mathbf{r}, t)$ into the component that describes its spatial behavior solely due to its wave nature, i.e., that given by $\exp(i\mathbf{r} \cdot \mathbf{k})$ where $\mathbf{k}$ is the wave vector in the direction of propagation, and the remaining component $U(\mathbf{r})$ that incorporates all the other spatial variations, i.e., those due to the random spatial variation of the permittivity and diffraction effects. Taking the direction of propagation to be along the x axis of an otherwise arbitrarily assigned rectangular coordinate system, one can write

$$E(\mathbf{r}) = U(\mathbf{r}) \exp(ikx) \tag{1.32}$$

Substituting this expression into Eq. (1.16), performing the required differentiations, and simplifying yields

$$\nabla_\rho^2 U(\mathbf{r}) + \frac{\partial^2 U(\mathbf{r})}{\partial x^2} + 2ik\,\frac{\partial U(\mathbf{r})}{\partial x} = -k^2\tilde{\varepsilon}(\mathbf{r})U(\mathbf{r}) \tag{1.33}$$

where $\nabla_\rho^2 \equiv \partial^2/\partial y^2 + \partial^2/\partial z^2$ is the transverse laplacian operator. Since $\lambda \ll l_0$ [this, of course, is *a priori* once Eq. (1.16) is used], spatial variation of the complex amplitude $U(\mathbf{r})$ due to diffraction will be dominated by the variation due to $\tilde{\varepsilon}$; thus, one has

$$\frac{\partial^2 U(\mathbf{r})}{\partial x^2} \sim \frac{U}{l_0^2} \qquad 2ik\,\frac{\partial U(\mathbf{r})}{\partial x} \sim \frac{U}{\lambda l_0}$$

and hence

$$\frac{\partial^2 U(\mathbf{r})}{\partial x^2} \ll 2ik\,\frac{\partial U(\mathbf{r})}{\partial x}$$

allowing Eq. (1.33) to be written

$$\nabla_\rho^2 U(\mathbf{r}) + 2ik\,\frac{\partial U(\mathbf{r})}{\partial x} = -k^2\tilde{\varepsilon}(\mathbf{r})U(\mathbf{r}) \tag{1.34}$$

Equation (1.34) is a parabolic wave equation with a stochastic source term on the right side and will form the basis for the remainder of this chapter. This *stochastic parabolic wave equation* is also known as the *diffusion approximation* because of the form of the left side, i.e., a diffusion equation with an imaginary factor in the diffusion coefficient (analogous to the Schrodinger equation of quantum mechanics). Application of this equation to the analysis of the statistics of the random wave field as will be formulated here was first noted by V. I. Tatarskii.[10] It is important to note that the condition $\lambda \ll l_0$ is a necessary one; there are, however, other constraints that prevail in the transition to and use of Eq. (1.34). These will now be briefly examined.

Including a unit source in the right side of Eq. (1.34) and writing the Green function solution to the resulting equation, one has

$$U(\mathbf{r}) = \int G_p(\mathbf{r}, \mathbf{r}')f(\mathbf{r}')\,d^3r' \tag{1.35}$$

where

$$\nabla_\rho^2 G_p(\mathbf{r}, \mathbf{r}') + 2ik \frac{\partial G_p(\mathbf{r}, \mathbf{r}')}{\partial x} = \delta(\mathbf{r} - \mathbf{r}') \tag{1.36}$$

defines the Green function $G_p(\mathbf{r}, \mathbf{r}')$ for the parabolic wave equation and

$$f(\mathbf{r}) = -k^2\tilde{\varepsilon}(\mathbf{r})U(\mathbf{r}) + \delta(\mathbf{r} - \mathbf{R}) \tag{1.37}$$

incorporates the source terms. It is now desired to obtain an expression for $G_p(\mathbf{r}, \mathbf{r}')$ via solution of Eq. (1.36) and compare it to the solution of the full stochastic wave equation, viz., Eqs. (1.20) and (1.22). To this end, let there be defined the Fourier transform pair

$$G_p(\mathbf{r}, \mathbf{r}') = G_p(x, \boldsymbol{\rho}; x', \boldsymbol{\rho}')$$

$$= \int_{-\infty}^{\infty} g_p(x, \boldsymbol{\kappa}; x', \boldsymbol{\kappa}') \exp(i\boldsymbol{\kappa} \cdot \boldsymbol{\rho})\, d^2\kappa \tag{1.38}$$

$$g_p(x, \boldsymbol{\kappa}; x', \boldsymbol{\kappa}') = \left(\frac{1}{2\pi}\right)^2 \int_{-\infty}^{\infty} G_p(x, \boldsymbol{\rho}; x', \boldsymbol{\rho}') \exp(-i\boldsymbol{\kappa} \cdot \boldsymbol{\rho})\, d^2\rho \tag{1.39}$$

where the three-dimensional position vector $\mathbf{r}$ is decomposed into its longitudinal coordinate x and its two-dimensional transverse vector $\boldsymbol{\rho}$, i.e., $\mathbf{r} = (x, \boldsymbol{\rho})$. Thus, Fourier transforming Eq. (1.36) yields the first-order differential equation

$$\frac{\partial g_p(\boldsymbol{\kappa})}{\partial x} + \frac{i\kappa^2}{2k} g_p(\boldsymbol{\kappa}) = \left(\frac{1}{2\pi}\right)^2 \frac{\delta(x - x') \exp(-i\boldsymbol{\kappa} \cdot \boldsymbol{\rho})}{2ik} \tag{1.40}$$

which has as a solution (using Neumann boundary conditions where the "surface" term at $x = 0$ is not specified)

$$g_p(\boldsymbol{\kappa}) = \frac{\exp[-i\kappa^2(x - x')/(2k)] \exp(-i\boldsymbol{\kappa} \cdot \boldsymbol{\rho})}{(2\pi)^2 2ik} \tag{1.41}$$

Substituting this result into Eq. (1.38) and transforming the two-dimensional integral into plane-polar coordinates yields

$$G_p(x, \boldsymbol{\rho}; x', \boldsymbol{\rho}') = \left(\frac{1}{2\pi}\right)^2 \left(\frac{1}{2ik}\right) \int_0^{\infty} \exp\left[\frac{-i\kappa^2(x - x')}{2k}\right]$$

$$\times \int_0^{2\pi} \exp[i\kappa(\rho - \rho')] \cos\theta\, d\theta\, \kappa\, d\kappa \tag{1.42}$$

Integration and rearrangement of the resulting factors finally yields for the Green function of the parabolic equation

$$G_p(x, \boldsymbol{\rho}; x', \boldsymbol{\rho}') = -\frac{1}{4\pi} \frac{\exp\{i[k(\rho - \rho')^2/2(x - x')]\}}{x - x'} \qquad (1.43)$$

which is the well-known Fresnel (or parabolic) approximation.

These results are now to be compared with those of Sec. 1.3; one must, however, remember the transformation of Eq. (1.32) when making this comparison. Thus, substituting Eq. (1.37) into Eq. (1.35) and employing Eq. (1.32) solved for $E(\mathbf{r})$ (so as to return to a description of the full electric field) gives

$$E(\mathbf{r}) = G_0(\mathbf{r} - \mathbf{R}) \exp(ikx) + \frac{k^2}{4\pi} \int \frac{\exp\{ik(\rho - \rho')^2/[2(x - x')] + ik(x - x')\}}{x - x'} \times \tilde{\varepsilon}(x', \rho')E(x', \rho')\, dx'\, d^2\rho' \qquad (1.44)$$

The conditions under which the kernel of Eq. (1.44) can be obtained from the kernel of the more general Eq. (1.22) are determined by expanding the Green function given in Eq. (1.20) in powers of the ratio of transverse distance $\boldsymbol{\rho} - \boldsymbol{\rho}'$ to longitudinal distance $x - x'$, where $\mathbf{r} - \mathbf{r}' = (x - x', \boldsymbol{\rho} - \boldsymbol{\rho}')$. One thus has the following development:

$$\begin{aligned} k|\mathbf{r} - \mathbf{r}'| &= k[(x - x')^2 + (\rho - \rho')^2]^{1/2} \\ &= k|x - x'| \left[1 + \frac{(\rho - \rho')^2}{(x - x')^2}\right]^{1/2} \\ &= k|x - x'| \left[1 + \frac{(\rho - \rho')^2}{2(x - x')^2} - \frac{(\rho - \rho')^4}{8(x - x')^4} + \cdots\right] \end{aligned}$$

One immediately sees that the phase term of Eq. (1.44) is obtained from this expansion so long as the condition

$$\frac{k(\rho - \rho')^4}{(x - x')^3} \ll 2\pi \qquad (1.45)$$

holds. (Of course, the denominator of the kernel of Eq. (1.44) need not satisfy such rigid requirements.) Letting $(x - x') \sim L$, where L is the maximum distance of propagation to the reception point, and $|\rho - \rho'| \sim \rho$, this condition becomes

$$\rho^4 \ll L^3\lambda \qquad (1.46)$$

It is important to distinguish between this condition, which places restrictions only on the use of the Green function in the parabolic approximation, and the other conditions that dominate over the use of the entire solution of Eq. (1.44), in particular, the regions of applicability within the integration volume in $(x', \boldsymbol{\rho})$ space. These considerations will now be briefly discussed and appended to the analysis above; this will lead to the complete specification of the region of applicability of the parabolic equation in addition to some further insights into the propagation process.

Consider now the following portion of the phase term in the Green function kernel in Eq. (1.44):

$$S = k\left[\frac{(\boldsymbol{\rho} - \boldsymbol{\rho}')^2}{2(x - x')}\right]$$

The expression that defines radii ρ_m of the constant phase contours in a plane transverse to the propagation axis, in particular, the mth phase surface S_m, at which the phase takes on the value $m\pi$, is simply found from this equation to be given by

$$\rho_m = \sqrt{m\lambda(x - x')} \tag{1.47}$$

The difference in the radii of two such adjacent contours, i.e., $\Delta\rho_m = \rho_{m+1} - \rho_m$, is thus given by

$$\begin{aligned}\Delta\rho_m &= \sqrt{(m + 1)\lambda(x - x')} - \rho_m \\ &= \sqrt{\rho_m^2 + \lambda(x - x')} - \rho_m\end{aligned}$$

Considering only those phase contours such that the condition

$$\lambda(x - x') \ll \rho_m^2 \tag{1.48}$$

holds, one can expand the radical in powers of the ratio $\lambda(x - x')/\rho_m^2$ and obtain the expression

$$\Delta\rho_m \approx \frac{1}{2}\sqrt{\frac{\lambda(x - x')}{m}} \tag{1.49}$$

Now, if the scale of the inhomogeneities l is such that $\sqrt{\lambda(x - x')} \gg l$, then one has from Eq. (1.49) that $\Delta\rho_m \gg l$ and the phase factor S will vary with ρ more slowly than will the quantity $\tilde{\varepsilon}(x', \rho')$, which can execute many variations on the scale of l. Using the condition $\Delta\rho_m \gg$

l in Eq. (1.49), solving for the quantity m, and substituting the resulting expression into Eq. (1.47) yields

$$\rho_m \equiv \rho < \frac{\lambda(x - x')}{l} \tag{1.50}$$

which defines the transverse region positioned at x' in which most of the scattering takes place that contributes to the field observed at the reception point x. In the opposite case, on the other hand, when $\Delta\rho_m \ll l$, one has that $\sqrt{\lambda(x - x')} \ll l$ and the phase factor S varies more rapidly with ρ than will $\tilde{\varepsilon}(x', \rho')$; thus, scattering in the transverse plane in the region defined by $\rho < \lambda(x - x')/l$ contributes little to the total scattered field at x.

Taking an equality to hold in Eq. (1.50) as the worst case, substituting the result into Eq. (1.46), and using the minimal value l_0 for the size of the inhomogeneities l yields the condition

$$\lambda^3 L \ll l_0^4 \tag{1.51}$$

that must hold for the use of the parabolic Green function in Eq. (1.44). This restriction has an interesting interpretation; rearranging factors, it can be made to yield

$$\frac{L\lambda}{l_0^2} \ll \frac{l_0^2}{\lambda^2}$$

which states that the ratio of the characteristic areas of the first Fresnel zone, i.e., $L\lambda$, and the smallest inhomogeneity, i.e., l_0^2, is much smaller than the quantity l_0^2/λ^2. At the other extreme, another region of applicability can be established; in particular, with the condition $\sqrt{\lambda(x - x')} \sim \sqrt{\lambda(L)} \gg l$ potentially prevailing in the strong scattering region, and taking l in this case to represent the maximal value L_0, one has the additional condition

$$\lambda L \gg L_0^2 \tag{1.52}$$

for which the parabolic wave equation can be applied. Alternatively, combining Eqs. (1.51) and (1.52) yields for this range of applicability of the solution of Eq. (1.44)

$$L_0^2 \ll \lambda L \ll \frac{l_0^4}{\lambda^2} \tag{1.53}$$

which defines for the distance L the Fraunhoffer diffraction zone for this stochastic scattering problem. It is interesting to note that this requirement also places an implicit restriction on the characteristic lengths λ, L_0, and l_0, viz.,

$$\lambda L_0 < l_0^2 \tag{1.54}$$

Equations (1.51) and (1.53) thus define the ranges of applicability of the parabolic wave equation, Eq. (1.34). With the condition $\lambda \ll l_0$ prevailing at the outset, one finds from these relations that Eq. (1.34) applies not only to distances L that lie in the *Fraunhoffer diffraction zone* [given by Eq. (1.53)], but also to those within the *Fresnel diffraction zone,* i.e., $\lambda L / l_0^2 \sim 1$, and the *geometrical optics zone,* i.e., $\lambda L \ll l_0^2$. This, however, still does not exhaust the restrictions that apply to the use of Eq. (1.34).

As can be seen from the integral solution of Eq. (1.44), the field $E(\mathbf{r})$ is determined by the inhomogeneities $\tilde{\varepsilon}(x', \rho')$ at the longitudinal positions $x' < x$, i.e., at points that are located *before* that at which $E(\mathbf{r})$ is evaluated. Therefore, as borne out in Eq. (1.46), backscattering phenomena are neglected. For this to be so, there must be some prevailing condition that incorporates the maximum level of fluctuations $\tilde{\varepsilon}$ that can be tolerated in this scenario. The analysis of this situation must await further developments of the theory and is therefore deferred to an examination in Appendix B. What is found is that

$$\tfrac{1}{4}k\langle\tilde{\varepsilon}^2\rangle \ll 1 \tag{1.55}$$

i.e., the level of permittivity fluctuations over distances on the order of a wavelength of the field must be small. The region of integration along the longitudinal x direction in Eq. (1.44) can thus be taken along the interval from the source to the point of reception, the total distance between which is L.

1.4.2 Application of the stochastic parabolic wave equation to optical propagation problems—field moments and the Markov approximation

At the outset, it will be useful to derive a Helmholtz-Kirchhoff integral relationship for the field $U(\mathbf{r})$. To this end, consider the Green function $G(\mathbf{r}, \mathbf{r}') = G(x, \rho; x', \rho')$ that is taken to satisfy Eq. (1.34) applied at the intermediate field point x', viz.,

$$\nabla_{\rho'}^2 G(x, \boldsymbol{\rho}; x', \boldsymbol{\rho}') - 2ik\,\frac{\partial G(x, \boldsymbol{\rho}; x', \boldsymbol{\rho}')}{\partial x'} + k^2\tilde{\varepsilon}(x', \boldsymbol{\rho}')G(x, \boldsymbol{\rho}; x', \boldsymbol{\rho}') = 0 \tag{1.56}$$

and the boundary condition

$$G(x, \boldsymbol{\rho}; x, \boldsymbol{\rho}') = \delta(\boldsymbol{\rho} - \boldsymbol{\rho}') \tag{1.57}$$

In writing Eq. (1.56), it is assumed that $G(\mathbf{r}, \mathbf{r}') = G(\mathbf{r} - \mathbf{r}')$, as with G_p above, thus generating a sign change upon changing variables in the derivative in the second term of Eq. (1.56) as compared to Eq. (1.34). This Green function $G(\mathbf{r}, \mathbf{r}')$ differs from $G_p(\mathbf{r}, \mathbf{r}')$ above and defined by Eq. (1.36) in that the former includes the stochastic factor $\tilde{\varepsilon}(x', \boldsymbol{\rho}')$ in its definition. Taking Eq. (1.34) to hold at $\mathbf{r}' = (x', \boldsymbol{\rho}')$ and multiplying it by G, and subtracting from this the result obtained by multiplying Eq. (1.56) by U yields

$$G\nabla^2_{\rho'}U - U\nabla^2_{\rho'}G + 2ik\frac{\partial(GU)}{\partial x'} = 0 \tag{1.58}$$

Integrating this equation within the semi-infinite volume contained between the planes defined at $x' = 0$ (i.e., the plane containing the sources of the field) and $x' = L$ (i.e., the plane containing the observation points) gives

$$\begin{aligned}2ik\int_{-\infty}^{\infty}&[G(L, \boldsymbol{\rho}; L, \boldsymbol{\rho}')U(L, \boldsymbol{\rho}') - G(L, \boldsymbol{\rho}; 0, \boldsymbol{\rho}')U(0, \boldsymbol{\rho}')]\,d^2\rho' \\ &+ \int_{-\infty}^{\infty}[G(L, \boldsymbol{\rho}; L, \boldsymbol{\rho}')\nabla^2_{\rho'}U(L, \boldsymbol{\rho}') - U(L, \boldsymbol{\rho}')\nabla^2_{\rho'}G(L, \boldsymbol{\rho}; L, \boldsymbol{\rho}') \\ &\qquad - G(L, \boldsymbol{\rho}; 0, \boldsymbol{\rho}')\nabla^2_{\rho'}U(0, \boldsymbol{\rho}') \\ &\qquad + U(0, \boldsymbol{\rho}')\nabla^2_{\rho'}G(L, \boldsymbol{\rho}; 0, \boldsymbol{\rho}')]\,d^2\rho' = 0\end{aligned} \tag{1.59}$$

Integrating by parts in the second member of Eq. (1.59), noting, as usual, that the radiation condition holds for U and G, thus allowing these field quantities to go to zero as $\rho \to \infty$, and using the boundary condition of Eq. (1.57) in the first term within the first member of Eq. (1.59) finally gives

$$U(L, \boldsymbol{\rho}) = \int_{-\infty}^{\infty} G(L, \boldsymbol{\rho}; 0, \boldsymbol{\rho}')U(0, \boldsymbol{\rho}')\,d^2\rho' \tag{1.60}$$

This integral relation connects a field source function $U(0, \rho') \equiv U_0(\boldsymbol{\rho}')$, which can be taken to be a deterministic field distribution (e.g., that of a two-dimensional image), to the random field $U(L, \boldsymbol{\rho})$ that results from propagation through the random medium described by the stochastic Green function $G(L, \boldsymbol{\rho}; 0, \boldsymbol{\rho}')$ that is given as the solution of the stochastic parabolic wave equation, Eq. (1.56). It should be noted that Eq. (1.56) is just the complex conjugate of the original parabolic wave equation, Eq. (1.34). Also, unless otherwise noted, the sources of the field, viz., the "object" plane, will be taken to occur at $x' = 0$.

The relevant statistical moments of the random propagating electric field are as defined in Eq. (1.25). Although one can define any moment of the electric field that contains an arbitrary number of complex conjugated factors, the most interesting and relevant ones from the point of view of measurements and applications are those *even* moments where $n = m$, i.e., ones that have equal numbers of unconjugated and conjugated field terms. The odd moments contain uncompensated phase factors and therefore oscillate rapidly with longitudinal distance and decay away quickly. Therefore, even though the treatment of the equations to follow will be kept general with the proviso that $n \geq m$, applications of these expressions will be specific to the even moments.

Substituting Eq. (1.32) into Eq. (1.25) yields

$$\Gamma_{nm}(\mathbf{r}_1, \mathbf{r}_2, \ldots, \mathbf{r}_n; \mathbf{r}_{n+1}, \ldots, \mathbf{r}_{n+m})$$

$$= \left\langle \prod_{i=1}^{n} U(\mathbf{r}_i) \exp(ikx_i) \prod_{j=n+1}^{n+m} U^*(\mathbf{r}_j) \exp(-ikx_j) \right\rangle$$

$$= \left\langle \prod_{i=1}^{n} U(\mathbf{r}_i) \prod_{j=n+1}^{n+m} U^*(\mathbf{r}_j) \right\rangle \exp(ikx_i) \exp(-ikx_j) \quad (1.61)$$

where the constant phase terms have been factored out of the averaging operation. In all situations considered in this work and in most of the cases in the applications, one has $x_i = x_j$. Thus, using Eq. (1.60), one can write for the moments of the electric field in the observation plane situated at $x = L$

$$\Gamma_{nm}(L; \rho_1, \rho_2, \ldots, \rho_n; \rho_{n+1}, \ldots, \rho_{n+m}) = \Gamma_{nm}(L, \{\boldsymbol{\rho}_i\}; \{\boldsymbol{\rho}_j\})$$

$$= \left\langle \prod_{i=1}^{n} U(L, \boldsymbol{\rho}_i) \prod_{j=n+1}^{n+m} U^*(L, \boldsymbol{\rho}_j) \right\rangle$$

$$= \int_{-\infty}^{\infty} \ldots (n) \ldots \int_{-\infty}^{\infty} \int_{-\infty}^{\infty} \ldots (m) \ldots \int_{-\infty}^{\infty}$$

$$\times\ \gamma_{nm}(L; \{\boldsymbol{\rho}_i - \boldsymbol{\rho}_i'\}, \{\boldsymbol{\rho}_j - \boldsymbol{\rho}_j'\}) \prod_{i=1}^{n} U_0(\boldsymbol{\rho}_i') \prod_{j=n+1}^{n+m} U_0^*(\boldsymbol{\rho}_j')$$

$$\times\ d^2\rho_1' \cdots d^2\rho_n'\, d^2\rho_{n+1}' \cdots d^2\rho_{n+m}' \quad (1.62)$$

where $\gamma_{nm}(x - x'; \{\boldsymbol{\rho}_i - \boldsymbol{\rho}_i'\}, \{\boldsymbol{\rho}_j - \boldsymbol{\rho}_j'\})$

$$\equiv \left\langle \prod_{i=1}^{n} G(x, \boldsymbol{\rho}_i; x', \boldsymbol{\rho}_i') \prod_{j=n+1}^{n+m} G^*(x, \boldsymbol{\rho}_j; x', \boldsymbol{\rho}_j') \right\rangle \quad (1.63)$$

is the nmth moment of the random electric field, and the notation $\{\boldsymbol{\rho}_i\}$ represents the set of transverse coordinates with $1 \le i \le n$ and $\{\boldsymbol{\rho}_j\}$ represents those with $n + 1 \le j \le n + m$. Equation (1.60) has been used to obtain the final result in Eq. (1.62). For example, one has for the mean field

$$\Gamma_{10} = \int_{-\infty}^{\infty} \langle G(L, \boldsymbol{\rho}_1; 0, \boldsymbol{\rho}_1')\rangle U_0(\boldsymbol{\rho}_1')\, d^2\rho_1' \tag{1.64}$$

and for the second-order moment of the field

$$\Gamma_{11} = \int_{-\infty}^{\infty}\int_{-\infty}^{\infty} \langle G(L, \boldsymbol{\rho}_1; 0, \boldsymbol{\rho}_1')G^*(L, \boldsymbol{\rho}_2; 0, \boldsymbol{\rho}_2')\rangle \times U_0(\boldsymbol{\rho}_1')U_0^*(\boldsymbol{\rho}_2')\, d^2\rho_1'\, d^2\rho_2' \tag{1.65}$$

As before, the problem reduces to one having to do with the evaluation of the moments γ_{nm} of the field $G(x, \boldsymbol{\rho}_i; x', \boldsymbol{\rho}_i')$ which in this instance is given by the differential equation Eq. (1.56). However, unlike the case considered earlier, a closed-form solution can be obtained for these moments with the aid of a palatable approximation. This will now be demonstrated.

Adopting the abbreviated notation $G(x', \boldsymbol{\rho}') = G(x, \boldsymbol{\rho}; x', \boldsymbol{\rho}')$ where only the intermediate field points are denoted for simplicity, dropping the primes, and defining the quantity g_{nm}

$$\begin{aligned} g_{nm} &\equiv g_{nm}(x - x'; \{\boldsymbol{\rho}_i - \boldsymbol{\rho}_i'\}; \{\boldsymbol{\rho}_j - \boldsymbol{\rho}_j'\}) \\ &= \prod_{i=1}^{n} G(x, \boldsymbol{\rho}_i) \prod_{j=n+1}^{m+n} G^*(x, \boldsymbol{\rho}_j) \end{aligned} \tag{1.66}$$

multiplying the identities

$$\frac{g_{nm}}{G(x, \rho_i)} = \prod_{k=1,k\neq i}^{n} G(x, \boldsymbol{\rho}_k) \prod_{j=n+1}^{n+m} G^*(x, \boldsymbol{\rho}_j)$$

$$\frac{g_{nm}}{G^*(x, \boldsymbol{\rho}_j)} = \prod_{i=1}^{n} G(x, \boldsymbol{\rho}_i) \prod_{l=n+1,l\neq j}^{n+m} G^*(x, \boldsymbol{\rho}_l)$$

respectively, by Eq. (1.56) and its complex conjugate, and adding the results gives

$$2ik\left[\sum_{i=1}^{n}\frac{g_{nm}}{G(x,\boldsymbol{\rho}_i)}\frac{\partial G(x,\boldsymbol{\rho}_i)}{\partial x}+\sum_{j=n+1}^{n+m}\frac{g_{nm}}{G^*(x,\boldsymbol{\rho}_j)}\frac{\partial G^*(x,\boldsymbol{\rho}_j)}{\partial x}\right]$$
$$-\left[\sum_{i=1}^{n}\frac{g_{nm}}{G(x,\boldsymbol{\rho}_i)}\nabla^2_{\rho_i}G(x,\boldsymbol{\rho}_i)-\sum_{j=n+1}^{n+m}\frac{g_{nm}}{G^*(x,\boldsymbol{\rho}_j)}\nabla^2_{\rho_j}G^*(x,\boldsymbol{\rho}_j)\right]$$
$$-k^2\left[\sum_{i=1}^{n}\tilde{\varepsilon}(x,\boldsymbol{\rho}_i)-\sum_{j=n+1}^{n+m}\tilde{\varepsilon}^*(x,\boldsymbol{\rho}_j)\right]g_{nm}=0 \quad (1.67)$$

Using the definition of Eq. (1.66), one finds that the first term on the left in Eq. (1.67) is simply $\partial g_{nm}/\partial x$ and that the second term is an expansion of $(\sum_{i=1}^{n}\nabla^2_{\rho_i}-\sum_{j=n+1}^{n+m}\nabla^2_{\rho_j})g_{nm}$, noting carefully that $\nabla^2_{\rho_i}$ operates only on those factors in g_{nm} that contain $\boldsymbol{\rho}_i$ and that $\nabla^2_{\rho_j}$ operates on those that contain $\boldsymbol{\rho}_j$. Hence, Eq. (1.67) can be written as

$$2ik\frac{\partial g_{nm}}{\partial x}-\left(\sum_{i=1}^{n}\nabla^2_{\rho_i}-\sum_{j=n+1}^{n+m}\nabla^2_{\rho_j}\right)g_{nm}-k^2Qg_{nm}=0 \quad (1.68)$$

where
$$Q=Q(x;\{\boldsymbol{\rho}_i\};\{\boldsymbol{\rho}_j\})$$
$$=\sum_{i=1}^{n}\tilde{\varepsilon}(x,\boldsymbol{\rho}_i)-\sum_{j=n+1}^{n+m}\tilde{\varepsilon}^*(x,\boldsymbol{\rho}_j) \quad (1.69)$$

is the random term. The associated boundary condition follows from that of Eq. (1.57) and is given by, after reinstating the full set of variable notation,

$$g_{nm}(x;\{\boldsymbol{\rho}_i\};\{\boldsymbol{\rho}_j\};x;\{\boldsymbol{\rho}_i'\};\{\boldsymbol{\rho}_j'\})=\prod_{i=1}^{n}\delta(\boldsymbol{\rho}_i-\boldsymbol{\rho}_i')\prod_{j=n+1}^{n+m}\delta(\boldsymbol{\rho}_j-\boldsymbol{\rho}_j') \quad (1.70)$$

At this point, it is useful to introduce the identity

$$2ik\frac{\partial g_{nm}}{\partial x}-k^2Q(x)g_{nm}=2ik\exp\left[-\frac{ik}{2}\int_0^x Q(x')\,dx'\right]\times\frac{\partial}{\partial x}\left\{\exp\left[\frac{ik}{2}\int_0^x Q(x')\,dx'\right]g_{nm}\right\}$$

into Eq. (1.68) and, upon rearranging terms, obtain

$$2ik\frac{\partial}{\partial x}\left\{\exp\left[\frac{ik}{2}\int_0^x Q(x')\,dx'\right]g_{nm}\right\}$$
$$-\exp\left[\frac{ik}{2}\int_0^x Q(x')\,dx'\right]\left(\sum_{i=1}^{n}\nabla^2_{\rho_i}-\sum_{j=n+1}^{n+m}\nabla^2_{\rho_j}\right)g_{nm}=0 \quad (1.71)$$

Integrating this equation with respect to x yields

$$2ik\left\{\exp\left[\frac{ik}{2}\int_0^x Q(x')\,dx'\right]g_{nm}(x) - g_{nm}(0)\right\}$$
$$-\int_0^x \exp\left[\frac{ik}{2}\int_0^{x_1} Q(x')\,dx'\right]\left(\sum_{i=1}^{n}\nabla_{\rho_i}^2 - \sum_{j=n+1}^{n+m}\nabla_{\rho_j}^2\right)g_{nm}(x_1)\,dx_1 = 0 \tag{1.72}$$

Finally, rearranging the exponential factors and averaging the resulting expression [remembering the definition of Eq. (1.63)] gives

$$2ik\gamma_{nm}(x) - 2ik\gamma_{nm}(0)\left\langle\exp\left[-\frac{ik}{2}\int_0^x Q(x')\,dx'\right]\right\rangle$$
$$-\int_0^x\left\langle\exp\left[-\frac{ik}{2}\int_{x_1}^x Q(x')\,dx'\right]\left(\sum_{i=1}^{n}\nabla_{\rho_i}^2 - \sum_{j=n+1}^{n+m}\nabla_{\rho_j}^2\right)\right.$$
$$\left.\times\, g_{nm}(x_1)\,dx_1\right\rangle = 0 \tag{1.73}$$

At this point, a very convenient and crucial approximation is made. The third term of Eq. (1.73) contains an ensemble average over two functions of x, viz., (1) the exponential factor containing the integral from an intermediate point x_1 to the point x, i.e., a functional of the random function $\tilde{\varepsilon}(x', \boldsymbol{\rho}')$ over the range $x' > x_1$, and (2) the factor $g_{nm}(x_1)$ evaluated at the intermediate point x_1, i.e., a functional of $\tilde{\varepsilon}(x'', \boldsymbol{\rho}')$ over the range $x'' \leq x_1$. [This latter fact can be substantiated by solving Eq. (1.72) for $g_{nm}(x)$ and forming a series solution by iteration of this equation.] There is, in the general case, a statistical correlation between the random functions $\tilde{\varepsilon}(x'', \boldsymbol{\rho}')$ for $x'' \leq x_1$ and $\tilde{\varepsilon}(x', \boldsymbol{\rho}')$ for $x' > x_1$; however, such longitudinal spatial correlations, like all types of spatial correlations, have a length $l_\parallel$ associated with them such that the dependence of the correlation is significant only for $|x' - x''| < l_\parallel$. Now what makes the longitudinal correlation different from, e.g., the transverse correlation is the fact that the maximum longitudinal distance encountered in the problem, i.e., the total propagation distance L, is such that $L \gg l_\parallel$. Thus, the exponential and $g_{nm}(x)$ functionals in the last member of Eq. (1.73) can be considered statistically independent for most of the longitudinal region of integration. This assumption is consistent with the conditions necessary for the application of the parabolic equation, in particular, the condition of Eq. (1.55) that allows the neglect of backscattered radiation along the propagation direction; here too the permittivity of the atmosphere in the region $x > x_1$ does not affect the field at the point x_1. This circumstance defines the property of *dynamic causality*. In fact, the parabolic equation of Eq. (1.56) is a *causal* relation.[11,12] These considerations then allow the last member of Eq. (1.73) to be written

$$\left\langle \exp\left[-\frac{ik}{2}\int_{x_1}^{x} Q(x')\,dx'\right]\left(\sum_{i=1}^{n}\nabla^2_{\rho_i} - \sum_{j=n+1}^{n+m}\nabla^2_{\rho_j}\right) g_{nm}(x_1)\right\rangle$$

$$= \left\langle \exp\left[-\frac{ik}{2}\int_{x_1}^{x} Q(x')\,dx'\right]\right\rangle\left(\sum_{i=1}^{n}\nabla^2_{\rho_i} - \sum_{j=n+1}^{n+m}\nabla^2_{\rho_j}\right)\gamma_{nm}(x_1) \quad (1.74)$$

thus letting Eq. (1.73) take the form of a *closed* equation in the function $\gamma_{nm}(x)$.

This development can be formally stated as follows. Consider the two-point spatial correlation function $B_\varepsilon(x', \boldsymbol{\rho}'; x'', \boldsymbol{\rho}'') = \langle\tilde{\varepsilon}(x', \boldsymbol{\rho}')\tilde{\varepsilon}(x'', \boldsymbol{\rho}'')\rangle$ of the atmospheric permittivity. The fact that $|x' - x''| < l_\parallel \ll L$ allows one to expand the correlation in the small parameters defined by the ratios $|\boldsymbol{\rho}'|/L$ and $|\boldsymbol{\rho}''|/L$. The first term of such a series expansion is formally obtained by equating to zero the longitudinal correlation length $l_\parallel$. If only this first expansion term is retained, then one formally obtains an effective correlation function $B_{\varepsilon\text{eff}}(x', \boldsymbol{\rho}'; x'', \boldsymbol{\rho}'')$ given by

$$B_{\varepsilon\text{eff}}(x', \boldsymbol{\rho}'; x'', \boldsymbol{\rho}'') = \delta(x' - x'')A(x'; \boldsymbol{\rho}', \boldsymbol{\rho}'') \quad (1.75)$$

where the two-dimensional spatial correlation function $A(x'; \boldsymbol{\rho}', \boldsymbol{\rho}'')$ is found from the requirement

$$\int_{-\infty}^{\infty} B_\varepsilon(x', \boldsymbol{\rho}'; x'', \boldsymbol{\rho}'')\,dx'' = \int_{-\infty}^{\infty} B_{\varepsilon\text{eff}}(x', \boldsymbol{\rho}'; x'', \boldsymbol{\rho}'')\,dx''$$
$$= A(x'; \boldsymbol{\rho}', \boldsymbol{\rho}'') \quad (1.76)$$

It should be noted that the two-dimensional correlation function can be a function of position x', thus accounting for any large-scale variations of the permittivity statistics along the propagation path.

In general, for an arbitrary n-point correlation function $B_\varepsilon(x', \boldsymbol{\rho}'; x'', \boldsymbol{\rho}''; \ldots; x^{(n)}, \boldsymbol{\rho}^{(n)})$, one has

$$B_\varepsilon(x', \boldsymbol{\rho}'; x'', \boldsymbol{\rho}''; \ldots; x^{(n)}, \boldsymbol{\rho}^{(n)}) = \delta(x_1 - x_2)\delta(x_1 - x_3)$$
$$\cdots \delta(x_1 - x_n)A(x_1; \boldsymbol{\rho}_1, \ldots, \boldsymbol{\rho}_n) \quad (1.77)$$

Thus, the transition from the left side to the right side of Eq. (1.74) is tantamount to modeling the statistical nature of the atmospheric permittivity field as a δ-function correlated process in the direction of propagation, leaving the meaningful properties to the transverse, two-dimensional correlation function $A(x; \boldsymbol{\rho}_1, \ldots, \boldsymbol{\rho}_n)$. Through the use of this *δ-function approximation,* the form of the set of equations for the moments $\gamma_{nm}(x)$ that result from the application of Eq. (1.74) in Eq. (1.73) is isomorphic to that of the second-order partial differential equations of the generalized parabolic type that resemble the Fokker-Planck-Kolmogorov diffusion equation for a Markov random process. For this reason, the δ-function approximation described above is also called the *Markov approximation* in stochastic wave propagation theory.[10] [For obvious reasons, it is also sometimes called the diffusion approximation, along with the use of Eq. (1.34).]

Substituting Eq. (1.74) into Eq. (1.73) yields

$$
\begin{aligned}
2ik\gamma_{nm}(x) - 2ik\gamma_{nm}(0)&\left\langle \exp\left[-\frac{ik}{2}\int_0^x Q(x')\,dx'\right]\right\rangle \\
&- \int_0^x \left\langle \exp\left[-\frac{ik}{2}\int_{x_1}^x Q(x')\,dx'\right]\right\rangle \\
&\times \left(\sum_{i=1}^{n} \nabla_{\rho_i}^2 - \sum_{j=n+1}^{n+m} \nabla_{\rho_j}^2\right) \gamma_{nm}(x_1)\,dx_1 \qquad (1.78)
\end{aligned}
$$

Noting that due to the δ-function correlation along the longitudinal x direction assumed for the permittivity field, one has

$$
\begin{aligned}
&\left\langle \exp\left[-\frac{ik}{2}\int_0^x Q(x')\,dx'\right]\right\rangle \\
&\quad = \left\langle \exp\left[-\frac{ik}{2}\int_0^{x_1} Q(x')\,dx'\right] \exp\left[-\frac{ik}{2}\int_{x_1}^x Q(x')\,dx'\right]\right\rangle \\
&\quad = \left\langle \exp\left[-\frac{ik}{2}\int_0^{x_1} Q(x')\,dx'\right]\right\rangle \left\langle \exp\left[-\frac{ik}{2}\int_{x_1}^x Q(x')\,dx'\right]\right\rangle \\
&\quad = M(x_1)M(x, x_1) \qquad (1.79)
\end{aligned}
$$

where
$$M(x, y) \equiv \left\langle \exp\left[-\frac{ik}{2}\int_y^x Q(x')\,dx'\right]\right\rangle$$

$$M(x) \equiv M(x, 0) \qquad (1.80)$$

Hence, using this notation and solving Eq. (1.79) for $M(x, x_1)$ allows Eq. (1.78) to be written as

$$2ik\gamma_{nm}(x) - 2ik\gamma_{nm}(0)M(x) - \int_0^x M(x)M^{-1}(x_1) \times \left(\sum_{i=1}^{n} \nabla^2_{\rho_i} - \sum_{j=n+1}^{n+m} \nabla^2_{\rho_j}\right) \gamma_{nm}(x_1)\, dx_1 = 0 \quad (1.81)$$

Dividing this equation through by the factor $M(x)$, noting that it can be taken out of the integrand, and differentiating the resulting expression with respect to x gives

$$2ik\left[M^{-1}(x)\frac{\partial \gamma_{nm}(x)}{\partial x} - \gamma_{nm}(x)M^{-2}(x)\frac{\partial M(x)}{\partial x}\right] - M^{-1}(x)\left(\sum_{i=1}^{n} \nabla^2_{\rho_i} - \sum_{j=n+1}^{n+m} \nabla^2_{\rho_j}\right)\gamma_{nm}(x) = 0 \quad (1.82)$$

Multiplying this result through by $M(x)$ and noting the well-known result

$$M^{-1}(x)\frac{\partial M(x)}{\partial x} = \frac{\partial \ln [M(x)]}{\partial x}$$

one finally has for the differential equation for the $(n + m)$th moment of the wave field, upon reinstating the primed variables,

$$\frac{\partial \gamma_{nm}}{\partial x'} = \gamma_{nm}\frac{\partial \ln [M(x')]}{\partial x'} - \frac{i}{2k}\left(\sum_{i=1}^{n} \nabla^2_{\rho'_i} - \sum_{j=n+1}^{n+m} \nabla^2_{\rho'_j}\right)\gamma_{nm} \quad (1.83)$$

where

$$M(x') = \left\langle \exp\left[-\frac{ik}{2}\int_0^{x'} Q(x'')\, dx''\right]\right\rangle \quad (1.84)$$

and, as defined earlier,

$$\gamma_{nm} \equiv \gamma_{nm}(x - x'; \{\boldsymbol{\rho}_i - \boldsymbol{\rho}'_i\}, \{\boldsymbol{\rho}_j - \boldsymbol{\rho}'_j\})$$

and

$$Q(x) = Q(x; \boldsymbol{\rho}_i; \boldsymbol{\rho}_j) \equiv \sum_{i=1}^{n} \tilde{\varepsilon}(x, \boldsymbol{\rho}_i) - \sum_{j=n+1}^{n+m} \tilde{\varepsilon}^*(x, \boldsymbol{\rho}_j)$$

Equations (1.83) and (1.84) are the most general form of the parabolic wave equation for arbitrary moments of the field since they are given in terms of the quantity $M(x)$, which can be related to the characteristic functional of the permittivity field.

Digression 1.2: Characteristic Functionals The concept of the characteristic functional of a random field is motivated by the simpler concept of the characteristic function of a random variable. Consider the general case of the probability density of n random variables, $p(u_1; u_2; \ldots; u_n)$. Since this is a function of n variables, one can define its Fourier transform, viz.,

$$\phi(\chi_1, \chi_2, \ldots, \chi_n) = \int \ldots (n) \ldots \int \exp\left(i \sum_{i=1}^{n} \chi_i u_i\right) \times p(u_1; u_2; \ldots; u_n)\, du_1\, du_2 \ldots du_n \quad \text{(D1.2.1)}$$

called the *characteristic function* of the variables u_i. Since one can analogously define the inverse transform to this relation, it becomes clear that there exists a unique connection between a probability density and its corresponding characteristic function. Thus, knowing the entire set of characteristic functions that corresponds to a set of random variables is equivalent to knowing the set of prevailing probability densities, which, analogous to the discussion of Digression 1.1, allows the associated random variables to be completely defined.

The form of the Fourier relation of Eq. (D1.2.1), which is such that one can interpret it as the mean of the exponential function that enters into it, i.e.,

$$\phi(\chi_1, \chi_2, \ldots, \chi_n) = \left\langle \exp\left(i \sum_{i=1}^{n} \chi_i u_i\right)\right\rangle \quad \text{(D.1.2.2)}$$

allows one to directly obtain the moments of the random variables by differentiation of ϕ with respect to the corresponding variables χ, called *conjugate variables* (conjugate to u). In particular, differentiation of Eq. (D1.2.2) with respect to any particular function χ_j out of the set of n such functions yields

$$\frac{\partial \phi(\chi_1, \chi_2, \ldots, \chi_n)}{\partial \chi_j} = \left\langle i u_j \exp\left(i \sum_{i=1}^{n} \chi_i u_i\right)\right\rangle$$

Evaluating this derivative by setting $\chi_i = 0$, $i = 1, \ldots, n$, gives

$$\left.\frac{\partial \phi(\chi_1, \chi_2, \ldots, \chi_n)}{\partial \chi_j}\right|_{\chi_1 = \chi_2 = \cdots = \chi_n = 0} = i\langle u_j \rangle \quad \text{(D1.2.3)}$$

for the first moment of the jth random variable. Similarly, for the second moment of any two random variables,

$$\left.\frac{\partial^2 \phi(\chi_1, \chi_2, \ldots, \chi_n)}{\partial \chi_j\, \partial \chi_k}\right|_{\chi_1 = \chi_2 = \cdots = \chi_n = 0} = -\langle u_j u_k \rangle \quad \text{(D1.2.4)}$$

The process can be continued for any arbitrary moment of the random variables. Thus, it is seen that the specification of the characteristic function of a set of random variables directly yields the statistical moments that prevail over their values. Since, in most applications, it is usually one or several of the moments that are desired, and not the probability densities, the use of characteristic functions simplifies the calculations.

In the case of random fields, where one is now concerned with random functions over space and time rather than with random variables, the characteristic function generalizes straightforwardly[7] rather than a sum over products of discrete values of χ_i and u_i appearing in the argument of the exponential of Eq. (D1.2.5), one now has an integral over space and time of the product of two functions $\chi(\mathbf{r}, t)$ and $u(\mathbf{r}, t)$, viz.,

$$\begin{aligned}\Phi[\chi(\mathbf{r}, t)] &= \left\langle \exp\left[i \int\int \chi(\mathbf{r}, t) u(\mathbf{r}, t)\, d\mathbf{r}\, dt \right] \right\rangle \\ &= \langle \exp\{iu[\chi(\mathbf{r}, t)]\}\rangle \end{aligned} \tag{D1.2.5}$$

where what was the characteristic function ϕ of the random variables u_i becomes the *characteristic functional* Φ of the random field $u(\mathbf{r}, t)$. The value of this "function of a function," i.e., the functional, when the *conjugate function* is such that $\chi(\mathbf{r}_0, t_0) = 1$ for some space-time point $(\mathbf{r}_0, t_0)$, is the value of the characteristic function of the random variable given by $u[\chi(\mathbf{r}_0, t_0) = 1]$. Thus, given any arbitrary function $\chi(\mathbf{r}, t)$, Eq. (D1.2.5) yields, in general, some complex number.

The partial derivatives that give the random variable moments in the case of characteristic functions now become functional derivatives that give the random field moments. Thus, for the first space-time moment, which holds at a particular spatial position $\mathbf{r}_1$ at a particular time t_1, one takes the functional derivative of Eq. (D1.2.5) with respect to the function $\chi(\mathbf{r}_1, t_1)$ which is evaluated at the space-time point $(\mathbf{r}_1, t_1)$, i.e.,

$$\frac{\delta\Phi[\chi(\mathbf{r}, t)]}{\delta\chi(\mathbf{r}_1, t_1)} = \left\langle iu(\mathbf{r}_1, t_1) \exp\left[i \int\int \chi(\mathbf{r}, t) u(\mathbf{r}, t)\, d\mathbf{r}\, dt \right] \right\rangle$$

and evaluate it in the case of $\chi(\mathbf{r}, t) = 0$, thus yielding

$$\left.\frac{\delta\Phi[\chi(\mathbf{r}, t)]}{\delta\chi(\mathbf{r}_1, t_1)}\right|_{\chi(\mathbf{r},t)=0} = i\langle u(\mathbf{r}_1, t_1)\rangle \tag{D1.2.6}$$

In the above, it is important that the point $(\mathbf{r}_1, t_1)$ at which the function χ is evaluated (with which the functional derivative is performed) is contained in the region over which the space-time integral of Eq. (D1.2.6) is defined. The second space-time moment of the field at the two points $(\mathbf{r}_1, t_1)$ and $(\mathbf{r}_2, t_2)$ follows via a double application of the prescription given for the first moment,

$$\left.\frac{\delta^2\Phi[\chi(\mathbf{r}, t)]}{\delta\chi(\mathbf{r}_1, t_1)\, \delta\chi(\mathbf{r}_2, t_2)}\right|_{\chi(\mathbf{r},t)=0} = -\langle u(\mathbf{r}_1, t_1) u(\mathbf{r}_2, t_2)\rangle \tag{D1.2.7}$$

Other moments follow similarly. In the general case, one has for the nth moment

$$B(\mathbf{r}_1, t_1; \mathbf{r}_2, t_2; \ldots; \mathbf{r}_n, t_n) = (-i)^n \left.\frac{\delta^n\Phi[\chi(\mathbf{r}, t)]}{\delta\chi(\mathbf{r}_1, t_1)\, \delta\chi(\mathbf{r}_2, t_2) \cdots \delta\chi(\mathbf{r}_n, t_n)}\right|_{\chi(\mathbf{r},t)=0} \tag{D1.2.8}$$

In many cases, like that in stochastic wave propagation under the quasi-stationary approximation, one simply has a spatial random field, independent of the time. In this case, the characteristic functional is given by

$$\Phi[\chi(\mathbf{r})] = \left\langle \exp\left[i \int \chi(\mathbf{r}) u(\mathbf{r})\, d\mathbf{r} \right] \right\rangle \tag{D1.2.9}$$

The first spatial moment is then given by

$$\left.\frac{\delta\Phi[\chi(\mathbf{r})]}{\delta\chi(\mathbf{r}_1)}\right|_{\chi(\mathbf{r})=0} = i\langle u(\mathbf{r}_1)\rangle \tag{D1.2.10}$$

the second moment by

$$\left.\frac{\delta^2\Phi[\chi(\mathbf{r})]}{\delta\chi(\mathbf{r}_1)\,\delta\chi(\mathbf{r}_2)}\right|_{\chi(\mathbf{r})=0} = -\langle u(\mathbf{r}_1)u(\mathbf{r}_2)\rangle \tag{D1.2.11}$$

and for the general nth-order spatial moment

$$B(\mathbf{r}_1, \mathbf{r}_2, \ldots, \mathbf{r}_n) = (-i)^n \left.\frac{\delta^n\Phi[\chi(\mathbf{r})]}{\delta\chi(\mathbf{r}_1)\,\delta\chi(\mathbf{r}_2)\ldots\delta\chi(\mathbf{r}_n)}\right|_{\chi(\mathbf{r})=0} \tag{D1.2.12}$$

As was indicated above, the functions χ can be arbitrary and usually appear as auxiliary, intermediate constructions that drop out of a final result, e.g., they are set to zero after functional differentiation to yield field moments in which they no longer appear. Sometimes, as will be done in what is to follow, their functional form is dictated by the particular problem to which a characteristic functional description is adapted.

Finally, it should be pointed out that in the case when the random field $u(\mathbf{r})$ in question is complex valued, it is in essence described by two real linearly independent functions, Re $[u(\mathbf{r})]$ and Im $[u(\mathbf{r})]$, which, in their characteristic functional, will be associated with two conjugate real linearly independent functions Re $[\chi(\mathbf{r})]$ and Im $[\chi(\mathbf{r})]$. However, this can be facilitated simply by introducing two independent conjugate functions $\chi_1(\mathbf{r})$ and $\chi_2(\mathbf{r})$ associated with $u(\mathbf{r})$ and $u^*(\mathbf{r})$ and writing for the prevailing characteristic functional

$$\begin{aligned}\Phi[\chi_1(\mathbf{r}), \chi_2(\mathbf{r})] &= \left\langle \exp\left\{ i \int [\chi_1(\mathbf{r})u(\mathbf{r}) + \chi_2(\mathbf{r})u^*(\mathbf{r})]\, d\mathbf{r} \right\}\right\rangle \\ &= \langle \exp\left(i\{u[\chi_1(\mathbf{r})] + u^*[\chi_2(\mathbf{r})]\}\right)\rangle \end{aligned} \tag{D1.2.13}$$

One therefore has

$$\begin{aligned}\left.\frac{\delta\Phi[\chi_1(\mathbf{r}), \chi_2(\mathbf{r})]}{\delta\chi_1(\mathbf{r}_1)}\right|_{\chi_1(\mathbf{r})=\chi_2(\mathbf{r})=0} &= i\langle u(\mathbf{r}_1)\rangle \\ \left.\frac{\delta\Phi[\chi_1(\mathbf{r}), \chi_2(\mathbf{r})]}{\delta\chi_2(\mathbf{r}_1)}\right|_{\chi_1(\mathbf{r})=\chi_2(\mathbf{r})=0} &= i\langle u^*(\mathbf{r}_1)\rangle\end{aligned} \tag{D1.2.14}$$

for the first-order moments, and for one of the three possible second-order moments

$$\left.\frac{\delta^2\Phi[\chi_1(\mathbf{r}), \chi_2(\mathbf{r})]}{\delta\chi_1(\mathbf{r}_1)\,\delta\chi_2(\mathbf{r}_2)}\right|_{\chi_1(\mathbf{r})=\chi_2(\mathbf{r})=0} = -\langle u(\mathbf{r}_1)u^*(\mathbf{r}_2)\rangle \tag{D1.2.15}$$

where the other two follow in an obvious fashion.

Defining the characteristic functional $\Phi_\varepsilon(\chi_1, \chi_2)$ for the atmospheric permittivity field as

$$\Phi_\varepsilon(\chi_1, \chi_2) = \left\langle \exp\left[i \int \tilde{\varepsilon}(\mathbf{r})\chi_1(\mathbf{r})\, d\mathbf{r} + i \int \tilde{\varepsilon}^*(\mathbf{r})\chi_2(\mathbf{r})\, d\mathbf{r} \right]\right\rangle \tag{1.85}$$

and letting the auxiliary functions χ_1, χ_2 be given by

$$\chi_1 = -\left(\frac{k}{2}\right)\sum_{i=1}^{n}\delta(\boldsymbol{\rho} - \boldsymbol{\rho}_i') \qquad \chi_2 = \left(\frac{k}{2}\right)\sum_{j=n+1}^{n+m}\delta(\boldsymbol{\rho} - \boldsymbol{\rho}_j') \tag{1.86}$$

one has, upon substituting Eq. (1.86) into Eq. (1.85) and performing the $\boldsymbol{\rho}$ integrations,

$$\Phi_\varepsilon(\chi_1, \chi_2) = \left\langle \exp\left[-\frac{ik}{2}\int_0^x \sum_{i=1}^{n} \tilde{\varepsilon}(x', \boldsymbol{\rho}_i')\,dx' + \frac{ik}{2}\int_0^x \sum_{j=n+1}^{n+m} \tilde{\varepsilon}^*(x', \boldsymbol{\rho}_j')\,dx' \right]\right\rangle$$

Remembering the definition of Eq. (1.69) for the quantity $Q(x)$, one obtains Eq. (1.84).

Specific forms of the characteristic functional will be considered in Chap. 2, where the statistical basis of atmospheric turbulence and turbidity will be considered.

1.5 The Green Function Solution of the Stochastic Parabolic Wave Equation

Although Eq. (1.62), which relates the arbitrary moment of a propagating wave field to the moments of the source of the field, is rigorous and of general form, it does not easily lend itself to interpretation in terms of the individual propagation components that are convolved within it. In particular, and especially if one wants to apply a "linear systems" interpretation to the random propagation process,[13] it is desired to obtain a form of solution for the moments of the field that incorporates the factors that are representative of the source distribution, the propagation of the wave field as if it were in "free space" devoid of any intervening random variations of its propagation characteristics, and of course the random action of the atmosphere.

To this end, consider once again the solution of Eq. (1.56) and require it to take the multiplicative form

$$G(x, \boldsymbol{\rho}; x', \boldsymbol{\rho}') = G_{p0}(x, \boldsymbol{\rho}; x', \boldsymbol{\rho}')G_{\tilde{\varepsilon}}(x, \boldsymbol{\rho}; x', \boldsymbol{\rho}') \tag{1.87}$$

where G_{p0} is taken to satisfy the complex conjugate of the parabolic wave equation in the case of $\tilde{\varepsilon} = 0$, i.e.,

$$\nabla^2_{\rho'} G_{p0}(x, \boldsymbol{\rho}; x', \boldsymbol{\rho}') - 2ik\,\frac{\partial G_{p0}(x, \boldsymbol{\rho}; x', \boldsymbol{\rho}')}{\partial x'} = 0 \tag{1.88}$$

with the Dirichlet boundary condition of Eq. (1.57), i.e., $G_{p0}(x', \boldsymbol{\rho}; x', \boldsymbol{\rho}') = \delta(\boldsymbol{\rho} - \boldsymbol{\rho}')$, and $G_{\tilde{\varepsilon}}$ satisfies the differential equation that results from substituting Eq. (1.87) into Eq. (1.56) and applying the requirement of Eq. (1.88). As in Sec. 1.4.1, Eq. (1.88) can be solved via Fourier transformation, making careful note that Eq. (1.88) is the complex conjugate of a simplified version of the equation considered in Sec. 1.4.1, viz., Eq. (1.36). Hence, using the conjugates of the transform relations of Eqs. (1.38) and (1.39), Eq. (1.88) and its associated boundary condition yield

$$\frac{\partial g_{p0}(\boldsymbol{\kappa}')}{\partial x'} - \frac{i\kappa'^2}{2k} g_{p0}(\boldsymbol{\kappa}') = 0$$

$$g_{p0}(x, \boldsymbol{\rho}; x, \boldsymbol{\kappa}') = \left(\frac{1}{2\pi}\right)^2 \exp\,(i\boldsymbol{\kappa}' \cdot \boldsymbol{\rho})$$

which give

$$g_{p0}(x, \boldsymbol{\rho}; x', \boldsymbol{\kappa}') = \left(\frac{1}{2\pi}\right)^2 \exp\left[\frac{i\kappa'^2}{2k}(x - x') + i\boldsymbol{\kappa}' \cdot \boldsymbol{\rho}\right]$$

Finally, inverse transforming this relation using the conjugate of Eq. (1.38) yields for the Green function of the free space parabolic wave equation

$$G_{p0}(x, \boldsymbol{\rho}; x', \boldsymbol{\rho}') = \frac{ik}{2\pi(x - x')} \exp\left[-i\frac{k(\boldsymbol{\rho} - \boldsymbol{\rho}')^2}{2(x - x')}\right] \tag{1.89}$$

One now substitutes Eq. (1.87) into Eq. (1.56) and, employing Eq. (1.88), obtains

$$-2ikG_{p0}\frac{\partial G_{\tilde{\varepsilon}}}{\partial x'} + 2\nabla_{\rho'}G_{\tilde{\varepsilon}}\nabla_{\rho'}G_{p0} + G_{p0}\nabla^2_{\rho'}G_{\tilde{\varepsilon}} + k^2\tilde{\varepsilon}G_{p0}G_{\tilde{\varepsilon}} = 0 \tag{1.90}$$

and upon using Eq. (1.89) to rid of the G_{p0} term,

$$-2ik\left[\frac{\partial G_{\tilde{\varepsilon}}}{\partial x'} + \frac{(\boldsymbol{\rho} - \boldsymbol{\rho}')}{x - x'} \cdot \nabla_{\rho'}G_{\tilde{\varepsilon}}\right] + \nabla^2_{\rho'}G_{\tilde{\varepsilon}} + k^2\tilde{\varepsilon}G_{\tilde{\varepsilon}} = 0 \tag{1.91}$$

where the prevailing boundary condition is now given by

$$G_{\tilde{\varepsilon}}(x, \boldsymbol{\rho}; 0, \boldsymbol{\rho}') = 1 \tag{1.92}$$

The structure of Eq. (1.91) is isomorphic to the parabolic wave equation Eq. (1.56). The first term within the brackets of Eq. (1.91) describes propagation along the x axis, and the second describes propagation in directions that deviate from the x axis. This apparent specialization of propagation directions is simply due to the requirement of the paraxial approximation of unperturbed wave propagation along the x axis by Eq. (1.88) and its solution, Eq. (1.89). This form of the parabolic wave equation was first considered in the analysis of the phenomena of "backscatter enhancement."[14]

The development given above allows one to write the following palatable solution for the random wave field: Substituting Eqs. (1.87) and (1.89) into Eq. (1.60) yields

$$U(L, \boldsymbol{\rho}) = \frac{ik}{2\pi L} \int_{-\infty}^{\infty} \exp\left[-i\,\frac{k(\boldsymbol{\rho} - \boldsymbol{\rho}')^2}{2L}\right] G_{\tilde{\varepsilon}}(L, \boldsymbol{\rho}; 0, \boldsymbol{\rho}') \times U_0(\boldsymbol{\rho}')\, d^2\rho' \qquad (1.93)$$

which can be interpreted as an extended version of the Huygens-Fresnel principle: The exponential function describes the propagation of the spherical "wavelets" from the source distribution $U_0(\boldsymbol{\rho}')$ positioned at $(0, \boldsymbol{\rho}')$ to the observation point $(L, \boldsymbol{\rho})$. The random function $G_{\tilde{\varepsilon}}$ accounts for the perturbation of the propagation of these spherical wavelets due to the random atmosphere. Equation (1.91), the solution of which gives $G_{\tilde{\varepsilon}}$, can be reduced to the form of the parabolic equation dealt with in Sec. 1.4.2, viz., Eq. (1.56), thus rendering applicable to this particular case the general expression for the moments of the field, Eq. (1.83). This is easily accomplished via the well-known method of characteristics. To this end, one can write a total differential for the first term on the left side of Eq. (1.91), viz.,

$$\frac{\partial}{\partial x'} + \frac{\partial \boldsymbol{\rho}'}{\partial x'} \cdot \nabla_{\rho'} = \frac{\partial}{\partial x'} + \frac{\boldsymbol{\rho} - \boldsymbol{\rho}'}{x - x'} \cdot \nabla_{\rho'}$$

so long as one admits the characteristic $\boldsymbol{\rho}' \equiv \boldsymbol{\rho}'(x')$ defined by the differential equation

$$\frac{\partial \boldsymbol{\rho}'}{\partial x'} = \frac{\boldsymbol{\rho} - \boldsymbol{\rho}'}{x - x'}$$

with the attendant boundary condition $\boldsymbol{\rho}'(0) = 0$. In this instance, one then has

$$\boldsymbol{\rho}'(x') = \boldsymbol{\rho}\left(\frac{x'}{x}\right) \qquad (1.94a)$$

and Eq. (1.91) becomes

$$-2ik\,\frac{\partial G_{\tilde{\varepsilon}}[x, \boldsymbol{\rho}; x, \boldsymbol{\rho}'(x')]}{\partial x'} + \nabla^2_{\rho'} G_{\tilde{\varepsilon}}[x, \boldsymbol{\rho}; x, \rho'(x')] + k^2 \tilde{\varepsilon} G_{\tilde{\varepsilon}}[x, \boldsymbol{\rho}; x', \boldsymbol{\rho}'(x')] = 0 \qquad (1.94b)$$

with its boundary condition

$$G_{\tilde{\varepsilon}}(x, \boldsymbol{\rho}; 0, 0) = 1 \qquad (1.94c)$$

Thus, the development of the last section [e.g., Eq. (1.83)] is applicable to the case considered here so long as one confines the solution to the characteristic defined by Eq. (1.94*a*).

This form of solution provides an intuitive interpretation to the atmospheric imaging process. The function $(ik/2\pi L) \exp[-ik(\boldsymbol{\rho} - \boldsymbol{\rho}')^2/2L]$ can be interpreted as the spatial impulse response of the random atmosphere in the absence of permittivity fluctuations, and $G_{\tilde{\varepsilon}}(x, \boldsymbol{\rho}; x', \boldsymbol{\rho}')$ is the additional perturbation to the impulse response due to the random variations of the atmospheric permittivity field. Such a modeling formalism will greatly facilitate the analysis of image degradation due to the atmosphere and its subsequent compensation.

The generation of arbitrary field moments Γ_{nm} from Eq. (1.93) is straightforward and exactly follows that which leads to Eq. (1.62). Similarly, the differential equation for the arbitrary moments of the point source Green function $\gamma_{nm} \equiv \langle G_{\tilde{\varepsilon}} G_{\tilde{\varepsilon}}^* \rangle$ follows that of Eqs. (1.83) and (1.84) on the characteristic curve given by Eq. (1.94*a*).

1.6 The Rytov Transformation and the Theory of Weak Fluctuations

As was shown in Sec. 1.4.1, the scalar stochastic wave equation, Eq. (1.16), can be put into the form of a parabolic differential equation that, like Eq. (1.16), is parametric in the random function $\tilde{\varepsilon}(\mathbf{r})$. However, by the use of the Markov approximation, a closed-form differential equation can be obtained that describes arbitrary field moments of the propagating wave field. In this section, another method of treatment will be discussed that was at the basis of early studies of wave propagation in the atmosphere. In particular, the method adopted here will be to attempt to obtain a nonparametric differential equation that describes the random wave field. The form of the theory to be developed below necessarily employs some mathematical simplifications which will place an additional restriction on the maximum level of intensity fluctuations that the theory is able to describe. Hence, what will be presented below is a "weak fluctuation" theory that finds application in short-range optical propagation problems. However, the form of the theory also allows separate descriptions to be given to amplitude and phase fluctuations of a wave rather than just its intensity. This fact, coupled with the relatively simple expressions that are obtained, tends to make weak fluctuation theory a viable approach in many atmospheric imaging problems.

Consider once again Eq. (1.34), which derives directly from the scalar stochastic wave equation, Eq. (1.16), under the suitable and realistic approximations given in Eqs. (1.54) and (1.55), and define a complex phase $\psi(\mathbf{r})$ such that one can write for the field $U(\mathbf{r})$

$$U(\mathbf{r}) = \exp[\psi(\mathbf{r})] \tag{1.95}$$

Substituting Eq. (1.95) into Eq. (1.34), performing the required differentiations, and simplifying yields the nonparametric relationship

$$2ik\frac{\partial\psi(\mathbf{r})}{\partial x} + \nabla_\rho^2\psi(\mathbf{r}) + [\nabla_\rho\psi(\mathbf{r})]^2 + k^2\tilde{\varepsilon}(\mathbf{r}) = 0 \tag{1.96}$$

which, however, is also a *nonlinear* Riccati equation. Thus, through the transformation of Eq. (1.95), the complexities associated with the parametric nature of the scalar stochastic wave equation and its approximate form of Eq. (1.34) have been transferred to the nonparametric but nonlinear equation Eq. (1.96). This transformation was first proposed in Ref. 15 for the deterministic solution of the diffraction of light by acoustical radiation and then later used for random wave propagation studies.[1] As with equations of the type of Eq. (1.96), one can proceed with a perturbation solution for $\psi(\mathbf{r})$; the perturbation parameter can be a nominal, representative value of the range of the random function $\tilde{\varepsilon}(\mathbf{r})$. In particular, it is useful to define the perturbation expansion parameter ν and an auxiliary random function $\xi(\mathbf{r})$ such that

$$\tilde{\varepsilon}(\mathbf{r}) = \nu\xi(\mathbf{r}) \qquad \nu = \sqrt{\langle\tilde{\varepsilon}^2\rangle} \tag{1.97}$$

Then, letting $\psi(\mathbf{r})$ be given by the expansion

$$\psi(\mathbf{r}) = \sum_{i=0}^{\infty}\psi_i(\mathbf{r}) \qquad \psi_i(\mathbf{r}) = \nu^i\Phi_i \tag{1.98}$$

substituting this and Eq. (1.97) into Eq. (1.96), and equating terms of corresponding perturbation order yields the following set of differential equations for the set of functions $\Phi_i(\mathbf{r})$:

$$\begin{aligned}
2ik\frac{\partial\Phi_0(\mathbf{r})}{\partial x} + \nabla_\rho^2\Phi_0(\mathbf{r}) + [\nabla_\rho\Phi_0(\mathbf{r})]^2 &= 0\\
2ik\frac{\partial\Phi_1(\mathbf{r})}{\partial x} + \nabla_\rho^2\Phi_1(\mathbf{r}) + 2\nabla_\rho\Phi_0(\mathbf{r})\cdot\nabla_\rho\Phi_1(\mathbf{r}) &= -k^2\xi(\mathbf{r})\\
2ik\frac{\partial\Phi_2(\mathbf{r})}{\partial x} + \nabla_\rho^2\Phi_2(\mathbf{r}) + 2\nabla_\rho\Phi_0(\mathbf{r})\cdot\nabla_\rho\Phi_2(\mathbf{r}) &= -[\nabla_\rho\Phi_1(\mathbf{r})]^2\\
&\ \vdots
\end{aligned} \tag{1.99}$$

One now multiplies the nth equation of this set by ν^n and, remembering the second relations in each of Eqs. (1.97) and (1.98), obtains

$$
\begin{aligned}
2ik\frac{\partial\psi_0(\mathbf{r})}{\partial x} + \nabla_\rho^2\psi_0(\mathbf{r}) + [\nabla_\rho\psi_0(\mathbf{r})]^2 &= 0\\
2ik\frac{\partial\psi_1(\mathbf{r})}{\partial x} + \nabla_\rho^2\psi_1(\mathbf{r}) + 2\nabla_\rho\psi_0(\mathbf{r})\cdot\nabla_\rho\psi_1(\mathbf{r}) &= -k^2\tilde{\varepsilon}(\mathbf{r})\\
2ik\frac{\partial\psi_2(\mathbf{r})}{\partial x} + \nabla_\rho^2\psi_2(\mathbf{r}) + 2\nabla_\rho\psi_0(\mathbf{r})\cdot\nabla_\rho\psi_2(\mathbf{r}) &= -[\nabla_\rho\psi_1(\mathbf{r})]^2\\
&\vdots
\end{aligned}
\tag{1.100}
$$

At this point, one defines a function $w_i(\mathbf{r})$ by the transformation

$$\psi_i(\mathbf{r}) = \exp\left[-\psi_0(\mathbf{r})\right]w_i(\mathbf{r}) \tag{1.101}$$

designed so that application of Eq. (1.101) to all but the first expression in the set of Eq. (1.100) will get rid of the $\nabla_\rho\psi_0(\mathbf{r})$ factors that appear. Carrying out this procedure yields a new set of equations, all of which take the form

$$2ik\frac{\partial w_i(\mathbf{r})}{\partial x} + \nabla_\rho^2 w_i(\mathbf{r}) = f_i(\mathbf{r})\exp\left[\psi_0(\mathbf{r})\right] \tag{1.102}$$

where

$$
\begin{aligned}
f_1(\mathbf{r}) &= -k^2\tilde{\varepsilon}(\mathbf{r})\\
f_2(\mathbf{r}) &= -[\nabla\psi_1(\mathbf{r})]^2\\
f_3(\mathbf{r}) &= -2\nabla_\rho\psi_1(\mathbf{r})\cdot\nabla_\rho\psi_2(\mathbf{r})\\
&\vdots
\end{aligned}
$$

As shown in Sec. 1.4.1, an equation of the type given in Eq. (1.102) has a solution given by

$$w_i(\mathbf{r}) = \int_0^x\int_{-\infty}^{\infty} G_p(x, \boldsymbol{\rho}; x', \boldsymbol{\rho}')f_i(\mathbf{x}', \boldsymbol{\rho}')\exp\left[\psi_0(\mathbf{x}', \boldsymbol{\rho}')\right]d^2\rho'\,dx' \tag{1.103}$$

Thus, transforming back to the functions $\psi_i(\mathbf{r})$ via Eq. (1.101) and using Eq. (1.95) to relate the "zeroth"-order perturbation term $\psi_0(\mathbf{r})$ to that of the field $U_0(\mathbf{r})$, one obtains

$$\psi_1(\mathbf{r}) = -k^2 \int_0^x \int_{-\infty}^{\infty} G_p(x, \boldsymbol{\rho}; x', \boldsymbol{\rho}')\tilde{\varepsilon}(x', \boldsymbol{\rho}') \frac{U_0(x', \boldsymbol{\rho}')}{U_0(x, \boldsymbol{\rho})} d^2\rho' \, dx' \tag{1.104}$$

$$\psi_2(\mathbf{r}) = -\int_0^x \int_{-\infty}^{\infty} G_p(x, \boldsymbol{\rho}; x', \boldsymbol{\rho}')[\nabla_\rho \psi_1(x', \boldsymbol{\rho}')]^2 \frac{U_0(x', \boldsymbol{\rho}')}{U_0(x, \boldsymbol{\rho})} d^2\rho' \, dx' \tag{1.105}$$

An expression for the unperturbed field $U_0(\mathbf{r})$ is simply found by transforming the first equation of Eq. (1.100) with the inverse of the relation $U_0(\mathbf{r}) = \exp[\psi_0(\mathbf{r})]$, which derives from Eq. (1.95) and relates the zeroth-order Rytov solution $\psi_0(\mathbf{r})$ to the field component $U_0(\mathbf{r})$. This gives

$$2ik \frac{\partial U_0}{\partial x} + \nabla_\rho^2 U_0 = 0 \tag{1.106}$$

Therefore, once the unperturbed field distribution $U_0(\mathbf{r})$ has been selected that satisfies Eq. (1.106), one continues to use this field function in the higher-order "corrective" Rytov solutions of Eqs. (1.104), (1.105), etc., for the field given by Eq. (1.95) propagating in the presence of random permittivity fluctuations $\tilde{\varepsilon}(\mathbf{r})$.

Since the complex phase is the central quantity in this formulation, it is desired to relate it to the more fundamental quantities of amplitude and phase that can appear in the applications or experiment. To this end, substituting the series expansion for $\psi(\mathbf{r})$ as given by Eq. (1.98) into the Rytov transformation, Eq. (1.95), and separating out the $\psi_0(\mathbf{r})$ term yields

$$U(\mathbf{r}) = U_0(\mathbf{r}) \exp\left[\sum_{i=1}^{\infty} \psi_i(\mathbf{r})\right] \tag{1.107}$$

Representing the fields $U(\mathbf{r})$ and $U_0(\mathbf{r})$ within this expression by their amplitude and phase, i.e., $U(\mathbf{r}) = A(\mathbf{r}) \exp[iS(\mathbf{r})]$ and $U_0(\mathbf{r}) = A_0(\mathbf{r}) \exp[iS_0(\mathbf{r})]$, one obtains, after rearranging terms,

$$\sum_{i=1}^{\infty} \psi_i(\mathbf{r}) = \chi(\mathbf{r}) + iS_1(\mathbf{r}) \tag{1.108}$$

where
$$\chi(\mathbf{r}) \equiv \ln\left[\frac{A(\mathbf{r})}{A_0(\mathbf{r})}\right] \tag{1.109}$$

is, for obvious reasons, the *log-amplitude* and

$$S_1(\mathbf{r}) \equiv S(\mathbf{r}) - S_0(\mathbf{r}) \tag{1.110}$$

is the *phase fluctuation*. From Eq. (1.108), one has

$$\chi(\mathbf{r}) = \mathrm{Re}\left[\sum_{i=1}^{\infty} \psi_i(\mathbf{r})\right] = \frac{1}{2}\sum_{i=1}^{\infty} [\psi_i(\mathbf{r}) + \psi_i^*(\mathbf{r})] \tag{1.111}$$

$$S_1(\mathbf{r}) = \mathrm{Im}\left[\sum_{i=1}^{\infty} \psi_i(\mathbf{r})\right] = \frac{1}{2i}\sum_{i=1}^{\infty} [\psi_i(\mathbf{r}) - \psi_i^*(\mathbf{r})] \tag{1.112}$$

Note that one can also write $\chi(\mathbf{r}) = \ln[A(\mathbf{r})] - \ln[A_0(\mathbf{r})]$, i.e., the log-amplitude is the variation of the logarithm of the fluctuating amplitude about that of the unperturbed "free-space" amplitude.

One can now form the various statistical parameters that characterize the random functions $\chi(\mathbf{r})$ and $S_1(\mathbf{r})$. For reasons that will be apparent shortly, it is of interest to consider only the first- and second-order initial moments (i.e., the mean and correlation function) of $\chi(\mathbf{r})$ and $S_1(\mathbf{r})$. The means are simply the means of the individual complex phase terms in Eqs. (1.111) and (1.112), and the correlation functions are, after expanding the required multiplications and simplifying,

$$B_\chi(\mathbf{r}_1, \mathbf{r}_2) \equiv \langle \chi(\mathbf{r}_1)\chi(\mathbf{r}_2)\rangle$$
$$= \frac{1}{2}\sum_{i=1}^{\infty}\sum_{j=1}^{\infty} \mathrm{Re}[\langle \psi_i(\mathbf{r}_1)\psi_j^*(\mathbf{r}_2)\rangle + \langle \psi_i(\mathbf{r}_1)\psi_j(\mathbf{r}_2)\rangle] \tag{1.113}$$

$$B_{S_1}(\mathbf{r}_1, \mathbf{r}_2) \equiv \langle S_1(\mathbf{r}_1)S_1(\mathbf{r}_2)\rangle$$
$$= \frac{1}{2}\sum_{i=1}^{\infty}\sum_{j=1}^{\infty} \mathrm{Re}[\langle \psi_i(\mathbf{r}_1)\psi_j^*(\mathbf{r}_2)\rangle - \langle \psi_i(\mathbf{r}_1)\psi_j(\mathbf{r}_2)\rangle] \tag{1.114}$$

where, as usual, Re[...] denotes the real part of [...]. Similarly, one can also consider the cross-correlation function

$$B_{\chi S_1}(\mathbf{r}_1, \mathbf{r}_2) \equiv \langle \chi(\mathbf{r}_1)S_1(\mathbf{r}_2)\rangle$$
$$= \frac{1}{2}\sum_{i=1}^{\infty}\sum_{j=1}^{\infty} \mathrm{Im}[\langle \psi_i(\mathbf{r}_1)\psi_j(\mathbf{r}_2)\rangle - \langle \psi_i(\mathbf{r}_1)\psi_j^*(\mathbf{r}_2)\rangle] \tag{1.115}$$

where Im[...] denotes the imaginary part of [...]. It is important to note that as defined above, $B_{\chi S_1}(\mathbf{r}_1, \mathbf{r}_2) \neq B_{\chi S_1}(\mathbf{r}_2, \mathbf{r}_1)$ in general; only in the case where the log-amplitude and phase fields are statistically homogeneous and isotropic will an equality hold. Thus, one need only consider linear combinations of $\langle \psi_i(\mathbf{r}_1)\psi_j(\mathbf{r}_2)\rangle$ and $\langle \psi_i(\mathbf{r}_1)\psi_j^*(\mathbf{r}_2)\rangle$. Just as in Sec. 1.3, one can perform a diagrammatic analysis of the summation of these series.[16]

Consider now the solution for the random wave field in the approximation that is good to the first Rytov solution $\psi_1(\mathbf{r})$. Thus one need only form expressions for the products $\psi_1(\mathbf{r}_1)\psi_1(\mathbf{r}_2)$ and $\psi_1(\mathbf{r}_1)\psi_1^*(\mathbf{r}_2)$ from Eq. (1.104). The formulation of weak fluctuation theory requires one to select the form of the initial wave field at the outset that satisfies Eq. (1.106). Because of the usefulness in applications in atmospheric imaging, spherical wave propagation will be considered; the propagation of plane waves can also be obtained as a special case. For a spherical wave in the paraxial approximation of Eq. (1.106), one has

$$U_0(\mathbf{r}') = U_0(x', \boldsymbol{\rho}') = \frac{\exp(ik\rho'^2/2x')}{x'} \tag{1.116}$$

Using this expression and that of Eq. (1.43) in Eq. (1.104) yields

$$\begin{aligned}\psi_1(x, \boldsymbol{\rho}) &= \frac{k^2}{4\pi}\int_0^x \int_{-\infty}^{\infty} \\ &\quad \times \frac{1}{x'(x-x')}\exp\left\{\frac{ik}{2}\left[\frac{(\boldsymbol{\rho}-\boldsymbol{\rho}')^2}{(x-x')} + \frac{\rho'^2}{x'} - \frac{\rho^2}{x}\right]\right\} \\ &\quad \times \tilde{\varepsilon}(x', \boldsymbol{\rho}')\, d^2\rho'\, dx' \\ &= \frac{k^2}{4\pi}\int_0^x \int_{-\infty}^{\infty} \frac{1}{\gamma(x-x')}\exp\left\{\frac{ik}{2}\left[\frac{(\rho-\gamma\rho')^2}{\gamma(x-x')}\right]\right\} \\ &\quad \times \tilde{\varepsilon}(x', \boldsymbol{\rho}')\, d^2\rho'\, dx'\end{aligned} \tag{1.117}$$

where $\gamma = \gamma(x') = x'/x$. Note that for the plane wave case, one simply has $U_0(\mathbf{r}')/U_0(\mathbf{r}) = 1$. Proceeding in the same way as above, one finds that this case is represented by Eq. (1.117) with $\gamma = 1$. In fact, for the general beam wave case,[17,18] where the initial radiation distribution is represented by a radius W_0 [the distance in the transverse plane where the amplitude decreases to $\exp(-1)$ of its value at the center] and a radius of curvature R_0 of the phase front, one has Eq. (1.117) where $\gamma = (1 + i\alpha x')/(1 + i\alpha x)$ with $\alpha \equiv 2/(kW_0^2) + i/R_0$. When $W_0 \to \infty$, $R_0 \to 0$, a plane wave obtains, i.e., $\gamma \to 1$, and when $W_0 \to 0$, $R_0 \to 0$, one has a spherical wave, i.e., $\gamma \to x'/x$. In this general case, the factor γ is a complex quantity. *In what is to follow, only spherical and plane wave propagation will be considered, and thus expressions will be derived where γ will always be taken to be real.*

The convolutional form of Eq. (1.117) suggests that subsequent analysis will be facilitated by employing the spectral expansion of the random function $\tilde{\varepsilon}(x', \boldsymbol{\rho}')$ in the variable $\boldsymbol{\rho}$. Implicit in such a selection is the fact that one is dealing with a statistically homogeneous random field which, as will be discussed in Chap. 2, models the atmospheric permittivity field.

Digression 1.3: The Spectral Representation of Random Functions Consider a random space-time field $f(\mathbf{r}, t)$ that is stationary and homogeneous; from the results of Digression 1.1, one can write for the correlation function

$$\langle (f(\mathbf{r}_1, t_1) f(\mathbf{r}_2, t_2)) \rangle = B_f(\mathbf{r}_1, t_1; \mathbf{r}_2, t_2) = B_f(\mathbf{q}, \tau) \tag{D1.3.1}$$

Such a random field can be represented by a Fourier-Stieltjes integral, viz.,

$$f(\mathbf{r}, t) = \int\int \exp\,[i(\mathbf{K} \cdot \mathbf{r} + \omega t)]\, dZ(\mathbf{K}, \omega) \tag{D1.3.2}$$

where $dZ(\mathbf{K}, \omega)$ is the spectral amplitude, which is a function of both the increments of the three-dimensional spatial frequency vector $\mathbf{K}$ and the temporal frequency ω. This spectral amplitude is defined by the requirements that reflect the statistics of the field. For the first moment of the spectral amplitude increment, one has from the requirement of stationarity and homogeneity

$$\langle f(\mathbf{r}, t) \rangle = \int\int \exp\,[i(\mathbf{K} \cdot \mathbf{r} + \omega t)] \langle dZ(\mathbf{K}, \omega) \rangle = 0$$

which yields the requirement

$$\langle dZ(\mathbf{K}, \omega) \rangle = 0 \tag{D1.3.3}$$

(Without loss of generality, the constant space-time average was set to zero.) Similarly, applying the same statistical conditions to the second-order moment, one must have

$$\langle dZ(\mathbf{K}, \omega)\, dZ(\mathbf{K}', \omega') \rangle = \delta(\mathbf{K} - \mathbf{K}')\delta(\omega + \omega')\Omega(\mathbf{K}, \omega)\, d\mathbf{K}\, d\mathbf{K}'\, d\omega\, d\omega' \tag{D1.3.4}$$

since this allows one to obtain, from applying Eq. (D1.3.2) to Eq. (D1.3.1), a relation that possess homogeneity and stationarity, viz.,

$$\begin{aligned} B_f(\mathbf{r}_1, t_1; \mathbf{r}_2, t_2) &= B_f(\mathbf{q}, \tau) \\ &= \int\int\int\int \exp\,[i(\mathbf{K} \cdot \mathbf{r}_1 + \omega t_1 - \mathbf{K}' \cdot \mathbf{r}_2 - \omega' t_2)] \\ &\quad \times \langle dZ(\mathbf{K}, \omega)\, dZ(\mathbf{K}', \omega') \rangle \\ &= \int\int \exp\,[i(\mathbf{K} \cdot \mathbf{q} + \omega\tau)]\Omega(\mathbf{K}, \omega)\, d\mathbf{K}\, d\omega \end{aligned} \tag{D1.3.5}$$

where $\Omega(\mathbf{K}, \omega)$ is the four-dimensional space-time spectral density of the correlation. In the case where $\tau = 0$, Eq. (D1.3.5) gives just the spatial correlation

$$B(\mathbf{q}) = \int \exp(i\mathbf{K} \cdot \mathbf{q})\Phi(\mathbf{K})\, d\mathbf{K} \tag{D1.3.6}$$

where the spatial spectral density $\Phi(k)$ is given by

$$\Phi(\mathbf{K}) = \int \Omega(\mathbf{K}, \omega)\, d\omega \tag{D1.3.7}$$

Hence, the spatial correlation function and the spectral density are Fourier transform pairs, the inverse transform being given by

$$\Phi(\mathbf{K}) = \left(\frac{1}{2\pi}\right)^3 \int \exp(-i\mathbf{K} \cdot \mathbf{q})B(\mathbf{q})\, d\mathbf{q} \tag{D1.3.8}$$

Similarly, setting $q = 0$ in Eq. (D1.3.5) yields just the temporal correlation

$$B(\tau) = \int \exp(i\omega\tau)W(\omega)\, d\omega \tag{D1.3.9}$$

where the temporal spectral density $W(\omega)$ is given by

$$W(\omega) = \int \Omega(\mathbf{K}, \omega)\, d\mathbf{K} \tag{D1.3.10}$$

As with the spatial correlation and spectral density above, the temporal correlation and associated spectrum are Fourier transform pairs where the inverse transform is

$$W(\omega) = \left(\frac{1}{2\pi}\right) \int \exp(-i\omega\tau)B(\tau)\, d\tau \tag{D1.3.11}$$

In what is to follow in this chapter, it will be found useful to reduce the three-dimensional spatial spectrum of Eq. (D1.3.6) into a two-dimensional spectrum in a plane that is transverse to a preferred direction, i.e., the direction of propagation along the x axis. To this end, expanding the three-dimensional displacement $\mathbf{q}$ into its longitudinal x and transverse $\boldsymbol{\rho}$ components and, similarly, the three-dimensional spatial frequency vector $\mathbf{K}$ into its longitudinal K_x and transverse $\boldsymbol{\kappa}$ components, one can write Eq. (D1.3.6) in the form

$$B(x, \boldsymbol{\rho}) = \iint \exp(i\boldsymbol{\kappa} \cdot \boldsymbol{\rho})F(x, \boldsymbol{\kappa})\, d\boldsymbol{\kappa} \tag{D1.3.12}$$

where $F(x, \kappa)$ is the two-dimensional spectral density given by

$$F(x, \boldsymbol{\kappa}) = \int \exp(iK_x x)\Phi(K_x, \boldsymbol{\kappa})\, dK_x \tag{D1.3.13}$$

In the particular case of δ-function correlation along the x axis, one has from Eq. (1.75)

$$B(x' - x'', \boldsymbol{\rho}' - \boldsymbol{\rho}'') = \delta(x' - x'')A(x', \boldsymbol{\rho}' - \boldsymbol{\rho}'') \qquad \text{(D1.3.14)}$$

with the associated requirement that

$$\int B(x, \boldsymbol{\rho})\, dx = A(x', \boldsymbol{\rho}) \qquad x \equiv x'', \boldsymbol{\rho} \equiv \boldsymbol{\rho}' - \boldsymbol{\rho}'' \qquad \text{(D1.3.15)}$$

Substituting Eqs. (D1.3.12) and (D1.3.13) into the left side of Eq. (D1.3.15), one has

$$\begin{aligned} A(x, \boldsymbol{\rho}) &= \int\int\int \exp(i\boldsymbol{\kappa} \cdot \boldsymbol{\rho}) \exp[iK_x(x' - x'')] \\ &\quad \times \Phi(K_x, \boldsymbol{\kappa})\, dK_x\, d\boldsymbol{\kappa}\, dx'' \\ &= 2\pi \int\int \exp(i\boldsymbol{\kappa} \cdot \boldsymbol{\rho})\delta(K_x)\Phi(K_x, \boldsymbol{\kappa})\, dK_x\, d\boldsymbol{\kappa} \\ &= 2\pi \int \exp(i\boldsymbol{\kappa} \cdot \boldsymbol{\rho})\Phi(\boldsymbol{\kappa})\, d\boldsymbol{\kappa} \qquad \text{(D1.3.16)} \end{aligned}$$

When one has that statistical isotropy prevails, Eqs. (D1.3.6) and (D1.3.8) become independent of the direction of the spatial and spatial frequency vectors and, upon converting them to spherical polar coordinates, respectively become

$$\begin{aligned} B(q) &= \int_0^{2\pi}\int_0^{\pi}\int_0^{\infty} \exp(ikq\cos\theta)\Phi(K)K^2 \sin\theta\, dK\, d\theta\, d\phi \\ &= 2\pi \int_0^{\infty} \left\{\int_{-1}^{1} \exp(iKqx)\, dx\right\} \Phi(K)K^2\, dK \\ &= 4\pi \int_0^{\infty} \frac{\sin(Kq)}{Kq}\, \Phi(K)\, K^2\, dK \qquad \text{(D1.3.17)} \end{aligned}$$

and

$$\begin{aligned} \Phi(K) &= \left(\frac{1}{2\pi}\right)^3 \int_0^{2\pi}\int_0^{\pi}\int_0^{\infty} \exp(-iKq\cos\theta)B(q)q^2 \sin\theta\, dq\, d\theta\, d\phi \\ &= \frac{1}{2\pi^2}\int_0^{\infty} \frac{\sin(Kq)}{Kq} B(q)q^2\, dq \qquad \text{(D1.3.18)} \end{aligned}$$

Similarly, the two-dimensional relation of Eq. (D1.3.16) becomes, in plane polar coordinates,

$$A(x, \boldsymbol{\rho}) = (2\pi)^2 \int_0^{\infty} J_0(\kappa\rho)\Phi(\kappa)\kappa\, d\kappa \qquad \text{(D1.3.19)}$$

upon expanding the exponential function in a Bessel series and having only the "zeroth" order remain after performing the θ integration.

Thus, one can write

$$\tilde{\varepsilon}(x', \boldsymbol{\rho}') = \int_{-\infty}^{\infty} \exp(i\boldsymbol{\kappa}' \cdot \boldsymbol{\rho}')\, d\omega(x', \boldsymbol{\kappa}') \tag{1.118}$$

where the spectral amplitude $d\nu(x', \boldsymbol{\kappa}')$ is taken to be endowed with statistical properties that reproduce those of $\tilde{\varepsilon}(x', \boldsymbol{\rho}')$ through Eq. (1.118). Hence, since $\tilde{\varepsilon}$ is a random function with a zero mean and is also a real function, one has, respectively, using Eq. (1.118),

$$\langle d\nu(x', \boldsymbol{\kappa}')\rangle = 0 \qquad d\nu(x', \boldsymbol{\kappa}') = d\nu^*(x', -\boldsymbol{\kappa}') \tag{1.119}$$

Homogeneity of the permittivity field allows one to write for the two-dimensional spatial correlation function [cf. Eq. (1.75)] $A(x; \boldsymbol{\rho}, \boldsymbol{\rho}') = A(x; \boldsymbol{\rho} - \boldsymbol{\rho}')$. Using Eq. (1.118) to form the three-dimensional correlation function of the δ-correlated permittivity field as given by Eq. (1.75), which is now also taken to be statistically homogeneous, one has

$$\begin{aligned}\langle \tilde{\varepsilon}(x, \boldsymbol{\rho})\tilde{\varepsilon}(x', \boldsymbol{\rho}')\rangle &= \delta(x - x')A(x, \boldsymbol{\rho} - \boldsymbol{\rho}') \\ &= \int_{-\infty}^{\infty}\int_{-\infty}^{\infty} \exp(i\boldsymbol{\kappa} \cdot \boldsymbol{\rho} + i\boldsymbol{\kappa}' \cdot \boldsymbol{\rho}')\langle d\nu(x, \boldsymbol{\kappa})\, d\nu(x', \boldsymbol{\kappa}')\rangle\end{aligned} \tag{1.120}$$

In order to get the integrand to reflect the δ-correlated homogeneous statistics of $\tilde{\varepsilon}$, the following condition must prevail:

$$\langle d\nu(x, \boldsymbol{\kappa})\, d\nu(x', \boldsymbol{\kappa}')\rangle = \delta(x - x')\delta(\boldsymbol{\kappa} + \boldsymbol{\kappa}')F_\varepsilon(x, \boldsymbol{\kappa})\, d\boldsymbol{\kappa}\, d\boldsymbol{\kappa}' \tag{1.121}$$

where $F_\varepsilon(x, \kappa)$ is the two-dimensional spectral density of the correlation. Using Eq. (1.121) in Eq. (1.120) and solving for the spectral density via the inverse Fourier transform, one has

$$F_\varepsilon(x, \boldsymbol{\kappa}) = \left(\frac{1}{2\pi}\right)^2 \int_{-\infty}^{\infty} A(x, \boldsymbol{\rho}_d) \exp(-i\boldsymbol{\kappa} \cdot \boldsymbol{\rho}_d)\, d^2\rho_d \tag{1.122}$$

Substituting Eq. (1.118) into Eq. (1.117) and performing the ρ' integration yields

$$\psi_1(x, \boldsymbol{\rho}) = \frac{ik}{2}\int_0^x \int_{-\infty}^{\infty} \exp(i\gamma\boldsymbol{\kappa}' \cdot \boldsymbol{\rho})H(x, x', \boldsymbol{\kappa}')\, d\nu(x', \boldsymbol{\kappa}')\, dx' \tag{1.123}$$

where $$H(x, x', \boldsymbol{\kappa}) \equiv \exp\left[-\frac{i\kappa^2}{2k}\gamma(x')(x - x')\right] \tag{1.124}$$

From application of Eq. (1.123), and employing Eq. (1.121), one has

$$\begin{aligned}
&\langle\psi_1(x, \boldsymbol{\rho}_1)\psi_1(x, \boldsymbol{\rho}_2)\rangle \\
&= -\frac{k^2}{4}\int_0^x\int_{-\infty}^{\infty}\int_0^x\int_{-\infty}^{\infty}\exp\{i[\gamma(x')\boldsymbol{\kappa}'\cdot\boldsymbol{\rho}_1 + \gamma(x'')\boldsymbol{\kappa}''\cdot\boldsymbol{\rho}_2]\} \\
&\quad\times H(x, x', \boldsymbol{\kappa}')H(x, x'', \boldsymbol{\kappa}'')\langle d\nu(x', \boldsymbol{\kappa}')\, d\nu(x'', \boldsymbol{\kappa}'')\rangle\, dx'\, dx'' \\
&= -\frac{k^2}{4}\int_0^x\int_{-\infty}^{\infty}\exp(i\boldsymbol{\kappa}'\cdot\mathbf{Q}) \\
&\quad\times H^2(x, x', \boldsymbol{\kappa}')F_\varepsilon(x', \boldsymbol{\kappa}')\, d\kappa'\, dx'
\end{aligned} \tag{1.125}$$

where $\mathbf{Q} \equiv \gamma(x')(\boldsymbol{\rho}_1 - \boldsymbol{\rho}_2)$. Similarly, one has

$$\begin{aligned}
\langle\psi_1(x, \boldsymbol{\rho}_1)\psi_1^*(x, \boldsymbol{\rho}_2)\rangle &= \frac{k^2}{4}\int_0^x\int_{-\infty}^{\infty}\int_0^x\int_{-\infty}^{\infty}\exp\{i[\gamma(x')\boldsymbol{\kappa}'\cdot\boldsymbol{\rho}_1 - \gamma(x'')\boldsymbol{\kappa}''\cdot\boldsymbol{\rho}_2]\} \\
&\quad\times H(x, x', \boldsymbol{\kappa}')H^*(x, x'', \boldsymbol{\kappa}'') \\
&\qquad\qquad\times\langle d\nu(x', \boldsymbol{\kappa}')\, d\nu^*(x'', \boldsymbol{\kappa}'')\rangle\, dx'\, dx'' \\
&= \frac{k^2}{4}\int_0^x\int_{-\infty}^{\infty}\exp(i\boldsymbol{\kappa}'\cdot\mathbf{Q}) \\
&\quad\times|H(x, x', \boldsymbol{\kappa}')|^2F_\varepsilon(x', \boldsymbol{\kappa}')\, d\kappa'\, dx'
\end{aligned} \tag{1.126}$$

where the second relation of Eq. (1.119) was used to obtain the form required for the use of Eq. (1.121) as well as the fact that $H(x, x', \boldsymbol{\kappa})$ is an even function of $\boldsymbol{\kappa}$.

One can finally write for the log-amplitude and phase correlation functions of Eqs. (1.113) and (1.114) applied in the observation plane located at $x = L$

$$\begin{aligned}
B_{\substack{\chi\\S}}(L, \boldsymbol{\rho}_1, \boldsymbol{\rho}_2) &= B_{\substack{\chi\\S}}(L, \boldsymbol{\rho}_1 - \boldsymbol{\rho}_2) \\
&= \frac{1}{8}k^2\int_0^L\int_{-\infty}^{\infty}\mathrm{Re}\,\{[|H(L, x', \boldsymbol{\kappa}')|^2 \\
&\quad\mp H^2(L, x', \boldsymbol{\kappa}')]\exp(i\boldsymbol{\kappa}'\cdot\mathbf{Q})\}F_\varepsilon(x', \boldsymbol{\kappa}')\, d^2\kappa'\, dx'
\end{aligned} \tag{1.127}$$

where the functional dependence of $\mathbf{Q}$ on $\boldsymbol{\rho}_1 - \boldsymbol{\rho}_2$ is noted. Hence, these correlation functions are solely functions of the difference $\boldsymbol{\rho}_1 - \boldsymbol{\rho}_2$, which, by definition, holds for a statistically homogeneous field. In the same way, the associated variances $\sigma^2(L) \equiv B(L, 0)$ are independent of position in the field.

An affectation of using the spectral expansion technique in the analysis that led to Eq. (1.127) is that there are implicit requirements placed on the spectral density $F_\varepsilon(x, \boldsymbol{\kappa})$. In particular, it is assumed that Eq. (1.127) converges for all values of spatial frequencies $\boldsymbol{\kappa}$. However, there are several situations, including that of the atmospheric permittivity field, where integrals involving the spectral density diverge at small spatial frequencies. This is simply a manifestation of the fact that statistical homogeneity does not hold at such correspondingly large spatial separations. However, the existence of such integrands at larger spatial frequencies implies that homogeneity prevails over the smaller spatial separations. Thus, even though "global" homogeneity does not exist for these situations, "local" homogeneity does prevail.

Digression 1.4: Locally Homogeneous Random Fields The concepts of stationarity and homogeneity introduced in Digressions 1.1 and 1.3 above, although allowing immense simplification of the expressions that result, are nevertheless oversimplifications for many of the random space-time fields that are encountered in nature. In particular, the troposphere of the earth is neither stationary nor homogeneous; diurnal fluctuations are reflected in every meteorological field that is defined (e.g., temperature, humidity, etc.), and the thermal boundary represented by the earth's irregular surface features induces spatial variations of these meteorological fields, particularly with respect to height above the surface. Thus, the "global" stationarity and homogeneity assumptions used in the digressions above are not realized when considering the random meteorological fields of the earth. This situation can present several inconsistencies in the modeling of such situations, especially those where one employs the ergodic hypothesis with its implicit requirement of stationarity and homogeneity.

To allow for this large-scale nonstationarity and inhomogeneity, one resorts to the use of the concept of space-time random fields that possess *stationary and homogeneous increments,* i.e., where the space-time gradient of the random field is a stationary and homogeneous random function, rather than the random field itself. Thus, the global stationarity and homogeneity considered earlier is a special case of this more general concept of incremental stationarity and homogeneity.

Consider a random function of position $u(\mathbf{r})$; from what was just said above, such a field has homogeneous increments (also, for reasons that will become obvious, it is said to be *locally* homogeneous), if for any two positions $\mathbf{r}$ and $\mathbf{q}$ within the field, one has

$$\langle u(\mathbf{r}') - u(\mathbf{r})\rangle = \langle u(\mathbf{r} + \mathbf{q}) - u(\mathbf{r})\rangle = m(\mathbf{q}) \tag{D1.4.1}$$

where $\mathbf{q} \equiv \mathbf{r}' - \mathbf{r}$ and the position-dependent mean m satisfies

$$\nabla_q m(\mathbf{q}) = 0 \tag{D1.4.2}$$

Thus, Eq. (D1.4.2) demands that, in general,

$$m(\mathbf{q}) = \mathbf{c} \cdot \mathbf{q} \tag{D1.4.3}$$

where c is a vector constant. This constant can be calculated in principle for a particular random field by forming the spatial averages of the difference

$$\nabla_{\mathbf{q}} u(\mathbf{r}) = u(\mathbf{r} + \mathbf{q}) - u(\mathbf{r}) \tag{D1.4.4}$$

with respect to r. Once the constant vector c has been determined, one can then subtract the function $\mathbf{c} \cdot \mathbf{r}$ from the field $u(\mathbf{r})$ to yield, if desired, a transformed field $u(\mathbf{r}) - \mathbf{c} \cdot \mathbf{r}$ with a zero mean. Since it is the increment and not the field itself that is subject to homogeneity conditions, one can proceed to consider the second moment of these increments, i.e.,

$$D_u(\mathbf{r}_1, \mathbf{q}_1, \mathbf{r}_2, \mathbf{q}_2) \equiv \langle [u(\mathbf{r}_1 + \mathbf{q}_1) - u(\mathbf{r}_1)][u(\mathbf{r}_2 + \mathbf{q}_2) - u(\mathbf{r}_2)] \rangle \tag{D1.4.5}$$

Simplifying this expression using the identity

$$(a - b)(c - d) = \tfrac{1}{2}[(a - d)^2 + (b - c)^2 + (a - c)^2 + (b - d)^2]$$

one obtains

$$\begin{aligned} D_u(\mathbf{r}_1, \mathbf{q}_1, \mathbf{r}_2, \mathbf{q}_2) = \tfrac{1}{2}[&D_u(\mathbf{r}_1 + \mathbf{q}_1 - \mathbf{r}_2) + D_u(\mathbf{r}_1 + \mathbf{q}_2 - \mathbf{r}_2) \\ &- D_u(\mathbf{r}_1 + \mathbf{q}_1 - \mathbf{r}_2 - \mathbf{q}_2) - D_u(\mathbf{r}_1 - \mathbf{r}_2)] \end{aligned} \tag{D1.4.6}$$

where the quantities

$$D_u(\mathbf{r}_1 + \mathbf{q}_1 - \mathbf{r}_2) \equiv \langle [u(\mathbf{r}_1 + \mathbf{q}_1) - u(\mathbf{r}_2)]^2 \rangle \tag{D1.4.7}$$

etc.

are called *structure functions* of the field $u(\mathbf{r})$ and, as per the discussion above, are functions of the differences of the coordinates at which the field is considered. By their very nature, structure functions contain less information about the random field than do correlation functions; in fact, in situations where global homogeneity prevails, one can express a structure function in terms of the correlation functions, viz.,

$$\begin{aligned} D_u(\mathbf{q}) &= \langle [u(\mathbf{r} + \mathbf{q}) - u(\mathbf{r})]^2 \rangle \\ &= B_u(\mathbf{r} + \mathbf{q}, \mathbf{r} + \mathbf{q}) + B_u(\mathbf{r}, \mathbf{r}) - 2B_u(\mathbf{r} + \mathbf{q}, \mathbf{r}) \\ &= 2B_u(0) - 2B_u(\mathbf{q}) \end{aligned} \tag{D1.4.8}$$

This relationship demonstrates that the structure function is bounded, $|D_u(\mathbf{q})| \leq 4B_u(0)$. More importantly, however, this indicates how, once the correlation function for the homogeneous random field is determined, one can find the corresponding structure function. Proceeding further, one can then lift the restriction of homogeneity and consider the same field to be locally homogeneous; the correlation functions are then undefined, but the structure function which corresponded to them and was calculated from them in the homogeneous case is still a valid statistical parameter. Thus, any statistical expressions that involve the correlation functions can be employed in Eq. (D1.4.8) to yield the corresponding expressions in the more general locally homogeneous case.

In particular, and most importantly, one can define and obtain expressions for the spectral density that corresponds to structure functions using the results of Digression 1.3. Using Eq. (D1.3.6) in Eq. (D1.4.8), one obtains

$$D_u(\mathbf{q}) = 2 \int [1 - \exp (i\mathbf{K} \cdot \mathbf{q})]\Phi(\mathbf{K})\, d\mathbf{K} \tag{D1.4.9}$$

which relates the spectral density $\Phi(\mathbf{K})$ to the structure function. Thus, one can still define a spectral density for a locally homogeneous field which is identical to that for the correlation function of a globally homogeneous random field. The determination of the equation for the spectral density in terms of the structure function commences by applying the laplacian operator to both sides of Eq. (D1.4.9), which yields a Fourier transform relationship that has as its inverse

$$\Phi(\mathbf{K}) = \frac{1}{16\pi^3\mathbf{K}^2} \int \exp (-i\mathbf{K} \cdot \mathbf{q})\nabla_q^2 D_u(\mathbf{q})\, d\mathbf{q} \tag{D1.4.10}$$

To make contact with the two-dimensional spectrum $F(x, \boldsymbol{\kappa})$ defined in Eqs. (D1.3.12) and (D1.3.13), one expands the coordinates in Eq. (D1.4.8) into longitudinal and transverse coordinates:

$$D_u(x, \boldsymbol{\rho}) = 2B_u(0, 0) - 2B_u(x, \boldsymbol{\rho}) \tag{D1.4.11}$$

It will be convenient to get rid of the first term on the right side of this relation and replace it with quantities that are functions of at least one of the coordinates. This is simply done by applying Eq. (D1.4.11) to the case where $\boldsymbol{\rho} = 0$, solving for $B_u(0, 0)$, and substituting the resulting expression back into Eq. (D1.4.11). This gives

$$D_u(x, \boldsymbol{\rho}) = D_u(x, 0) + 2[B_u(x, 0) - B_u(x, \boldsymbol{\rho})]$$

Substituting Eq. (D1.3.12) into this expression yields

$$D_u(x, \boldsymbol{\rho}) = D_u(x, 0) + 2 \int [1 - \exp (i\boldsymbol{\kappa} \cdot \boldsymbol{\rho})]F(x, \boldsymbol{\kappa})\, d\boldsymbol{\kappa} \tag{D1.4.12}$$

relating the structure function to the corresponding two-dimensional spectrum.

In the case of isotropic field statistics (i.e., *locally isotropic* field statistics), one obtains from Eqs. (D1.4.9) and (D1.4.10), using the same procedures as those used in the similar consideration in Digression 1.3,

$$D_u(q) = 8\pi \int_0^\infty \left[1 - \frac{\sin (Kq)}{(Kq)}\right] \Phi(K)\, K^2\, dK \tag{D1.4.13}$$

and

$$\Phi(K) = \frac{1}{4\pi^2K^2} \int_0^\infty \frac{\sin (Kq)}{(Kq)} \frac{d}{dq} \left[q^2 \frac{d}{dq} D_u(q)\right] dq \tag{D1.4.14}$$

where the radial part of the laplacian in spherical coordinates was employed to obtain the last relation. Similarly, Eq. (D1.4.12) reduces to

$$D_u(x, \boldsymbol{\rho}) = D_u(x, 0) + 4\pi \int_0^\infty [1 - J_0(\kappa\rho)]F(x, \boldsymbol{\kappa})\kappa\, d\kappa \tag{D1.4.15}$$

It is straightforward to adopt this concept to two or more such random fields, e.g.,

$$D_{uv}(\mathbf{r}_1, \mathbf{q}_1, \mathbf{r}_2, \mathbf{q}_2) \equiv \langle [u(\mathbf{r}_1 + \mathbf{q}_1) - u(\mathbf{r}_1)][v(\mathbf{r}_2 + \mathbf{q}_2) - v(\mathbf{r}_2)] \rangle$$

giving rise to structure functions of the type

$$D_{uv}(\mathbf{r}_1 + \mathbf{q}_1 - \mathbf{r}_2) \equiv \langle [u(\mathbf{r}_1 + \mathbf{q}_1) - v(\mathbf{r}_2)]^2 \rangle$$

etc.

Taking $\tilde{\varepsilon}(x, \boldsymbol{\rho})$ [and thus, subsequently, $\chi(x, \boldsymbol{\rho})$ and $S(x, \boldsymbol{\rho})$] to be a locally homogeneous random field, one can define log-amplitude and phase structure functions given by

$$\begin{aligned} D_\chi(L, \boldsymbol{\rho}_1, \boldsymbol{\rho}_2) &= \langle [\chi(L, \boldsymbol{\rho}_1) - \chi(L, \boldsymbol{\rho}_2)]^2 \rangle \\ &= 2\sigma_\chi^2(L) - 2B_\chi(L, \boldsymbol{\rho}_1, \boldsymbol{\rho}_2) \end{aligned} \tag{1.128}$$

and similarly for $D_S(L, \boldsymbol{\rho}_1, \boldsymbol{\rho}_2)$. One thus obtains, using Eq. (1.128),

$$\begin{aligned} D_{\substack{\chi \\ S}}(L, \boldsymbol{\rho}_1, \boldsymbol{\rho}_2) &= D_{\substack{\chi \\ S}}(L, \boldsymbol{\rho}_1 - \boldsymbol{\rho}_2) \\ &= \frac{k^2}{4} \int_0^L \int_{-\infty}^{\infty} \mathrm{Re}\{[|H(L, x', \boldsymbol{\kappa}')|^2 \mp H^2(L, x', \boldsymbol{\kappa}')] \\ &\quad \times [1 - \exp(i\boldsymbol{\kappa}' \cdot \mathbf{Q})]\} F_\varepsilon(x', \boldsymbol{\kappa}')\, d^2\kappa'\, dx' \end{aligned} \tag{1.129}$$

As will be shown in the next section, the use of these results coupled with the result of Eq. (1.93) of the last section leads to a very transparent and well-known theoretical formulation of random wave propagation. However, there is a very important aspect of the use of the Rytov approximation that deserves some mention, as it has led to considerable conceptual difficulty in the past. This has to do with the derivation of the second-order moment of the electric field using the first Rytov approximation. Consider the expression for the second-order moment obtained from using Eq. (1.107), written for the first Rytov approximation, in Eq. (1.61), written for the second moment:

$$\Gamma_{11}(L; \boldsymbol{\rho}_1, \boldsymbol{\rho}_2) = U_0(L, \boldsymbol{\rho}_1) U_0^*(L, \boldsymbol{\rho}_2) \langle \exp[\psi_1(L, \boldsymbol{\rho}_1) + \psi_1^*(L, \boldsymbol{\rho}_2)] \rangle \tag{1.130}$$

At this point, one needs to know, or at least assume, the probability distribution that governs the complex phase so that the ensemble average of the exponential may be evaluated. [One could, in principle, write Eq. (1.130) in terms of characteristic functionals (see, e.g., Ref. 16), although this approach has never been used in the applications of this method.] Instead, the statistics that govern the statistics of

$\psi_1(L, \boldsymbol{\rho})$ are assumed to be gaussian based on the following plausibility argument. As stated briefly in Sec. 1.4.1 and covered more thoroughly in Chap. 2, the largest spatial scale over which the random field is correlated is characterized by a length L_0. The fluctuating permittivity field is then essentially uncorrelated over distances that exceed L_0. Now if the length L of the total propagation path is such that $L \gg L_0$, so that the propagation path traverses many such "correlation cells," i.e., $L/L_0 \gg 1$, then the longitudinal integration appearing in Eq. (1.104) will subtend several regions of uncorrelated fluctuations of $\tilde{\varepsilon}$. By the well-known central limit theorem, the resulting derived random variable will be governed by gaussian statistics. (It is crucial to note that such an assumption only applies to the complex phase which results from integration over a large spatial interval in the permittivity field; this statement is independent of any assumption regarding the statistics governing the spatial or temporal variation of the permittivity field itself.)

Thus, integrating the exponential in Eq. (1.130) over a gaussian probability density yields

$$\langle \exp(\zeta) \rangle = \exp\left[\tfrac{1}{2}\langle(\zeta - \langle\zeta\rangle)^2\rangle + \langle\zeta\rangle\right] \tag{1.131}$$

where $\zeta \equiv \psi_1(L, \boldsymbol{\rho}_1) + \psi_1^*(L, \boldsymbol{\rho}_2)$. Using Eq. (1.108) for the special case of $i = 1$, assuming that local homogeneity prevails over the transverse lengths that are considered (this allows the means $\langle\chi\rangle$ and $\langle S\rangle$ to be constants), and further, assuming that the statistics of $\psi_1(L, \boldsymbol{\rho})$ are also locally isotropic (this allows the cross-correlations $\langle\chi S\rangle$ to vanish[19]), one can write

$$\begin{aligned}\Gamma_{11}(L; \boldsymbol{\rho}_1, \boldsymbol{\rho}_2) &= \Gamma_{11}(L; \boldsymbol{\rho}_1 - \boldsymbol{\rho}_2) \\ &= U_0(L, \boldsymbol{\rho}_1)U_0^*(L, \boldsymbol{\rho}_2) \exp\{-\tfrac{1}{2}D_\chi(L, \boldsymbol{\rho}_1 - \boldsymbol{\rho}_2) \\ &\quad - \tfrac{1}{2}D_S(L, \boldsymbol{\rho}_1 - \boldsymbol{\rho}_2) + 2[\sigma_\chi^2(L) - \langle\chi\rangle^2 + \langle\chi\rangle]\}\end{aligned} \tag{1.132}$$

At this point, it is now noted that to within the accuracy of the first Rytov approximation [i.e., Eq. (1.104)], one should have $\langle\chi\rangle = 0$, since all odd moments of a gaussian random function vanish; a nonzero average log-amplitude can only be obtained from the solution of the second Rytov approximation, Eq. (1.105), since this can be seen to employ the second moment of $\tilde{\varepsilon}$. However, if one considers the special case of Eq. (1.132) where $\boldsymbol{\rho}_1 = \boldsymbol{\rho}_2 \equiv \boldsymbol{\rho}$, i.e., considers the intensity $I(L) = \Gamma_{11}(L)$ of the field, and applies the conservation of energy (noting that no absorption has been admitted into the problem), one obtains the requirement that $\exp[2\sigma_\chi^2(L)] = 1$, which demands the inconsistent condition $\sigma_\chi^2(L) = 0$. The usual way out of this difficulty is to assume that higher-order Rytov approximations are needed to substantiate the fact that $\langle\chi\rangle \neq 0$ and indeed that $\sigma_\chi^2(L) - \langle\chi\rangle^2 + \langle\chi\rangle = 0$. However, as has been previously pointed out,[20] the perturbation expansion applied to obtain Eqs. (1.104), (1.105), etc., is done from the point of view of the "order of smallness" of the complex amplitude $\psi(L, \boldsymbol{\rho})$ with respect to the perturbation parameter $\tilde{\varepsilon}$ and not from the point of view of the statistical parameters that can be formed from such complex amplitudes. From the point of view of the statistical parameters, the quantity $\langle\chi\rangle$, which must be found by solution of the second Rytov approximation ψ_2, is of the same order of magnitude, i.e., $\sim|\tilde{\varepsilon}|^2$, as σ_χ^2, which can be obtained from the first Rytov approximation ψ_1. By the same reasoning, $\langle\chi\rangle^2$ is of the order of magnitude $\sim|\tilde{\varepsilon}|^4$. Thus, to within an approximation of accuracy $\sim|\tilde{\varepsilon}|^2$, rather than to within the first Rytov approximation, one must retain $\langle\chi\rangle$ in a solution but can dispense with $\langle\chi\rangle^2$. Hence, the abovementioned condition imposed by the conservation of energy becomes, with accuracy $\sim|\tilde{\varepsilon}|^2$, $\sigma_\chi^2(L) + \langle\chi\rangle = 0$, with the obvious proviso that $\langle\chi\rangle^2 \ll \langle\chi\rangle$, i.e., $|\langle\chi\rangle| \ll 1$. Indeed, an analytical solution obtained for the average log-amplitude[20] yields the fact that, for plane and spherical wave propagation, $\sigma_\chi^2(L) = -\langle\chi\rangle$. Also, since the condition $|\langle\chi\rangle| \ll 1$ is assumed to prevail, one is immediately provided with the well-known restriction

$$|\langle\chi\rangle| = \sigma_\chi^2(L) \ll 1 \tag{1.133}$$

governing the use of the first *and second* Rytov solutions to the accuracy $\sim|\tilde{\varepsilon}|^2$. These circumstances allow Eq. (1.132) to be written

$$\Gamma_{11}(L; \boldsymbol{\rho}_1 - \boldsymbol{\rho}_2) = U_0(L, \boldsymbol{\rho}_1)U_0^*(L, \boldsymbol{\rho}_2) \exp\left[-\tfrac{1}{2}D_\phi(L, \boldsymbol{\rho}_1 - \boldsymbol{\rho}_2)\right] \tag{1.134}$$

where $D_\phi(L, \boldsymbol{\rho}_1 - \boldsymbol{\rho}_2)$ is the *wave structure function* defined and given by

$$\begin{aligned} D_\phi(L, \boldsymbol{\rho}_1 - \boldsymbol{\rho}_2) &\equiv D_\chi(L, \boldsymbol{\rho}_1 - \boldsymbol{\rho}_2) + D_S(L, \boldsymbol{\rho}_1 - \boldsymbol{\rho}_2) \\ &= \frac{k^2}{4}\int_0^L\int_{-\infty}^{\infty} \mathrm{Re}\,\{|H(L, x', \boldsymbol{\kappa}')|^2[1 - \exp(i\boldsymbol{\kappa}' \cdot \mathbf{Q})]\} \\ &\qquad \times F_\varepsilon(x', \boldsymbol{\kappa}')\, d^2\kappa'\, dx' \end{aligned} \tag{1.135}$$

where use was made of Eq. (1.129).

Thus, in addition to the constraints that prevail over the use of the parabolic wave equation from which Eq. (1.96) derives through the transformation of Eq. (1.95), the use of Eqs. (1.127), (1.129), and (1.135) is limited by the level of the log-amplitude variance as given by Eq. (1.133). Hence, this formulation is limited to weak fluctuation phenomena in which Eq. (1.133) obtains; it is this restriction, along with the complexity attendant in obtaining higher-order approximations and/or higher-order moments of the field, that limits the application of this method in favor of that using the parabolic equation, where no restriction is placed directly on the level of field fluctuations.

1.7 The Extended Huygens-Fresnel Principle and Its Phase Approximation

As was mentioned in the discussion following it, Eq. (1.93) is of the form that allows a transparent interpretation of the wave propagation process through random media and takes the form of a modified version of the well-known Huygens-Fresnel principle which is useful in understanding propagation (i.e., diffraction propagation) in deterministic media. In fact, the application of the Huygens-Fresnel principle to stochastic wave propagation problems was first heuristicly formulated by Z. I. Feizulin and Yu. A. Kravtsov in the Soviet Union[21,22] and later "rediscovered" by R. F. Lutomirski and H. T. Yura in the United States.[23] In what is to follow, various versions of this "extended" Huygens-Fresnel principle will be directly derived from the foregoing. In particular, the Rytov transformation and the associated developments of Sec. 1.6 will be used with the solution of Eq. (1.93). It will then be shown how a suggestive approximation that can be employed, i.e., the "phase" approximation where the log-amplitude is neglected and all fluctuations of the wave field are relegated to phase fluctuations, will lead to a formulation that lends itself as a basis for development of adaptive compensation methods but, at the same time, can give erroneous results if not properly applied to the calculation of the statistical moments of the field.

1.7.1 Variations of the extended Huygens-Fresnel principle

Applying the Rytov transformation Eq. (1.95) to the field $G_{\tilde{\varepsilon}}$ given by Eq. (1.94), noting that this latter equation is the conjugate of the original relation [i.e., Eq. (1.34)] to which Eq. (1.95) was applied in Sec. 1.6, one obtains from Eq. (1.93)

$$U(L, \boldsymbol{\rho}) = \frac{ik}{2\pi L} \int_{-\infty}^{\infty} \exp\left[-i \frac{k(\boldsymbol{\rho} - \boldsymbol{\rho}')^2}{2L} + \sum_{i=1}^{\infty} \psi_i^* \right] U_0(\boldsymbol{\rho}')\, d^2\rho' \tag{1.136}$$

where, when applying the perturbation expansion to the conjugate of the complex phase ψ^*, it was noted that $G_{\tilde{\varepsilon}}$ is taken to describe contributions to propagation due *only* to the random variations of the permittivity, thus necessitating the exclusion of the ψ^* term. Equation

(1.136) is the same result that obtains from application of the Huygens-Fresnel principle *extended* to include variations in the propagation of the spherical wavelets induced by the random permittivity field through the appearance of the ψ_i^* terms.

The use in Eq. (1.136) of just the first Rytov approximation ψ_i^*, given by the conjugate of Eq. (1.104), results in the classical form of the extended Huygens-Fresnel principle that was taken as the heuristic starting point of several investigations dealing with stochastic wave propagation.[21–23] The accuracy of the extended Huygens-Fresnel principle using the first Rytov approximation is the same as that of the first approximation itself, i.e., it can be applied to situations where $\sigma_\chi^2 \ll 1$.

Consider for the moment this limitation on the log-amplitude due to the use of only the first Rytov approximation. Because the limitation is placed only on the level of variations of the log-amplitude and *not* on the phase fluctuation S_1, one is motivated to consider the possibility that the phase statistics calculated via the first approximation hold for a range which exceeds that over which $\sigma_\chi^2 \ll 1$. This is in fact the case, as has been demonstrated by theory[24,25] and experiment.[26] Given this development, it is then suggested that one should simply consider using in Eq. (1.136) the approximation

$$\psi_1^* = \chi - iS_1 \approx -iS_1 \tag{1.137}$$

which is also rather appealing since it gives the equation the essence that the Huygens-Fresnel principle conveys. However, the use of the *phase approximation,* Eq. (1.137), in Eq. (1.136) is not entirely justified on the basis of the derivation of this latter relation from the results of Sec. 1.5. In fact, as will be discussed below, this circumstance can, and has, lead to erroneous results in calculation of the field moments.

1.7.2 The phase approximation of the extended Huygens-Fresnel principle

Implicit in the use of the phase approximation is the stipulation that $\chi \approx 0$. The ramifications of this can be gleaned from substituting $\psi(\mathbf{r}) = \chi + iS_1$ into Eq. (1.96), performing the required operations, and separating out the real and imaginary parts. This yields

$$2k\frac{\partial \chi}{\partial x} + \nabla_\rho^2 S + 2(\nabla_\rho \chi)\cdot(\nabla_\rho S) = 0 \tag{1.138a}$$

$$2k\frac{\partial S}{\partial x} + \nabla_\rho^2 \chi + (\nabla_\rho \chi)^2 - (\nabla_\rho S)^2 + k^2\tilde{\varepsilon} = 0 \tag{1.138b}$$

Considering the phase approximation as discussed above, where one implicitly has $\chi \approx \nabla_\rho \chi \approx \nabla_\rho^2 \chi \approx 0$, these relations become

$$\nabla_\rho^2 S = 0 \tag{1.139a}$$

$$2k \frac{\partial S}{\partial x} - (\nabla_\rho S)^2 + k^2 \tilde{\varepsilon} = 0 \tag{1.139b}$$

However, noting from Eq. (1.139*a*) that $\nabla_\rho S$ is a constant, one can, without loss of generality, set it to zero, thus allowing Eq. (1.139*b*) to be written

$$2k \frac{\partial S}{\partial x} + k^2 \tilde{\varepsilon} = 0 \tag{1.140}$$

Hence, by the neglect of the log-amplitude, one implicitly has in this phase approximation Eqs. (1.139*a*) and (1.140) governing the phase field and not the general equations of Eq. (1.138).

The same situation is obtained if one uses the representation $G_{\tilde{\varepsilon}} = \chi - iS$ in the parabolic equation of Eq. (1.94) with the stipulation that

$$\nabla_{\rho'}^2 G_{\tilde{\varepsilon}} = 0 \tag{1.141}$$

The real and imaginary parts of this requirement yield the relations

$$\nabla_{\rho'}^2 \chi + (\nabla_{\rho'} \chi)^2 - (\nabla_{\rho'} S)^2 = 0$$

$$\nabla_{\rho'}^2 S + 2(\nabla_{\rho'} \chi) \cdot (\nabla_{\rho'} S) = 0$$

and the remaining portion of Eq. (1.94) gives Eq. (1.140). Hence, the use of the condition of Eq. (1.141) with Eqs. (1.93) and (1.94) is analogous to the use of the phase approximation, Eq. (1.137), in the extended Huygens-Fresnel principle of Eq. (1.136), viz.,

$$U(L, \boldsymbol{\rho}) = \frac{ik}{2\pi L} \int_{-\infty}^{\infty} \exp \left[-i \frac{k(\boldsymbol{\rho} - \boldsymbol{\rho}')^2}{2L} - iS_1(\mathbf{L}, \boldsymbol{\rho}; 0, \boldsymbol{\rho}') \right] U_0(\boldsymbol{\rho}')\, d^2\rho' \tag{1.142}$$

One can now perform the instructive task of ascertaining the impact of the approximate relation Eq. (1.142) on the results derived from Eqs. (1.93) and (1.94) for arbitrary moments of the wave field.[27–29]

To this end, it will be expedient to employ the general representation of Eqs. (1.56) and (1.60) at the outset of the analysis which is to follow, since this deals with the "full" form of the Green function $G(x, \boldsymbol{\rho}; x',$

$\boldsymbol{\rho}'$) and not the specialized form represented by Eqs. (1.93) and (1.94). In particular, consider Eq. (1.68), which describes the general product g_{nm} of the random wave field, i.e.,

$$-2ik\,\frac{\partial g_{nm}}{\partial x'} + \left(\sum_{i=1}^{n} \nabla^2_{\rho_i'} - \sum_{j=n+1}^{n+m} \nabla^2_{\rho_j'}\right) g_{nm} + k^2 Q g_{nm} = 0 \quad (1.143)$$

where $g_{nm}(x - x'; \{\boldsymbol{\rho}_i - \boldsymbol{\rho}_i'\}, \{\boldsymbol{\rho}_j - \boldsymbol{\rho}_j'\})$

$$= \prod_{i=1}^{n} G(x, \boldsymbol{\rho}_i; x', \boldsymbol{\rho}_i') \prod_{j=n+1}^{m+n} G^*(x, \boldsymbol{\rho}_j; x', \boldsymbol{\rho}_j')$$

as originally defined by Eq. (1.66) in a slightly condensed notation. The extended Huygens-Fresnel principle for the *nm*th moment, as would be obtained from the application of Eq. (1.93), can be obtained from Eq. (1.143) by commencing with the prescription

$$g_{nm} = g_{0_{nm}} g_{\varepsilon_{nm}} \quad (1.144)$$

where the product of the unperturbed field moments is simply given by Eq. (1.143) with $Q = 0$, i.e.,

$$-2ik\,\frac{\partial g_{0_{nm}}}{\partial x'} + \left(\sum_{i=1}^{n} \nabla^2_{\rho_i'} - \sum_{j=n+1}^{n+m} \nabla^2_{\rho_j'}\right) g_{0_{nm}} = 0 \quad (1.145)$$

and where $g_{\varepsilon_{nm}}$ is the spherical wave solution of Eq. (1.143). However, the phase approximation $g_{\varepsilon S_{nm}}$ of this formulation of the extended Huygens-Fresnel principle is, as per the discussion in the last paragraph, given by Eq. (1.143) under the condition that

$$\left(\sum_{i=1}^{n} \nabla^2_{\rho_i'} - \sum_{j=n+1}^{n+m} \nabla^2_{\rho_j'}\right) g_{\varepsilon S_{nm}} = 0 \quad (1.146)$$

Substituting the phase approximation of Eq. (1.144) into Eq. (1.143) and using Eq. (1.146) yields, after rearrangement of terms,

$$-2ik\,\frac{\partial (g_{0_{nm}} g_{\varepsilon S_{nm}})}{\partial x'} + g_{\varepsilon S_{nm}} \left(\sum_{i=1}^{n} \nabla^2_{\rho_i'} - \sum_{j=n+1}^{n+m} \nabla^2_{\rho_j'}\right) g_{0_{nm}}$$
$$+ 2 \sum_{i=1}^{n} (\nabla_{\rho_i'} g_{\varepsilon S_{nm}}) \cdot (\nabla_{\rho_i'} g_{0_{nm}}) - 2 \sum_{j=n+1}^{n+m} (\nabla_{\rho_j'} g_{\varepsilon S_{nm}}) \cdot (\nabla_{\rho_j'} g_{0_{nm}})$$
$$+ k^2 Q (g_{0_{nm}} g_{\varepsilon S_{nm}}) = 0 \quad (1.147)$$

As was done in the derivation leading to Eq. (1.94), the third and fourth terms of Eq. (1.147) can be neglected in terms of their physical significance. However, their presence is required for the structural integrity of Eq. (1.147) that will be required in what is to follow in the next step of the analysis. This next step entails ascertaining the difference between the exact function g_{nm} and its extended Huygens-Fresnel form $g_{0_{nm}} g_{\varepsilon S_{nm}}$. Thus, subtracting Eq. (1.147) from the exact relation of Eq. (1.143), "readding" the term

$$-g_{0_{nm}} \left(\sum_{i=1}^{n} \nabla^2_{\rho'_i} - \sum_{j=n+1}^{n+m} \nabla^2_{\rho'_j} \right) g_{\varepsilon S_{nm}}$$

to both sides of the result, and, once again, rearranging terms, one finally obtains the relationship

$$-2ik \frac{\partial g_{d_{nm}}}{\partial x'} + \left(\sum_{i=1}^{n} \nabla^2_{\rho'_i} - \sum_{j=n+1}^{n+m} \nabla^2_{\rho'_j} \right) g_{d_{nm}} + k^2 Q g_{d_{nm}}$$
$$= -g_{0_{nm}} \left(\sum_{i=1}^{n} \nabla^2_{\rho'_i} - \sum_{j=n+1}^{n+m} \nabla^2_{\rho'_j} \right) g_{\varepsilon S_{nm}} \quad (1.148)$$

where $$g_{d_{nm}} = g_{d_{nm}}(x - x'; \{\boldsymbol{\rho}_i - \boldsymbol{\rho}'_i\}, \{\boldsymbol{\rho}_j - \boldsymbol{\rho}'_j\})$$
$$\equiv g_{nm} - g_{0_{nm}} g_{\varepsilon S_{nm}} \quad (1.149)$$

is the difference between the exact product of the wave field at the set of points $\{\boldsymbol{\rho}_i\}$, $\{\boldsymbol{\rho}_j\}$ in the observation plane from that which is obtained in the phase approximation. The boundary condition on this quantity is

$$g_{d_{nm}}(0; \{\boldsymbol{\rho}_i - \boldsymbol{\rho}'_i\}; \{\boldsymbol{\rho}_j - \boldsymbol{\rho}'_j\}) = 0 \quad (1.150)$$

For the Green function solution of Eq. (1.148), one can employ that defined by Eq. (1.143); following the procedure established in the beginning of Sec. 1.4.2, one rewrites Eq. (1.143) to hold at an intermediate observation point, fixing the field source points. This induces a sign change of the first term of Eq. (1.143), giving

$$2ik \frac{\partial g_{nm}}{\partial x'} + \left(\sum_{i=1}^{n} \nabla^2_{\rho'_i} - \sum_{j=n+1}^{n+m} \nabla^2_{\rho'_j} \right) g_{nm} + k^2 Q g_{nm} = 0 \quad (1.151)$$

where, in this instance,

$$g_{nm} = g_{nm}(x'; \{\boldsymbol{\rho}'_i - \boldsymbol{\rho}_{0i}\}, \{\boldsymbol{\rho}'_j - \boldsymbol{\rho}_{0j}\})$$

Multiplying Eq. (1.148) by g_{nm} as defined here, subtracting this result from Eq. (1.151) multiplied by $g_{d_{nm}}$, integrating over the region defined by $0 \le x' \le L$, $-\infty \le \{\boldsymbol{\rho}_i\}, \{\boldsymbol{\rho}_j\} \le \infty$, using the boundary conditions given by Eqs. (1.70) and (1.150), and noting that the "surface terms" that result from the laplacians approach zero in the limit yields, upon using the definition of Eq. (1.149),

$$g_{nm}(L; \{\boldsymbol{\rho}_i - \boldsymbol{\rho}_{0i}\}, \{\boldsymbol{\rho}_j - \boldsymbol{\rho}_{0j}\}) = g_{0_{nm}} g_{\varepsilon S_{nm}}$$
$$+ \frac{i}{2k} \int_0^L \int_{-\infty}^{\infty} \ldots (n) \ldots \int_{-\infty}^{\infty} \ldots (m) \ldots \int_{-\infty}^{\infty} g_{nm}(x'; \{\boldsymbol{\rho}_i' - \boldsymbol{\rho}_{0i}'\},$$
$$\{\boldsymbol{\rho}_j' - \boldsymbol{\rho}_{0j}\}) g_{0_{nm}}(L - x'; \{\boldsymbol{\rho}_i - \boldsymbol{\rho}_i'\}, \{\boldsymbol{\rho}_j - \boldsymbol{\rho}_j'\})$$
$$\times \left(\sum_{i=1}^{n} \nabla_{\rho_i'}^2 - \sum_{j=n+1}^{n+m} \nabla_{\rho_j'}^2 \right) g_{\varepsilon S_{nm}}(L - x'; \{\boldsymbol{\rho}_i - \boldsymbol{\rho}_i'\}, \{\boldsymbol{\rho}_j - \boldsymbol{\rho}_j'\})$$
$$\times d^2\rho_1' \cdots d^2\rho_n' \, d^2\rho_{n+1}' \cdots d^2\rho_{n+m}' \, dx' \quad (1.152)$$

Noting that, within the integrand, the random g_{nm} factor is dependent on longitudinal positions in the interval $0 \le x'' \le x'$ and that the random $g_{\varepsilon S_{nm}}$ factor is a function of longitudinal positions in the interval $x' \le x'' \le L$, one can evoke the Markov approximation discussed in Sec. 1.4.2 when ensemble averaging Eq. (1.152), thus obtaining separate averages of these quantities. Thus, upon averaging Eq. (1.152), remembering the definitions of Eqs. (1.63) and (1.66), and substituting the result into Eq. (1.62) gives

$$\Gamma_{nm}(L, \{\boldsymbol{\rho}_i - \boldsymbol{\rho}_j\}) = \Gamma_{S_{nm}}(L, \{\boldsymbol{\rho}_i - \boldsymbol{\rho}_j\}) + \frac{i}{2k} \int_0^L \int_{-\infty}^{\infty}$$
$$\ldots (n) \ldots \int_{-\infty}^{\infty} \ldots (m) \ldots \int_{-\infty}^{\infty} \Gamma_{nm}(x'; \{\boldsymbol{\rho}_i' - \boldsymbol{\rho}_j'\})$$
$$\times g_{0_{nm}}(L - x'; \{\boldsymbol{\rho}_i - \boldsymbol{\rho}_i'\}, \{\boldsymbol{\rho}_j - \boldsymbol{\rho}_j'\}) \left(\sum_{i=1}^{n} \nabla_{\rho_i'}^2 - \sum_{j=n+1}^{n+m} \nabla_{\rho_j'}^2 \right)$$
$$\times \gamma_{\varepsilon S_{nm}}(L - x'; \{\boldsymbol{\rho}_i - \boldsymbol{\rho}_i'\}, \{\boldsymbol{\rho}_j - \boldsymbol{\rho}_j'\}) \, d^2\rho_1' \cdots d^2\rho_n'$$
$$\times d^2\rho_{n+1}' \cdots d^2\rho_{n+m}' \, dx' \quad (1.153)$$

where

$$\Gamma_{nm}(x', \{\boldsymbol{\rho}_i' - \boldsymbol{\rho}_j'\})$$
$$= \int_{-\infty}^{\infty} \ldots (n) \ldots \int_{-\infty}^{\infty} \ldots (m) \ldots \int_{-\infty}^{\infty} \gamma_{nm}(x', \{\boldsymbol{\rho}_i' - \boldsymbol{\rho}_{0i}\},$$
$$\{\boldsymbol{\rho}_j' - \boldsymbol{\rho}_{0j}\}) \prod_{i=1}^{n} U_0(\boldsymbol{\rho}_{0i}) \prod_{j=n+1}^{n+m} U_0^*(\boldsymbol{\rho}_{0j})$$
$$\times d^2\rho_{0_1} \cdots d^2\rho_{0_n} \, d^2\rho_{0_{n+1}} \cdots d^2\rho_{0_{n+m}} \quad (1.154)$$

and $\Gamma_{S_{nm}}$ is given by Eq. (1.154) where γ_{nm} is replaced with $\gamma_{\varepsilon S_{nm}}$. Here, this latter quantity is just the phase approximation of Eq. (1.63), i.e.,

$$\gamma_{\varepsilon S_{nm}}(L - x'; \{\boldsymbol{\rho}_i - \boldsymbol{\rho}_i'\}, \{\boldsymbol{\rho}_j - \boldsymbol{\rho}_j'\})$$
$$= \left\langle \prod_{i=1}^{n} \exp\left[-iS(L, \boldsymbol{\rho}_i; x', \boldsymbol{\rho}_i')\right] \prod_{j=n+1}^{n+m} \exp\left[iS(L, \boldsymbol{\rho}_j; x', \boldsymbol{\rho}_j')\right] \right\rangle \quad (1.155)$$

Equations (1.153) to (1.155) are exact results (at least to the level of correctness of the parabolic wave equation) that, by design, are in the form that elucidates the phase approximation of the extended Huygens-Fresnel principle. The first term of Eq. (1.153) is what would result if one were to apply the phase approximation to the calculation of the *nm*th moment of the wave field. The second term of this equation is the "correction term" that results from the implicit restriction of Eq. (1.146) that prevails when one uses the phase approximation. To be sure, if one imposes the condition of Eq. (1.146) on Eq. (1.153), one simply recovers the statistical moments of the wave field within the phase approximation. Thus Eq. (1.153) indicates that, in general, the use of the phase approximation can only be approximate, depending, of course, on the relative contribution of the second term. As is easily seen, Eq. (1.153) admits an iterative solution that results in a Neumann series for Γ_{nm}.[27]

There are, however, two instances when the phase approximation can give results that agree with those of the parabolic equation method. This occurs just for the first- and second-order field moments and only when one assumes locally homogeneous and isotropic gaussian statistics that govern the phase of the field. In particular, one has for the first-order moment from Eqs. (1.153) and (1.155)

$$\Gamma_{10}(L, \boldsymbol{\rho}_1) = \Gamma_{S_{10}}(L, \boldsymbol{\rho}_1) + \frac{i}{2k}\int_0^L \int_{-\infty}^{\infty} \Gamma_{10}(x', \boldsymbol{\rho}_1')$$
$$\times\, g_{0_{10}}(L - x'; \boldsymbol{\rho}_1 - \boldsymbol{\rho}_1')\nabla^2_{\rho_1'}\langle \exp\left[-iS(L, \rho_1; x', \boldsymbol{\rho}_1')\right]\rangle\, d^2\boldsymbol{\rho}_1'\, dx' \quad (1.156)$$

Since the phase S is assumed to be a gaussian random function, one can use Eq. (1.131) and, in addition to local homogeneity and isotropy (i.e., where the statistics of S are not dependent on any direction or separation of spatial position), one obtains the fact that

$$\langle \exp\left[-iS(L, \boldsymbol{\rho}_1; x', \boldsymbol{\rho}_1')\right]\rangle = \exp\left(-\tfrac{1}{2}\langle S^2\rangle\right) = \text{const.}$$

thus rendering the second term in Eq. (1.156) equal to zero. Similarly, for the second-order moment,

$$
\begin{aligned}
\Gamma_{11}(L, \boldsymbol{\rho}_1 - \boldsymbol{\rho}_2) &= \Gamma_{S_{11}}(L, \boldsymbol{\rho}_1 - \boldsymbol{\rho}_2) \\
&+ \frac{i}{2k} \int_0^L \int_{-\infty}^{\infty} \Gamma_{11}(x', \boldsymbol{\rho}_1' - \boldsymbol{\rho}_2') g_{0_{11}}(L - x'; \boldsymbol{\rho}_1' - \boldsymbol{\rho}_2')(\nabla^2_{\rho_1'} - \nabla^2_{\rho_2'}) \\
&\quad \times \langle \exp\{-i[S(L, \boldsymbol{\rho}_1; x', \boldsymbol{\rho}_1') - S(L, \boldsymbol{\rho}_2; x', \boldsymbol{\rho}_2')]\}\rangle \\
&\quad \times d^2\rho_1'\, d^2\rho_2'\, dx' \qquad (1.157)
\end{aligned}
$$

Again, making use of Eq. (1.131) and the homogeneity and isotropy of the phase field [cf. Eq. (D1.4.5)], one has that the average over the exponential in Eq. (1.157) becomes, in terms of the primed coordinates,

$$
\langle \ldots \rangle = \exp[-\tfrac{1}{2} D_S(L, |\boldsymbol{\rho}_1' - \boldsymbol{\rho}_2'|)]
$$

Substituting this into Eq. (1.157), converting the difference of the laplacian operators over each coordinate into a laplacian operator over the difference of the coordinates, and performing the required operations gives a zero result. Hence, the second term of Eq. (1.157) vanishes. However, when one considers the fourth-order moment, these terms do not collectively vanish, thus requiring a general iterative treatment.

Thus, when the conditions of locally homogeneous and isotropic gaussian statistics govern the random phase field of the atmosphere (which is a realistic requirement), the first- and second-order moments of the propagating wave field, as calculated using the phase approximation of the extended Huygens-Fresnel principle, agree with those obtained via the rigorous parabolic wave equation method. Fourth- and higher-order moments, however, cannot be properly represented by the phase approximation. Attempts to do so have given rise to erroneous and paradoxical conclusions in the applications.

References

1. V. I. Tatarskii, *Wave Propagation in a Turbulent Medium*. McGraw-Hill, New York, 1961. Chap. 6.
2. V. I. Tatarskii, *The Effects of the Turbulent Atmosphere on Wave Propagation*. U.S. Dept. of Commerce, TT-68-50464, Springfield, Va., 1971. Chap. 2.
3. J. W. Strohbehn and S. F. Clifford, "Polarization and Angle-of-Arrival Fluctuations for a Plane Wave Propagated Through a Turbulent Medium," *IEEE Trans. Antennas Propagat.* **AP-15**, pp. 416–421 (1967).
4. J. W. Strohbehn, "Optical Propagation Through the Turbulent Atmosphere," *Progr. Opt.* **9**, pp. 75–122 (1971).
5. V. I. Tatarskii, "Propagation of Electromagnetic Waves in a Medium with Strong Dielectric Constant Fluctuations," *Sov. Phys. JETP* **19** (4), pp. 946–953 (1964).
6. A. M. Yaglom, *Stationary Random Functions*. Prentice-Hall, Englewood Cliffs, N.J., 1962.
7. A. S. Monin and A. M. Yaglom, *Statistical Fluid Mechanics*. The MIT Press, Cambridge, Mass., 1973. Chap. 2.
8. Ref. 2, Section 61.
9. Yu. N. Barabanenkov, "Multiple Scattering of Waves by Ensembles of Particles and the Theory of Radiation Transport," *Sov. Phys. Usp.* **18** (9), pp. 673–689 (1976), and references therein.

10. V. I. Tatarskii, "Light Propagation in a Medium with Random Refractive Index Inhomogeneities in the Markov Random Process Approximation," *Sov. Phys. JETP* **29** (6), pp. 1133–1138 (1969).
11. V. I. Klyatskin, "Applicability of the Approximation of a Markov Random Process in Problems Relating to the Propagation of Light in a Medium with Random Inhomogeneities," *Sov. Phys. JETP* **30** (3), pp. 520–523 (1970).
12. V. I. Tatarskii and V. U. Zavorotnyi, "Strong Fluctuations in Light Propagation in a Randomly Inhomogeneous Medium," in E. Wolf, ed., *Progress in Optics,* **XVIII**, pp. 205–256 (1980).
13. R. L. Fante, "Electromagnetic Beam Propagation in Turbulent Media," *Proc. IEEE* **63** (12), pp. 1669–1692 (1975).
14. K. S. Gochelashvili and V. I. Shishov, "Propagation of Reflected Radiation in a Randomly Inhomogeneous Medium," *Sov. J. Quantum Electron* **11** (9), pp. 1182–1184 (1981).
15. S. M. Rytov, "Diffraction of Light by Ultrasonic Waves," *Izv. Akad. Nauk SSSR, Seriya Fizicheskaya,* No. 2, pp. 223–259 (1937). In Russian.
16. V. I. Klyatskin and V. I. Tatarskii, "Strong Fluctuations of a Plane Light Wave Moving in a Medium with Weak Random Inhomogeneities," *Sov. Phys. JETP* **28** (2), pp. 346–353 (1969).
17. A. Ishimaru, "Fluctuations of a Beam Wave Propagating Through a Locally Homogeneous Medium," *Radio Sci.* **4** (4), pp. 295–305 (1969).
18. A. Ishimaru, *Wave Propagation and Scattering in Random Media,* Vol. 2. Academic Press, New York, 1978. Chap. 18.
19. D. L. Fried, "Optical Heterodyne Detection of an Atmospherically Distorted Signal Wave Front," *Proc. IEEE* **55** (1), pp. 57–67 (1967).
20. R. M. Manning, "Theoretical Investigation of Millimeter Wave Propagation Through a Clear Atmosphere," *Proc. SPIE* **410**, *Laser Beam Propagation in the Atmosphere,* J. C. Leader, ed., pp. 119–136 (1983).
21. Z. I. Feizulin and Yu. A. Kravtsov, "Broadening of a Laser Beam in a Turbulent Medium," *Radiophys. Quantum Electron.* **10** (1), pp. 33–35 (1967).
22. Yu. A. Kravtsov and Z. I. Feizulin, "Some Consequences of the Huygens-Kirchhoff Principle for a Smoothly Nonuniform Medium," *Radiophys. Quantum Electron.* **12** (6), pp. 706–711 (1969).
23. R. F. Lutomirski and H. T. Yura, "Propagation of a Finite Optical Beam in an Inhomogeneous Medium," *Applied Optics* **10** (7), pp. 1652–1658 (1971).
24. V. I. Klyatskin, "Variance of the Angle of Arrival of a Plane Light Wave Propagating in a Medium with Weak Random Irregularities," *Radiophys. Quantum Electron.* **12** (5), pp. 578–580 (1969).
25. Yu. A. Kravtsov, "Strong Fluctuations of the Amplitude of a Light Wave and Probability of Formation of Caustics," *Sov. Phys. JETP* **28** (3), pp. 413–414 (1969).
26. A. S. Gurvich and M. A. Kallistratova, "Experimental Study of the Fluctuations in Angle of Incidence of a Light Beam Under Conditions of Strong Intensity Fluctuations," *Radiophys. Quantum Electron.* **11** (1), pp. 37–40 (1968).
27. V. A. Banakh and V. L. Mironov, "Phase Approximation of the Huygens-Kirchhoff Method in Problems of Laser-Beam Propagation in the Turbulent Atmosphere," *Optics Letters* **1** (5), pp. 172–174 (1977).
28. V. P. Aksenov and V. L. Mironov, "Spectral Expansion Method in Problems of Laser-Beam Propagation in the Turbulent Atmosphere," *Optics Letters* **3** (5), pp. 184–186 (1978).
29. V. P. Aksenov and V. L. Mironov, "Phase Approximation of the Huygens-Kirchhoff Method in Problems of Reflections of Optical Waves in the Turbulent Atmosphere," *J. Opt. Soc. Am.* **69** (11), pp. 1609–1614 (1979).

10. [illegible] "Light Propagation in a Medium with [illegible] Inhomogeneities [illegible] Approximation" [illegible] Phys. [illegible]
11. [illegible] Approximation [illegible] Markov [illegible] Problems Relating to the [illegible] in a Medium with Random [illegible] [illegible] 1970.
12. V. I. [illegible] and [illegible] "Strong Fluctuations in Light Propagation in a Randomly Inhomogeneous Medium" [illegible] Progress in Optics, XVI, pp. 205–256, 1978.
13. [illegible] "[illegible] Beam [illegible] Turbulent Media" [illegible] 1992.
14. K. [illegible] Gochelashvili and V. I. Shishov, "Propagation of Reflected Radiation in a Randomly Inhomogeneous Medium" [illegible] pp. [illegible]
15. [illegible] Brodie [illegible] 1997 [illegible]
16. [illegible] and V. I. [illegible] "Strong [illegible] of a [illegible] Light Wave [illegible] with [illegible] Random Inhomogeneities" [illegible] 42, pp. [illegible] 1976.
17. [illegible] "[illegible] Beam Wave Propagation Through a Locally Homogeneous Medium" [illegible] 1970.
18. A. [illegible] Wave Propagation and Scattering in Random Media, [illegible] Academic Press, [illegible] 1978 [illegible]
19. [illegible] "Optical Heterodyne Detection of an Atmospherically Distorted Signal Wave Front" [illegible] 1967.
20. [illegible] "Theory of [illegible] Detection [illegible] Through [illegible] Turbulence" [illegible] Atmospheric [illegible] pp. 130–139 [illegible]
21. [illegible] and [illegible] Tatarskii [illegible] Through [illegible] Media, Radiophys. Quantum Electron. [illegible] 1970.
22. [illegible] Kravtsov [illegible] "[illegible] Physics, Kirchhoff Principle for a Smoothly Nonuniform Medium" [illegible] 12, pp. 778–791, 1969.
23. [illegible] Tatarskii and [illegible] "[illegible] Propagation [illegible] Beam [illegible] Inhomogeneous Media" [illegible] 1971.
24. [illegible] "[illegible] and Angle-of-Arrival [illegible] Phase [illegible] Wave [illegible] Random Inhomogeneities" [illegible] 1978.
25. [illegible] "[illegible] of a Light Wave and Focused [illegible] Random Inhomogeneities" [illegible] 1978.
26. A. [illegible] and [illegible] "[illegible] Phase [illegible] Focused [illegible] Beam under [illegible] Strong [illegible] Fluctuations" [illegible] 1983.
27. [illegible] and [illegible] "[illegible] Properties of [illegible] Wave Propagation in the [illegible] Atmosphere" [illegible] pp. [illegible]–[illegible]
28. [illegible] and [illegible] "[illegible] from the Atmosphere" [illegible] pp. 147–150.
29. [illegible] "Phase Compensation of the [illegible] Kirchhoff Method in the [illegible] of Inhomogeneities in the [illegible] Atmosphere" [illegible] Opt. [illegible] 1987.

Chapter

2

The Statistical Modeling of the Atmospheric Permittivity Field

2.1 Introduction

The goal of this chapter will be a model that describes the statistics of the atmospheric permittivity field within the troposphere of the Earth. These results will allow one to connect the descriptions of the stochastic electromagnetic field that were derived in Chap. 1 with those of the basic components of the atmosphere.

The treatment will commence with a brief description in terms of the refractive index of the atmosphere at visible wavelengths, considering as additive components the dry gaseous continuum, the water vapor continuum, and aerosols and hydrometeors. The equations that describe the atmospheric refractivity and those that govern descriptive statistical aspects of atmospheric aerosols such as fogs and rain are established in Sec. 2.2. Section 2.3 then considers the statistical mechanical behavior of the fluctuations of the fundamental parameters of temperature and humidity that induce corresponding fluctuations in the continuum component of the refractive index. This is where the well-known Kolmogorov-Obukhov spectrum and its various modified forms are derived. The similar effects that result from the presence of aerosols are then considered, starting with general considerations of electromagnetic interactions with a typical aerosol on the basis of the fundamental Mie scattering theory. This is the subject of Sec. 2.4, where the necessary connection is established between the fundamental parameters that describe an aerosol and the "induced" or "effective" refractive index that prevails. Finally, Sec. 2.5 brings together all of the previous developments into a unified statistical model of the atmospheric propagation environment suitable for the application of the results of Chap. 1 to propagation problems at visible wavelengths.

2.2 The Description of the Refractive Index Field at Visible Wavelengths

The parameter known as the electric permittivity ε, along with the attendant magnetic permeability and the conductivity, provides the link between the mechanical properties of a medium and the electromagnetic field that can exist and propagate within that medium. In the case when the medium is the atmosphere that exists within the first 25 km above the earth's surface, it is only ε that is the relevant parameter, especially at infrared and visible wavelengths. The atmospheric permittivity is a measure of the effect of the atmosphere on the electrical properties of a prevailing electromagnetic field; it acts on the description of such a field through the form of a complex phase addition to the field. Since the source of the phenomenon that the permittivity field describes occurs at the molecular level with the interaction of the electromagnetic field with the gaseous components of the atmosphere, one must necessarily consider the spectroscopic description of the atmosphere throughout the range of optical wavelengths. Thus, as a first step in the statistical characterization of the atmospheric permittivity field, one must secure a description of the overall permittivity field and how it relates to the more fundamental parameters that characterize the atmosphere and, at the same time, are responsible for the random nature of the permittivity.

At the outset, the following analysis will be facilitated by considering, instead of the electric permittivity ε, the associated refractive index n, where

$$n^2 = \varepsilon \tag{2.1}$$

A complete thermodynamic description of the atmosphere allows one to relate n to the temperature, pressure, and concentrations of the constituent gases. Table 2.1 lists the major gaseous components that exist in the atmosphere that collectively contribute to the refractive index. Special note should be taken concerning water vapor, which is not included in Table 2.1. This particular component stands out from the rest and is given special treatment, since it has the largest space-time variability in concentration of any of the constituents and, more importantly, it can exist simultaneously in the gas and liquid state. The latter can result from either aqueous droplet formation that results in fog or the presence of hydrometeors such as rain, sleet, and snow.

TABLE 2.1 Gaseous Composition of Dry Air

Gaseous component	Tropospheric volumetric concentration, %	Evolution in atmosphere
N_2	78.08	Dissociates in ionosphere
O_2	20.95	Dissociates above 95 km
Ar	9.3×10^{-1}	Separates via diffusion above 110 km
CO_2	3.3×10^{-2}	Insignificant variations; dissociates above 100 km
Ne	1.8×10^{-3}	Separates via diffusion above 100 km
He	5.2×10^{-4}	Separates via diffusion above 110 km
CH_4	1.6×10^{-4}	Oxidizes and dissociates above troposphere
Kr	1.1×10^{-4}	Separates via diffusion above 110 km
H_2	5.0×10^{-5}	Dissociates above stratosphere
N_2O	3.5×10^{-5}	Insignificant variations; dissociates in stratosphere

The determination of the refractive index thus incorporates several aspects of the problem: the determination of (1) that portion which results from the dry gaseous continuum (i.e., void of water vapor); (2) the contribution from the water vapor continuum; and (3) the effect of fog droplets and hydrometeors. These will now be briefly reviewed; for more detailed treatments, one is directed to the references quoted during the development. In what follows, a scaled refractive index N referenced to the value of $n = 1$ will be employed in order to simplify the equations. This is given by

$$N = (n - 1) \times 10^6 \tag{2.2}$$

This quantity is called the refractivity associated with the refractive index n.

2.2.1 The dry gaseous continuum

With the exception of the water vapor, the atmospheric constituents of Table 2.1 define the "dry air" portion of the atmosphere that results in the dry air refractivity N_d given by[1,2]

$$N_d = \frac{0.3789P}{T} N_0[1 + (5.337 - 0.0157T) \times 10^{-6}]$$

$$\approx \frac{0.3789P}{T} N_0 \tag{2.3}$$

where P is the *total* atmospheric pressure with units of torr, T is the atmospheric temperature in kelvins, and N_0 is a "reference" refractivity specific to a pressure $P = 760$ torr and a temperature $T = 315$ K which is given by the wavelength-dependent relation

$$N_0 = 64.328 + \frac{29{,}498.1}{(146 - 0.0109\lambda^{-2})} + \frac{255.4}{(41 - \lambda^{-2})} \tag{2.4}$$

where the wavelength λ is expressed in micrometers. The transition to the approximate expression in Eq. (2.3) is justified by the fact that at normal atmospheric temperatures and pressures, the second term within the brackets of the first line of Eq. (2.3) is always $\ll 1$.

2.2.2 The water vapor continuum

The component of the refractivity N_w that is due to water vapor but devoid of the anomalous dispersion lines is given by[1,2]

$$\begin{aligned} N_w &= -1.765 \times 10^{-18}(1 - 0.0109\lambda^{-2})Q \\ &\approx -1.765 \times 10^{-18}Q \end{aligned} \tag{2.5}$$

where Q is the concentration of water vapor, expressed here in units of molecules/cm^3. The approximate result of the second line of Eq. (2.5) stems from the weak wavelength dependence at visible wavelengths ($0.4 \le \lambda \le 1.0$ μm).

At this point, one has characterized the entire refractivity field N_c due to the continuum, which is defined by (carefully noting the functional dependence on each of the fundamental atmospheric parameters)

$$N_c(P, T, Q) = N_d(P, T) + N_w(Q) \tag{2.6}$$

It is the variation in this quantity that is induced by variations in P, T, and Q that results in the random refractivity (or refractivity index) field that necessitates the use of the results of Chap. 1. Having established the functional dependence of Eq. (2.6) that, strictly speaking, only holds for wavelengths in the visible range, one can now obtain relations between the random refractivity field and the P, T, and Q fields. This will be the subject of Sec. 2.3.

2.2.3 Aerosols and hydrometeors

Aerosols are essentially any type of discrete particle that is dispersed within the gaseous medium of the atmosphere. In terms of the refractive index, they are distinguished by the fact that their presence causes discrete changes in the refractive index field; these changes are from the nominal refractive index of the gaseous atmosphere to that of the particular aerosol. However, the problem is extraordinarily complicated, especially in the case of hygroscopic particles, by the fact that the size as well as the refractive index of the aerosols is affected by the prevailing water vapor content of the atmosphere through the processes of condensation and evaporation.[3] In addition, aerosol properties are highly dependent on the aerosols' type and source, e.g., smoke, dust, sea salt particles near the ocean, etc.

Hydrometeors such as rain or snow exhibit the same variability; although the refractive index of a raindrop is relatively stable and approximately equal to that of water at the frequency of interest (with the possible exception of added impurities, acid content, etc., that can slightly change the value from that of condensed water), raindrop sizes are known to be functions of the rate at which the rain falls.[4]

The calculation of the overall refractive index of a suspension of particles can, in principle, be carried out by employing Mie scattering theory provided that one knows the refractive index of the particles. However, unless the particles are all the same and are assumed to be spherical, the calculation will be unwieldy. Even if such a calculation can be executed, if it is to be useful in realistic situations, the solution must also address the connection between the particle size and composition, and the water vapor content of the intervening gaseous atmosphere.

One can, however, approach the problem differently. At the outset, one can consider a level of approximation whereby the space and time scales that are admitted into the electromagnetic propagation and imaging problem (i.e., meters and/or seconds) are such that the properties of the aerosols can be taken to be constant relative to the gaseous water vapor fluctuations. In fact, the magnitude of the random fluctuations of the P, T, and Q fields that are responsible for the fluctuations in the refractive index field (which will be discussed later) are not of the spatial or temporal scale level to induce sufficient fluctuations in the aerosol properties; only large-scale (i.e., on the order of kilometers and/or hours) variations of the gaseous components of the atmosphere can induce variations on the same scale in the aerosol properties. This allows one to ignore the relationship between the characteristics of the gaseous atmosphere and those of the aerosols and to introduce as *a priori* basic modeling parameters the properties that characterize atmospheric aerosols at visible wavelengths, such as concentration, size distribution, and refractive index. A further and much more significant simplification can be realized if one notes that the refractive index of the aerosol particle n_p, that of the gaseous continuum component of the atmosphere n_c, and the minimum radius $a_{\min}$ of most atmospheric aerosols are such that $2ka_{\min}|n_p - n_c| \gg 1$. As will be discussed below, this combination of factors prominently figures in the result of a Mie scattering calculation for the scattered electric field from an isolated aerosol. Once such scattering behavior is established, one can model the aerosol component of the refractive index field so long as one establishes the significant statistical parameters that govern a typical ensemble of atmospheric aerosols, such as concentration and size distribution.

To this end, one considers an arbitrary volume V. The deposition of aerosols within this volume is an important consideration. One assumes, for simplicity, that the aerosols are uniformly distributed within this arbitrary volume, thus giving for the probability density $p(\mathbf{r}_i)$ of finding the ith particle at a position $\mathbf{r}_i$

$$p(\mathbf{r}_i) = \frac{1}{V} \tag{2.7}$$

Further, these particles can be taken to be statistically independent, thus allowing one to write for the joint probability distribution for m such aerosols in the volume

$$p(\mathbf{r}_1, \mathbf{r}_2, \ldots, \mathbf{r}_m) = \frac{1}{V^m} \tag{2.8}$$

The concentrations of aerosols can vary quite widely and are functions of the height above the earth.[5] However, for the purposes of a tractable model, the aerosol concentration ϑ_0 will be taken to be constant along whatever propagation path is chosen. The total number N of such aerosols within V will also be probabilistically described and is taken, to a very good approximation, to be given by a Poisson distribution, viz.,

$$p(N) = \frac{\langle N \rangle^N}{N!} \exp\left(-\langle N \rangle\right) \tag{2.9}$$

where $p(N)$ is the probability density of finding N aerosols in V characterized by an average number $\langle N \rangle$ which is connected to the concentration by the relation $\vartheta_0 = \langle N \rangle / V$.

Although there exist several models that purport to describe the size distribution of aerosols,[5,6] i.e., the probability density $p(a)$ that an aerosol particle has a radius a, a large majority of naturally occurring aerosols and hydrometeors seem to possess a universal characteristic whereby their radii are described by a single peak, asymmetrical distribution. A general representation for such size distributions is given by the two-parameter distribution[5,6]

$$p(a) = A a^{\nu - 1} \exp\left(-\mu a\right) \tag{2.10}$$

also known as the Deirmendjian modified gamma distribution, where ν and μ are two arbitrary parameters that are specific to the particular aerosol. The normalization constant A is such that the condition

$$\int_0^\infty p(a)\, da = 1$$

holds. Evaluation of this integral using Eq. (2.10) yields

$$A = \frac{\mu^\nu}{\Gamma(\nu)} \tag{2.11}$$

It is convenient, however, to get rid of one of the parameters that appears in Eq. (2.10) and replace it with the average radius $\langle a \rangle$ that governs the aerosol ensemble. This is simply given by

$$\langle a \rangle = \int_0^\infty a p(a)\, da = \frac{\nu}{\mu} \tag{2.12}$$

the use of which along with Eq. (2.11) in Eq. (2.10) gives for the aerosol size distribution of average radius $\langle a \rangle$

$$p(a) = \frac{\nu^\nu}{\Gamma(\nu)} \frac{a^{\nu-1}}{\langle a \rangle^\nu} \exp\left(-\nu \frac{a}{\langle a \rangle}\right) \tag{2.13}$$

Table 2.2 shows the most probable values that one will find for radiative and advective fogs.[5] It is important to note that the values for the concentration ϑ_0 are derived here from the corresponding values for water content w by the obvious relational constraint

$$w = \tfrac{4}{3}\, \pi \vartheta_0 \rho_w \int_0^\infty a^3 p(a)\, da \tag{2.14}$$

where ρ_w is the density of water.

Similarly, Table 2.3 shows parametric values for three types of rain size distributions, i.e., the Marshal-Palmer (MP), Joss thunderstorm

TABLE 2.2 Most Probable Values for Fogs

Fog type	$\langle a \rangle$, m	ν	Water content, g/m^3	ϑ_0, #/m^3
Radiative	5×10^{-6}	6	0.1	1.22×10^{11}
Advective	5×10^{-6}	3	0.1	8.59×10^{10}

TABLE 2.3 Most Probable Values for Rain Types

Rain type	$\langle a \rangle$, m	ν	ϑ_0, #/m^3
MP	$1.22 \times 10^{-4}R^{0.21}$	1	$1.42 \times 10^{3}R^{0.265}$
J-T	$1.67 \times 10^{-4}R^{0.21}$	1	$4.72 \times 10^{2}R^{0.265}$
J-D	$8.77 \times 10^{-5}R^{0.21}$	1	$4.50 \times 10^{3}R^{0.265}$

(J-T), and Joss drizzle (J-D) distributions.[4] In each case, the average raindrop radius is a function of the rain rate R; typically, R is measured in millimeters per hour. The applicability of these various rain models depends on what range of rain rates is being considered—one usually has for the MP distribution, 1 mm/h $< R <$ 50 mm/h; for the J-T distribution, 25 mm/h $< R <$ 150 mm/h; and for the J-D distribution, 0.25 mm/h $< R <$ 5 mm/h. As in the case for fogs, the values for ϑ_0 are obtained from the constraining relationship

$$R = \tfrac{4}{3}\pi\vartheta_0 \int_0^{\infty} a^3 p(a) v(a)\, da \tag{2.15}$$

where $v(a)$ is the terminal velocity of raindrops falling in the atmosphere. The model for the fall velocity used here is that of Gunn and Kinzer,[7] viz.,

$$v(a) = 7.229 \times 10^8 \sqrt{a}$$

where a is in meters and $v(a)$ is in millimeters per hour.

The above prescription for aerosol size distributions can be extended to any other type of aerosol, e.g., dust, smoke, etc., that is described by the general size distribution of Eq. (2.13). In fact, even if one encounters a polydispersion that is characterized by multiple peak distributions, one can form a sum of such expressions given by Eq. (2.13), taking care that the required prevailing constraints are satisfied. It must also be reiterated that the electromagnetic constraint $2ka_{\min,i}|n_{p,i} - n_c| \gg 1$ must always apply for the particular aerosol species i characterized by the set of values $n_{p,i}$, $a_{\min,i}$.

It should be noted that the presence and effect of discrete aerosols within the atmosphere is on the same level as the fluctuations in the P, T, and Q fields in that these inclusions, by their very presence, cause deviations in the nominal refractivity field. Thus, the specification of the aerosol field as given above, coupled with considerations concerning the scattering of an electromagnetic field by a single aerosol particle [one can consider single aerosol scattering because of the assumption leading to Eq. (2.8) that the scatterers are independent] will allow a sufficient basis from which to derive a general model to be used in conjunction with the results of Chap. 1 for the aerosol component of the permittivity field. It now remains to establish a descriptive model for the fluctuations in the gaseous refractivity field.

2.3 Fluctuations and Statistics of the Gaseous Refractivity Field

This section will consider the connection between the random fluctuations of the atmospheric temperature and humidity fields and the random fluctuations that result in the associated refractive index (or permittivity) field. What follows will provide the necessary link between the fundamental atmospheric turbulence statistics and the developments given in Chap. 1 that describe propagation of an electromagnetic wave field through such an environment.

2.3.1 The functional dependence of refractive index fluctuations on those of temperature and humidity

Having established the functional dependence of the atmospheric permittivity field, through use of the refractivity, on the fundamental parameters P, T, and Q that characterize the atmosphere at visible wavelengths, one can now derive the relations that govern the fluctuations that are induced in the refractivity field by prevailing fluctuations in these fundamental parameters. To this end, one can write

$$T = \langle T \rangle + \tilde{T} \qquad P = \langle P \rangle + \tilde{P} \qquad Q = \langle Q \rangle + \tilde{Q} \tag{2.16}$$

where the mean values $\langle \ldots \rangle$ are taken to be constant over the spatial and temporal scales that are considered, and $\tilde{\ldots}$ are the random perturbations of the fields over the same space and time scales, and where $\langle \ldots \rangle \gg \tilde{\ldots}$. These random fields give rise, through Eq. (2.6), to random fluctuations in the N_c field. For the small variations that are admitted into the otherwise nominal fields, one has from Eq. (2.6), to within first order of the fluctuating quantities,

$$\tilde{N}_c(P, T, Q) = \left(\frac{\partial N_d}{\partial P}\right)_{\langle T \rangle} \tilde{P} + \left(\frac{\partial N_d}{\partial T}\right)_{\langle P \rangle} \tilde{T} + \left(\frac{\partial N_w}{\partial Q}\right) \tilde{Q} \tag{2.17}$$

Using Eqs. (2.3) and (2.5) to evaluate the required partial derivatives, one obtains

$$\tilde{N}_c(P, T, Q) = \langle N_d \rangle \frac{\tilde{P}}{\langle P \rangle} - \langle N_d \rangle \frac{\tilde{T}}{\langle T \rangle} + \langle N_w \rangle \frac{\tilde{Q}}{\langle Q \rangle} \tag{2.18}$$

In terms of the refractive index due to the continuum, one has from Eq. (2.2) that $\tilde{n}_c = \tilde{N}_c \times 10^{-6}$, and upon using Eq. (2.18), one obtains

$$\tilde{n}_c = A_P \frac{\tilde{P}}{\langle P \rangle} + A_T \frac{\tilde{T}}{\langle T \rangle} + A_Q \frac{\tilde{Q}}{\langle Q \rangle} \tag{2.19}$$

$$\text{where} \quad A_P \equiv \langle N_d \rangle \times 10^{-6} \qquad A_T \equiv -\langle N_d \rangle \times 10^{-6} \qquad A_Q \equiv \langle N_w \rangle \times 10^{-6} \tag{2.20}$$

Before proceeding further, the relative magnitudes of the quantities that enter into Eqs. (2.19) and (2.20) will be examined for a possible simplification. At the outset, it should be noted that, at least within the troposphere, one has[8] $|\tilde{P}|/\langle P \rangle \approx 100|\tilde{T}|/\langle T \rangle$, and since $A_P = -A_T$, the pressure fluctuation contribution is negligible relative to that of the temperature. Other observations indicate that[9] one can have $|\tilde{T}| \sim 0.1$ K and $|\tilde{Q}| \sim 0.5\ \mu\text{g/cm}^3$. Taking as nominal values for the temperature and humidity $\langle T \rangle = 300$ K and $\langle Q \rangle = 6.82 \times 10^{17}$ molecules/cm^3, one has that $|\tilde{T}|/\langle T \rangle \approx 0.033$ percent and $|\tilde{Q}|/\langle Q \rangle \approx 2.2$ percent. However, using these same nominal values and considering a wavelength $\lambda = 0.63\ \mu$m, one finds, via Eqs. (2.3) to (2.5) and (2.20), $A_T = -2.655 \times 10^{-4}$ and $A_Q = -1.204 \times 10^{-6}$. Thus, the twofold difference in the magnitudes of fluctuations in the temperature and humidity fields is offset by the opposite contribution of their nominal values.

Hence, one can neglect the pressure fluctuation contribution in Eq. (2.19) relative to the combined effects of the temperature and humidity terms and write for the refractive index fluctuations, related to those of the temperature and humidity fields,

$$\tilde{n}_c = A_T \frac{\tilde{T}}{\langle T \rangle} + A_Q \frac{\tilde{Q}}{\langle Q \rangle} \tag{2.21}$$

One can now relate any statistical parameter involving the random refractivity field with those that govern the initiating random fields of temperature and humidity. Thus, one can write, assuming homogeneity and isotropy, the equation that connects the spatial correlation functions

$$B_n(r) = \left(\frac{A_T}{\langle T \rangle}\right)^2 B_T(r) + 2\left(\frac{A_T A_Q}{\langle T \rangle \langle Q \rangle}\right) B_{TQ}(r) + \left(\frac{A_Q}{\langle Q \rangle}\right)^2 B_Q(r) \tag{2.22}$$

or, in the case of local homogeneity and isotropy, the similar equation that relates the structure functions

$$D_n(r) = \left(\frac{A_T}{\langle T\rangle}\right)^2 D_T(r) + 2\left(\frac{A_T A_Q}{\langle T\rangle\langle Q\rangle}\right) D_{TQ}(r) + \left(\frac{A_Q}{\langle Q\rangle}\right)^2 D_Q(r) \quad (2.23)$$

where $B_n(r) \equiv \langle n_c(\mathbf{r}_1) n_c(\mathbf{r}_2)\rangle$, etc.

$$D_n(r) \equiv 2B_n(0) + 2B_n(r), \text{ etc.}$$

$$r = |\mathbf{r}_1 - \mathbf{r}_2|$$

It now remains to examine the statistical fluid mechanics of the atmosphere to determine the statistics of the T and Q fields required in Eqs. (2.22) and (2.23).

2.3.2 The fundamental statistical fluid mechanics governing atmospheric temperature and humidity fluctuations

Having related the statistics of the refractive index field of the atmosphere to those of the temperature and humidity fields, one now needs to establish a model of these fundamental atmospheric quantities.

At the outset, it is necessary to consider the equations that govern temperature and humidity (i.e., water vapor content) fluctuations that result from the random, turbulent nature of the atmospheric velocity field. It is necessary to first assume that the magnitudes of the temperature and humidity fields within the atmosphere are such that they do not affect the equations of motion of the atmosphere. In other words, temperature and water vapor are taken to be *passive additives* to the atmosphere. (Strictly speaking, temperature is not a passive additive to the atmospheric velocity field because of buoyancy forces that are generated by temperature fluctuations in the presence of the gravitational field; however, as was mentioned earlier, spatial and temporal scales considered in most atmospheric imaging problems are such that this process can be neglected.) Further, it is assumed that the fluctuations that enter into the temperature and humidity fields decay away only by the process of diffusion and not via phase transitions, chemical reactions, etc. In this way, the temperature "content" and water vapor content are conserved and are therefore taken as conservative additives to the atmospheric velocity field. (Such is not the case, for example, if one considers aerosol growth through condensation; however, as mentioned in Sec. 2.2.3, relatively small time scales of the problem are considered, and so this effect can be neglected.) Taking the temperature and humidity fields as conservative passive additives allows one to write for the equations that describe the evolution of these fields in the presence of velocity fluctuations[10]

$$\frac{\partial T}{\partial t} + \mathbf{v} \cdot \nabla T = \chi \nabla^2 T \tag{2.24}$$

$$\frac{\partial Q}{\partial t} + \mathbf{v} \cdot \nabla Q = D_Q \nabla^2 Q \tag{2.25}$$

where χ is the coefficient of thermal diffusivity and D_Q is the coefficient of molecular diffusion. Because of the isomorphism of these equations, and to simplify the notation in the development below, the generic quantity F will be taken to represent either the temperature or the humidity fields. Also, a generic designation D will be given to the diffusion coefficient, where $D = \chi$ in the case of $F = T$ and $D = D_Q$ in the case of $F = Q$. Note that with the exception of temperature, quantities represented by F are concentrations of some additive. However, to maintain generality, even in the case of temperature, one can loosely designate the temperature field as the "concentration" of temperature; of course, the dimensionality of D accounts for the actual units of measurement. Thus, Eqs. (2.24) and (2.25) take the form

$$\frac{\partial F}{\partial t} + \mathbf{v} \cdot \nabla F = D \nabla^2 F \tag{2.26}$$

Writing the additive concentration field in the form $F = \langle F \rangle + \tilde{F}$, $\langle F \rangle \gg |\tilde{F}|$ and the velocity field similarly, introducing these designations into Eq. (2.26), and averaging the resulting expression yields

$$\frac{\partial \langle F \rangle}{\partial t} + \nabla \cdot (\langle \mathbf{v} \rangle \langle F \rangle + \langle \tilde{\mathbf{v}} \tilde{F} \rangle - D \nabla \langle F \rangle) = 0 \tag{2.27}$$

The first term within the parentheses of this equation describes the transport of F due to the mean flow field, the second term describes the transport of F due to the random (turbulent) flow field, and the third term gives the reduction of F due to molecular (or, in the case of temperature, thermal) diffusion. Subtracting Eq. (2.27) from Eq. (2.26) then gives the equation that governs the random part of the additive field, viz.,

$$\frac{\partial \tilde{F}}{\partial t} + \nabla \cdot (\mathbf{v} F - \langle \mathbf{v} \rangle \langle F \rangle - \langle \tilde{\mathbf{v}} \tilde{F} \rangle - D \nabla \tilde{F}) = 0 \tag{2.28}$$

Writing the additive field that appears in the first term within the parentheses in terms of its mean and random components and simplifying gives

$$\frac{\partial \tilde{F}}{\partial t} + \nabla \cdot (\tilde{\mathbf{v}}\langle F\rangle + \mathbf{v}\tilde{F} - \langle \tilde{\mathbf{v}}\tilde{F}\rangle - D\nabla\tilde{F}) = 0 \tag{2.29}$$

Multiplying this equation by the random component of the additive field taken at the same point as Eq. (2.29) and rewriting relevant terms using the fluid mechanical condition of incompressibility, i.e., $\nabla \cdot \mathbf{v} = 0$, one obtains

$$\frac{\partial}{\partial t}\left(\frac{1}{2}\tilde{F}^2\right) + \nabla \cdot \left[\mathbf{v}\left(\frac{1}{2}\tilde{F}^2\right) - D\tilde{F}\nabla\tilde{F}\right] + \tilde{F}\tilde{\mathbf{v}}\nabla\langle F\rangle$$
$$+ D(\nabla\tilde{F})^2 - \tilde{F}\nabla\cdot\langle\tilde{\mathbf{v}}\tilde{F}\rangle = 0 \tag{2.30}$$

Finally, taking the ensemble average of this equation, integrating it over an arbitrary volume, and applying the divergence theorem to the second member of Eq. (2.30) gives

$$\frac{\partial H}{\partial t} = -\int_S \left(\left\langle \mathbf{v}\frac{\tilde{F}^2}{2}\right\rangle - D\langle\tilde{F}\nabla\tilde{F}\rangle\right)\cdot \hat{n}\, dS$$
$$- \int_V [\langle\tilde{F}\tilde{\mathbf{v}}\rangle \cdot \nabla\langle F\rangle + D\langle(\nabla\tilde{F})^2\rangle]\, dV \tag{2.31}$$

where

$$H \equiv \int_V \frac{\langle\tilde{F}^2\rangle}{2}\, dV$$

is a measure of the variance of the amount of passive additive in the turbulent mixture. It should be noted that the last term on the left side of Eq. (2.30) vanishes upon averaging, since $\langle\tilde{F}\nabla \cdot \langle\tilde{\mathbf{v}}\tilde{F}\rangle\rangle = \langle\tilde{F}\rangle\nabla \cdot \langle\tilde{\mathbf{v}}\tilde{F}\rangle = 0$ and $\langle\tilde{F}\rangle = 0$.

The surface integral on the right side of Eq. (2.31) represents the convective transfer of F through the bounding surface. The first term under the volume integral is the random increment of additive that appears in the field F due to random fluctuations in the velocity field, and the second such term is the amount of additive which dissipates as the result of diffusion. Equation (2.31) describes the scenario that is of interest here; velocity fluctuations, the evolution of which is governed by the fluid mechanical parameters of the atmosphere such as kinematic viscosity (to be discussed below), drive the fluctuations of the additive field, which is governed by the diffusion coefficient D.

However, such equations that endeavor to describe the statistical evolution of additives in the presence of random fluid flow, like the fundamental Navier-Stokes equation for statistical fluid motion itself, meet with severe difficulties when one attempts to obtain a closed set of equations. One needs to augment these equations with additional assumptions, as was done in Sec. 1.4.2 of Chap. 1 with the introduction of the Markov approximation into the equation that represents the electromagnetic field moments in a random medium. Furthermore, because of the high degree of coupling that exists between the quantities that enter into the equations of fluid dynamics, the exact relations are hardly ever used. Instead, it has been found to be useful to employ the dimensional analysis of similarity theory; relations are written based on the general concepts of what physical parameters can influence the flow and concentration fields. This approach was originally applied as far back as 1894 in the celebrated work of O. Reynolds.[11]

Using this approach, consider first the random velocity field. The transition from laminar (i.e., deterministic) flow to turbulent (i.e., random) flow is characterized by a quantity known as the Reynolds number, denoted Re, which is essentially the ratio of the inertial forces of the fluid to those due to friction within the fluid; this quantity is a measure of the stability of fluid flow against turbulent fragmentation of the velocity field. Letting L_0 be the characteristic scale length of a particular fluid flow, and letting v be the associated characteristic velocity, the condition that assures turbulent flow in a fluid with kinematic viscosity ν is given by

$$\mathrm{Re} \equiv \frac{Lv}{\nu} > \mathrm{Re}_{cr} \tag{2.32}$$

where Re_{cr} is the critical Reynolds number, a universal characteristic of every fluid in a given flow environment, which corresponds to the onset of turbulent, random irregularities in the motion of a fluid. Let $\tilde{v}$ be the random velocity fluctuation that exists in a turbulent flow, and let l be the characteristic distance within the velocity field of this fluctuation. From these characteristic parameters, one can define a corresponding characteristic time $t = l/\tilde{v}$. Therefore, the amount of energy per unit mass per unit time that this velocity fluctuation represents is simply given by

$$E = \frac{1}{2}\frac{\tilde{v}^2}{t} \sim \frac{\tilde{v}^3}{l} \tag{2.33}$$

However, at the same time, because of internal friction within the fluid, energy is dissipated as heat per unit mass per unit time; this is given by the well-known term that appears in the Navier-Stokes equations, viz.,

$$\varepsilon = \nu\nabla^2\tilde{v} \sim \nu\frac{\tilde{v}^2}{l^2} \tag{2.34}$$

(Since, for the remainder of this section, there will be no reference to electric permittivity, which is denoted by the same variable, no confusion should result from this double use.) If $E > \varepsilon$, i.e.,

$$\frac{\tilde{v}^3}{l} > \nu\frac{\tilde{v}^2}{l^2} \tag{2.35}$$

the inhomogeneity in the velocity field will continue to exist but decrease in size and velocity because of the relatively small expenditure of energy from viscosity. To be sure, especially if $\varepsilon \ll E$, then E must be approximately the same for the next phase of existence for the inhomogeneity characterized by a smaller velocity $\tilde{v}'$ and smaller scale l'; this latter fact stems from the requirement that

$$\frac{\tilde{v}^3}{l} \approx \frac{\tilde{v}'^3}{l'} \qquad |\tilde{v}| > |\tilde{v}'| \tag{2.36}$$

In fact, one can define a "local" Reynolds number Re$'$ that characterizes the stability of the inhomogeneities. The condition $\varepsilon < E$ allows one to write

$$\frac{E}{\varepsilon} = \frac{\tilde{v}^3/l}{\nu(\tilde{v}^2/l^2)} = \frac{\tilde{v}l}{\nu} \equiv \mathrm{Re}' > 1 \tag{2.37}$$

Thus, so long as Re$' > 1$, the inhomogeneities within the velocity field will continue to "fragment" into smaller and smaller inhomogeneities until the point is reached where the characteristic energy of the inhomogeneities is on the order of that dissipated into heat by viscosity, after which the inhomogeneity will cease to exist. However, if this turbulent cascading process is to continue to exist, the energy ε that is dissipated must be balanced by the average energy input by the average flow. Hence, this constant average energy input is exchanged throughout the hierarchy of inhomogeneity sizes; each inhomogeneity exchanges the amount ε of energy per unit mass per unit time which is converted into heat at the end of the fragmentation process where the velocity gradients are such that the equality in Eq. (2.34) is realized. In this way, the parameter ε is as fundamental as that of ν.

The scenario just qualitatively described forms the basis of motivation for two important assumptions introduced into statistical fluid mechanics to deal with such situations.[12] To begin with, it is noted that the stochastic nature of the inhomogeneity fragmentation or cascading process is such that the influence of the properties that govern the original flow field, i.e., properties characterized by scales on the order of those appearing in Eq. (2.32), are rapidly damped out during the first few orders of the cascading process. Thus, the statistics of the entire cascading inhomogeneity regime (at least those after the first few orders) should exhibit local homogeneity and isotropy, completely independent of the sources of such instability. Also, the temporal periods of these fluctuations, especially those of high order in the cascading process, will be much shorter than the time in which the average properties of the flow change. This allows one to further assign stationarity (i.e., time independence) to the prevailing statistics. Hence, there is the need to deal with stationary, local statistics parameters such as structure functions. Given this and the plausibility development in the last paragraph, one can hypothesize the following: The nature of all the small-scale fluctuations in the velocity field, i.e., those with characteristic sizes $l < L_0$, in the presence of sufficiently large Reynolds numbers (so that there is a large, well-established, and stationary turbulent field) exhibit locally homogeneous, isotropic, and stationary statistics and, furthermore, are determined only by the prevailing significant parameters ν and ε. This statement is known as *Kolmogorov's first similarity hypothesis.*

The fact that the turbulent field is now taken to be characterized only by ν and ε allows one, via dimensionality considerations, to define a length scale l_0 and a velocity scale v_0 given by

$$l_0 = \sqrt[4]{\frac{\nu^3}{\varepsilon}} \qquad v_0 = \sqrt[4]{\varepsilon\nu} \tag{2.38}$$

Since these parameters are determined by the viscosity, which governs the conversion of the energy entailed in the turbulent motion to heat, they are necessarily characteristic of the smallest inhomogeneities that exist in the flow. It therefore follows that for scale sizes $r > l_0$, the flow statistics must be independent of the viscosity and only a function of the quantity ε. Taken in conjunction with the hypothesis stated above, the locally homogeneous and isotropic turbulent field exists in the scale range $L_0 > r > l_0$, called the *inertial subrange,* and is determined only by ε. This is the content of *Kolmogorov's second similarity hypothesis.* The scale sizes that determine this subrange are called, respectively, the outer and inner scales of turbulence.

Specific quantitative results concerning the statistics of the turbulence field can be gleaned from the application of these two hypotheses and form the subject of the theory of locally isotropic turbulence.[13]

However, the goal here is to establish the statistics of the additive fields described by Eq. (2.31) in the presence of the turbulence field just described.

Just as the relative stability of the velocity flow field is reflected by the Reynolds number Re, the relative stability of the conservative passive additive field in the presence of such flow is indicated by two other such dimensionless measures. First, one has the Peclet number Pe, defined analogous to Re,

$$\mathrm{Pe} \equiv \frac{Lv}{D} \tag{2.39}$$

for the flow of an additive of diffusion coefficient D with a characteristic velocity v in a region with characteristic size L_0. One can also define another such dimensionless ratio that involves only the properties of the fluid and the additive, and not details of their motion; this is the Schmidt number Sc,

$$\mathrm{Sc} \equiv \frac{\nu}{D} \tag{2.40}$$

In the particular case where $D = \chi$, Eq. (2.40) defines the Prandtl number Pr. From these definitions, one has the obvious connection $\mathrm{Pe} = \mathrm{Re}\ \mathrm{Sc}$.

One can now construct an exact analogy between the evolution of the inhomogeneities within the additive field F and that of the velocity field above, and apply the two Kolmogorov hypotheses to obtain a statistical description of this random field, carefully noting that along with the viscosity of the air ν, one must also incorporate the diffusion coefficient D; both these parameters can be collectively introduced through the quantity Sc. Also, one has for the inner scale l_{F0} of this field that given by Eq. (2.38) with ν replaced by D, viz.,

$$l_{F0} = \sqrt[4]{\frac{D^3}{\varepsilon}}$$

However, even though it is within the spirit of the similarity approach not to deal directly with exact relations, one must nevertheless heed the relationships that they exhibit. In particular, Eq. (2.31) gives, to within the approximations made, an exact relationship between the temporal evolution of the concentration of the passive additive and the joint statistics of the velocity and concentration fields. Given the intervening developments, one can now analyze this equation further and obtain another quantity that must also be incorporated into the application of the Kolmogorov hypotheses.

To this end, consider the application of Eq. (2.31) to the assumed (via the Kolmogorov hypotheses) locally isotropic random concentration field of the additive. Assuming that $\mathrm{Pe} \gg 1$ prevails over this flow, one can extend the surface over which the integral of the first term on the left side of Eq. (2.31) is evaluated out to the diameter L_0. From the assumption that the details of the inhomogeneity fragmentation process in the additive field are independent of the average flows that occur on the scale L_0, this term is rendered negligible relative to the others. Furthermore, since statistical stationarity of the field is assumed, $\partial H/\partial t = 0$, and, additionally, Eq. (2.31) will apply over arbitrarily small volumes, thus allowing one to write

$$\langle \tilde{F}\tilde{\mathbf{v}} \rangle \cdot \nabla \langle F \rangle + D\langle (\nabla \tilde{F})^2 \rangle = 0 \tag{2.41}$$

Remembering that the quantity $\langle \tilde{F}\tilde{\mathbf{v}} \rangle$ represents the transport of the additive field F via the turbulent velocity field, i.e., the amount of inhomogeneity that appears as a result of the turbulent convective transport, one sees that the second member on the left side of Eq. (2.41) describes the dissipation of these fluctuations as a result of molecular diffusion. Thus, the quantity

$$N_F \equiv D\langle (\nabla \tilde{F})^2 \rangle \tag{2.42}$$

plays the analogous role in the dissipation of the inhomogeneities of the additive field to that of ε in the dissipation of the velocity inhomogeneities. The expansion of the theory of locally isotropic turbulence to incorporate the attendant statistics of conservative passive additives, as well as the significance of the fundamental quantity N_F, was first noted in Ref. 14.

One can now append the Kolmogorov hypotheses for the case of flow of a conservative passive additive, viz., (1) the nature of all the small-scale fluctuations in the concentration field F of an additive, i.e., those with characteristic sizes $r < L_0$, in the presence of sufficiently large Reynolds and Peclet numbers exhibit locally homogeneous, isotropic, and stationary statistics and, furthermore, are determined only by the prevailing significant parameters Sc, ε, and N_F; and (2) the locally homogeneous and isotropic turbulent field exists in the scale range $L_0 > r > \max(l_0, l_{F0})$, called the *inertial-convective subrange*, and is determined only by the values of ε and N_F. (It should be noted at this point that the inner scale values of l_0 and l_{F0} are approximately equal for the passive additive fields considered here; to be sure, they differ only in the value of ν or D, where one has $\nu = 0.15\ \mathrm{cm^2/s}$ for air, $D_T = \chi = 0.19\ \mathrm{cm^2/s}$ for the atmospheric temperature conductivity coefficient, and $D_Q = 0.20\ \mathrm{cm^2/s}$ for the atmospheric water vapor diffusion coefficient. Also, $30 < \varepsilon < 1000\ \mathrm{cm^2/s^3}$ for light to moderate turbulence.[15] Thus, one has inner scale values in the range from 0.05 to 0.1 cm; the nominal value of 0.1 cm is usually adopted for all inner scales.) Having motivated and finally established these basic postulates, one can now, by simple application of similarity theory, obtain relationships that describe the statistics of the additive field in the atmospheric turbulence field.

Consider the structure function of the concentration field F at two points, $\mathbf{x}$ and $\mathbf{x} + \mathbf{r}$; since, by the first hypothesis, within the scale range $r < L_0$ the spatial structure function is only a function of the scalar separation r of these points (due to local isotropy), i.e.,

$$D_F(r) = \langle [F(x + r) - F(x)]^2 \rangle$$

one has that the following functional relationship, constructed solely on dimensional considerations, must exist:

$$D_F(r) = N_F \varepsilon^{-1/2} D^{1/2} \beta \left(\frac{r}{l_{F0}}, \mathrm{Sc} \right) \qquad r < L_0 \tag{2.43}$$

The quantity $\beta(x, y)$ is a dimensionless function. The form of this function is dictated by the application of the second hypothesis. In particular, for the inertial-convective range $L_0 > r > l_{F0}$, Eq. (2.43) must become independent of Sc and D. Thus $\beta(x, y) = \beta(x)$, ridding Eq. (2.43) of Sc. To make this equation independent of D_F now requires one to assign an explicit functional form to $\beta(x)$. With $x = r/l_{F0}$ and l_{F0} given by the first relation of Eq. (2.38) (with the replacement $\nu = D$), Eq. (2.43) becomes independent of D provided $\beta(x) = Cx^{2/3}$, where C is a constant. Thus, Eq. (2.43) becomes, for the inertial-convective range,

$$D_F(r) = C_F^2 r^{2/3} \qquad l_{F0} < r < L_0 \tag{2.44}$$

where the constant

$$C_F^2 \equiv N_F \varepsilon^{-1/3} C$$

is called the structure parameter associated with the random concentration field F. Equation (2.44) is the well-known Kolmogorov-Obukhov "2/3 law" for a conservative passive additive in a turbulent field. Although the structure parameter, which is specific to each possible type of additive (e.g., temperature), can be related to more fundamental quantities that characterize the flow field (e.g., lapse rates, etc.),[13,16] its value for particular situations can be found simply by determining the structure function experimentally as a function of separation r and applying Eq. (2.44). This quantity is a measure of the relative strength of the turbulent fluctuations of the particular field. Thus, one has for the atmospheric temperature field

$$D_T(r) = C_T^2 r^{2/3} \qquad l_{T0} < r < L_0 \tag{2.45}$$

where C_T^2 is the temperature structure parameter, and similarly for the humidity field,

$$D_Q(r) = C_Q^2 r^{2/3} \qquad l_{Q0} < r < L_0 \tag{2.46}$$

where C_Q^2 is the humidity structure parameter. One can also consider the temperature-humidity structure function defined by

$$D_{TQ}(r) = \langle [T(x + r) - T(x)][Q(x + r) - Q(x)] \rangle$$

Applying the Kolmogorov hypotheses and following the reasoning leading to Eq. (2.44), obtain the relationship

$$D_{TQ}(r) = C_{TQ}^2 r^{2/3} \qquad \max\,(l_{T0}, l_{Q0}) < r < L_0 \tag{2.47}$$

where C_{TQ}^2 is the temperature-humidity structure parameter. Even though each inner scale measure has been separately denoted for clarity, they are all taken to be on the order of 0.1 cm as per the estimates made earlier.

Finally, it will prove useful later to consider the spectral density of the locally isotropic structure functions in Eqs. (2.45) to (2.47). The spectral density associated with a spatial structure function $D_{uv}(\mathbf{r})$ of, in the general case, two scalar fields follows from Eq. (D1.4.14) of Digression 1.4, viz.,

$$\Phi_{uv}(K) = \frac{1}{4\pi^2 K^2} \int_0^\infty \frac{\sin\,(Kr)}{Kr} \frac{d}{dr} \left[r^2 \frac{d}{dr} D_{uv}(r) \right] dr \tag{2.48}$$

Substituting the obvious generic form of the 2/3 law into this equation and performing the required operations within the integrand yields

$$\Phi_{uv}(K) = \frac{C_{uv}}{4\pi^2 K^2} \left(\frac{2}{3}\right)\left(\frac{5}{3}\right) \int_0^\infty \sin\,(Kr) r^{-1/3}\, dr \tag{2.49}$$

where

$$C_{uv} = \begin{cases} C_T^2 & \text{for temperature, } u = v = T \\ C_Q^2 & \text{for humidity, } u = v = Q \\ C_{TQ} & \text{for temperature/humidity, } u = T,\ v = Q \end{cases}$$

Evaluation of the integral gives

$$\int_0^\infty \sin(Kr)r^{-1/3}\,dr = K^{-2/3}\Gamma\left(\frac{2}{3}\right)\cos\left(-\frac{\pi}{6}\right)$$

Substituting this into Eq. (2.49) and simplifying results in the locally isotropic spectral density

$$\Phi_{uv}(K) = 0.033C_{uv}K^{-11/3} \tag{2.50}$$

which is the well-known Kolmogorov-Obukhov spectrum for fluctuations of passive additives within the inertial-convective subrange in the atmosphere.

The spatial length limitations imposed by the inertial-convective subrange translate into corresponding limitations of this spectral form; in particular, one has that Eq. (2.50) holds for frequencies $K = 2\pi/r$ in the range $2\pi/L_0 < K < 2\pi/l_0$. These finite limitations that prevail on the use of Eq. (2.50), particularly in integrals over spatial frequencies, give rise to equations that cannot always be analytically evaluated. Thus, for purposes of mathematical amenability, it is often desired to extend the range of Eq. (2.50) across an infinite range of frequency values. In the case of frequency values where $K > 2\pi/l_0$, i.e., length scales $r < l_0$, one is in the range called the dissipation subrange for the obvious reason that, as per the discussion above, the evolution of inhomogeneities of passive additives are governed solely by diffusion of the relevant concentrations. This, of course, is entailed within the first Kolmogorov similarity hypothesis. For sufficiently small separations r such that $r \ll l_0$, one has that the difference in the concentration field $F(r + x) - F(x)$ can be expanded in a power series in r. Retaining only the first term of such an expansion, one has that

$$D_F(r) \sim r^2 \qquad r \ll l_0$$

for the dissipation subrange. Noting that Sc $\sim$ 1 for the gaseous atmospheric components that are considered, one has from Eq. (2.43) that the function β must assume the form $\beta(x, y) = \beta(x) = C_1x^2$, where the constant $C_1 \sim C$. Thus, retaining the use of the constant C_F^2 in Eq. (2.43) and remembering the fact that

$$l_{F0} \sim l_0 = \sqrt[4]{\frac{D^3}{\varepsilon}}$$

one has for the dissipation subrange

$$D_F(r) = \frac{C_F^2}{C}\,\varepsilon^{-1/6}D^{1/2}\beta\left(\frac{r}{l_{F0}}, \mathrm{Sc}\right)$$

$$= C_F^2 l_0^{2/3}\left(\frac{r^2}{l_0^2}\right) \tag{2.51}$$

Unlike that for the inertial-convective subrange, Eq. (2.51) is a function (through l_0) of the diffusion coefficient D, a situation that is within the conditions specified by the first similarity hypothesis.

The discontinuous situation that exists between Eq. (2.44) for the inertial-convective subrange and Eq. (2.51) for the dissipation subrange is remedied by modifying the form of the spatial spectrum and requiring it to cover both subranges, thereby increasing its mathematical usefulness. To this end, one expects that the spectral density will rapidly decay toward zero within the dissipation subrange. Taking this decay to be exponential, one can assume the following modified form for Eq. (2.50):

$$\Phi_F(K) = 0.033C_F^2K^{-11/3}\exp(-\alpha^2K^2) \qquad \frac{2\pi}{l_0} < K < \infty \tag{2.52}$$

where the constant α is determined by the requirement that Eq. (2.52) yield the relation Eq. (2.51) through the use of Eq. (D1.4.13). Hence, using Eq. (2.52) in Eq. (D1.4.13) and considering the case where $Kr \ll 1$, such that

$$\left[1 - \frac{\sin(Kr)}{Kr}\right] \approx \frac{K^2r^2}{6} \qquad Kr \ll 1$$

one has

$$D_F(r) = 8\pi(0.033C_F^2)\int_0^\infty \left(\frac{K^2r^2}{6}\right)K^{-5/3}\exp(-\alpha^2K^2)\,dK$$

$$= 8\pi(0.033C_F^2)\left(\frac{r^2}{6}\right)\int_0^\infty K^{1/3}\exp(-\alpha^2K^2)\,dk \tag{2.53}$$

Evaluation of the integral in Eq. (2.53) gives

$$\int_0^\infty K^{1/3}\exp(-\alpha^2K^2)\,dK = (2\alpha^2)^{-2/3}\Gamma\left(\frac{4}{3}\right)D_{-4/3}(0)$$

where $D_{-4/3}(0)$ is a parabolic cylinder function of order $-4/3$. Substituting this result into Eq. (2.53) and simplifying [note: $D_{-4/3}(0) = 1.2036$] results in

$$D_F(r) = 0.09359\alpha^{-4/3}C_F^2r^2 \tag{2.54}$$

for the dissipation range. Equating this expression with that of Eq. (2.51) and solving for α gives

$$\alpha = 0.169l_0 \tag{2.55}$$

Thus, defining the transitional spatial frequency $K_m \equiv 1/\alpha$, one can write Eq. (2.52), in the general case of two different isotropic random fields u and v, as

$$\Phi_{uv}(K) = 0.033C_{uv}K^{-11/3}\exp\left(\frac{-K^2}{K_m^2}\right) \tag{2.56}$$

where, from Eq. (2.55), K_m and l_0 are related by

$$K_ml_0 = \frac{1}{0.169} = 5.92 \tag{2.57}$$

Equation (2.56), which holds for both the inertial-convective and dissipation subranges, is usually sufficient for use in most applications. However, Eq. (2.56) is still bounded at small frequencies by the value $K_0 \equiv 2\pi/L_0$. Because of this, some integrals in which Eq. (2.56) is employed may diverge as the spatial frequencies are taken to zero. This too can be circumvented by further modifying Eq. (2.56). The region where $K < K_0$, called the *input subrange,* cannot be universally described as are the inertial-convective and dissipation subranges since, by its very definition, it is dependent on the particular details of the processes that are responsible for the production of the additive concentration fields in question, the overall geometry of the flow region, etc. However, one can assume that the largest inhomogeneity encountered, i.e., the smallest spatial frequency, does not exceed that determined by the outer scale L_0. Hence, at small K, the spectral density saturates at the value K_0, viz.

$$\Phi_{uv}(K) = 0.033C_{uv}K_0^{-11/3} \qquad K < K_0 \equiv \frac{2\pi}{L_0}$$

For this situation to obtain but yet reduce to Eq. (2.56) at moderate K, one can modify Eq. (2.56) as follows:

$$\Phi_{uv}(K) = \frac{0.033C_{uv}}{(K^2 + K_0^2)^{11/6}}\exp\left(\frac{-K^2}{K_m^2}\right) \tag{2.58}$$

This is the most general modified form of the Kolmogorov-Obukhov spectrum for locally homogeneous and isotropic turbulent fluctuations of the concentration field of conservative passive additives such as temperature and humidity. Since $l_0 \sim 0.1$ cm and remembering Eq. (2.57), $K_m \sim 59.2\ \text{cm}^{-1}$. As per the discussion above, L_0 is determined by the maximum characteristic length of the atmospheric flow process, which, in the case of propagation in the open atmosphere, is essentially the height above the ground. However, the value does seem to saturate at about 100 m. Using this maximum value, one has $K_0 \sim 6.28 \times 10^{-4}\ \text{cm}^{-1}$.

2.3.3 The refractive index structure function, structure parameter, and associated spectral density

Substituting Eqs. (2.45) to (2.47) into Eq. (2.23) yields

$$D_n(r) = C_n^2 r^{2/3} \tag{2.59}$$

where

$$C_n^2 \equiv \left(\frac{A_T}{\langle T \rangle}\right)^2 C_T^2 + 2\left(\frac{A_T A_Q}{\langle T \rangle \langle Q \rangle}\right) C_{TQ} + \left(\frac{A_Q}{\langle Q \rangle}\right)^2 C_Q^2 \tag{2.60}$$

is the refractive index structure parameter and, just like those of the temperature and humidity fields, is a measure of the level of local isotropic fluctuations of the refractive index field. Similarly, using Eqs. (2.48) and (2.58), one has from Eq. (2.59)

$$\Phi_n(K) = \frac{0.033 C_n^2}{(K^2 + K_0^2)^{11/6}} \exp\left(\frac{-K^2}{K_m^2}\right) \tag{2.61}$$

for the spectral density of the refractive index field fluctuations.

Equation (2.60) provides the necessary contact between the statistics of the fluid dynamic and electromagnetic fields of the atmosphere. Hence, given measured or tabulated values of C_T^2, C_{TQ}, and C_Q^2 and the corresponding nominal values of T and Q, one can calculate the prevailing values of C_n^2 at any visible wavelength with application of Eqs. (2.3) to (2.5), (2.20), and (2.60). By implicit virtue of the passive nature of the additives represented by atmospheric temperature and humidity, the spectrum of the refractive index fluctuations then follows from the Kolmogorov-Obukhov form of Eq. (2.61).

From the development given in Sec. 2.3.2, in particular Eq. (2.42) and the parameters convolved in the definition of the structure parameter, it is seen that the values of structure parameters of the additive fields are a direct function of the field gradients. Thus, one can expect values of the structure parameters to be highly variable in a typical propagation scenario that exists over diverse terrain. For example, local solar heating effects of different ground covers and the subsequent thermal reradiation back into the atmosphere will give rise to thermal gradients that can vastly differ in magnitude. Such is also the case for water vapor content in the presence of oceans, lakes, or otherwise moist areas. This dependence is also the source of the rather paradoxical situation where these structure parameters can be relatively small, in some cases approaching zero, in extremely windy and turbulent (from the point of view of atmospheric velocity fluctuations) conditions; so long as the gradients of the temperature and/or humidity fields are small, so too will be the corresponding structure parameters. This is particularly the case with the temperature field at sunset or sunrise; at these times, the temperature field of the atmosphere is almost at thermal equilibrium with that of the earth's surface, resulting in little or no field gradient, even though the attendant velocity field is turbulent. In view of this wide range of diversity that can exist in the value of C_n^2, one cannot quote absolute values for structure parameters. However, some general but yet rather representative assessments can be made. Table 2.4 summarizes some typical values that can be assumed by C_n^2 (units: $m^{-2/3}$) at optical wavelengths at various times of the day over a transition in seasonal weather conditions.[17]

These data were compiled from observations in northern New York at a height of 3 m. Here, these representative values range over two

TABLE 2.4 Typical Values for C_n^2 at Various Times ($C_n^2 = 10^X$, $m^{-2/3}$)

Month	Diurnal period	X
February	Night	−14.13
	Day	−12.76
	Dusk	−13.59
	Dawn	−14.26
March	Night	−14.33
	Day	−12.70
	Dusk	−13.86
	Dawn	−14.34
April	Night	−13.67
	Day	−12.79
	Dusk	−13.59
	Dawn	−13.79
May	Night	−13.49
	Day	−12.56
	Dusk	−13.53
	Dawn	−13.66

orders of magnitude. Even though there is a small seasonal variation, most of the variation is realized over a diurnal cycle. It was also found that the variation of C_n^2 with height h (in meters) above the ground exhibited the functional dependence

$$C_n^2(h) = C_n^2(3) \times \begin{cases} h^{-4/3} & \text{daytime} \\ h^{-2/3} & \text{nighttime} \end{cases} \tag{2.62}$$

This model of the height dependence of C_n^2 is sufficient for line of sight image propagation across quasi-horizontal paths through the atmosphere, not exceeding a height of about 1 km. For the case of propagation paths that extend above 1 km or slant paths that can extend from the surface into space, a general model has been formulated[18] which encompasses a well-known "surge" phenomenon[19] in C_n^2 path profiles that occurs at a height of about 1.4 to 2.0 km because of a sharp drop in the value of the dissipation rate ε that prevails in a layer that is about 0.5 km thick. This general path profile model is given by

$$C_n^2(h) = C_n^2(0) \exp\left(-\frac{h}{H_2}\right)\left\{h^{-m} + K \exp\left[-n\left(1 - \frac{h}{H_1}\right)^2\right]\right\} \tag{2.63}$$

where $C_n^2(0)$ is the value of the structure parameter at (i.e., within a few meters of) the surface, $H_1 \sim 1.5$ to 2.0 km, $H_2 \sim 25$ km (i.e., the distance through the troposphere), and $⅓ < m < ⁴⁄₃$. The constant K represents the relative strength of the surge layer, $0 < K < 1$, and the constant n determines the width of this layer (i.e., the distances $h \pm \Delta h$ at which the argument of the exponential is $= -1$).

2.4 The Effective Refractive Index of Atmospheric Aerosols at Optical Wavelengths

The results of the foregoing provide sufficient material for the construction of a complete statistical description of the refractive index field, taking into account the effects both of the continuum and of aerosols. There is, however, one further development that must be made regarding the description of the effects of an aerosol on a propagating electromagnetic wave. That is, a link is required between the fundamental aerosol properties introduced in Sec. 2.2.3 and the quantities that enter into the equations of wave propagation; in essence, one needs to obtain an "effective refractive index," within the approximations suitable for visible wavelengths, that prevails in a suspension of aerosols.

To this end, consider once again reverting back to the permittivity description, and writing for the random permittivity field $\tilde{\varepsilon}(\mathbf{r})$ of the atmosphere

$$\tilde{\varepsilon}(\mathbf{r}) = \tilde{\varepsilon}_c(\mathbf{r}) + \tilde{\varepsilon}_p(\mathbf{r}) \tag{2.64}$$

where the first member on the right is the contribution due to fluctuations in the gaseous continuum, and the second term is the contribution due to the aerosols; the stochastic nature of the aerosol properties such as position, size, etc., makes this latter member a random function too. Since aerosols, by their very nature, are characterized by discrete properties, one can consider an arbitrary volume V containing a random number N of aerosols (as was done in Sec. 2.2.3) and write

$$\tilde{\varepsilon}_p(\mathbf{r}) = \sum_{i=1}^{N} V_0(\mathbf{r} - \mathbf{R}_i;\, \xi_i) \tag{2.65}$$

where $V_0(\mathbf{r} - \mathbf{R}_i;\, \xi_i)$ is the effective permittivity of the ith aerosol, located at the position $\mathbf{R}_i$ and characterized by properties such as radius and shape that are convolved in the descriptor ξ_i. Substituting Eq. (2.65) into Eq. (2.64) and this intermediate result into Eq. (1.16) yields

$$\left[\nabla^2 + k^2 + k^2\tilde{\varepsilon}_c(\mathbf{r}) + \sum_{i=1}^{N} V_0(\mathbf{r} - \mathbf{R}_i;\, \xi_i)\right] E(\mathbf{r}) = 0 \tag{2.66}$$

Using a phase representation for the random electric field similar to, but in a context different from, that used in the Rytov approach discussed in Chap. 1, one can write

$$E(\mathbf{r}) = \exp\left[\psi_c(\mathbf{r}) + \psi_p(\mathbf{r})\right] \tag{2.67}$$

where $\psi_c(\mathbf{r})$ is the random complex phase due to the continuum fluctuations and $\psi_p(\mathbf{r})$ is that due to random locations, sizes, and shapes of the aerosols. Substituting Eq. (2.67) into Eq. (2.66), simplifying, and separating the resulting expression into continuum-dependent and aerosol-dependent components gives

$$\nabla^2\psi_c(\mathbf{r}) + [\nabla\psi_c(\mathbf{r})]^2 + k^2 = -k^2\tilde{\varepsilon}(\mathbf{r}) \tag{2.68}$$

$$\nabla^2\psi_p(\mathbf{r}) + 2\nabla\psi_c(\mathbf{r}) \cdot \nabla\psi_p(\mathbf{r}) + [\nabla\psi_p(\mathbf{r})]^2 = -k^2 \sum_{i=1}^{n} V_0(\mathbf{r}) \tag{2.69}$$

Equation (2.68) is just the Rytov transformed version of the stochastic wave equation that describes propagation only in the presence of the random continuum component of the permittivity field. Equation (2.69), on the other hand, describes the development of the complex phase that is perturbed by interactions with aerosols. This particular relation can be reduced further by expanding the aerosol-dependent complex phase into components $\psi_{p,s}(\mathbf{r})$ which represent contributions to this random phase due to sth-order interactions of the electric field with the N aerosols, i.e.,

$$\psi_p(\mathbf{r}) = \sum_{i=1}^{N} \psi_{p,1}(\mathbf{r} - \mathbf{R}_i; \psi_i) + \sum_{i \neq j}^{N} \sum^{N} \psi_{p,2}(\mathbf{r}, \mathbf{R}_i, \mathbf{R}_j; \xi_i, \xi_j) + \cdots \tag{2.70}$$

where each term under the first summation on the right is a function of the properties of a single aerosol, since this group of terms represents single interactions with the aerosols, each term under the second group of summations is a function of the properties of any two distinct aerosols, since this group of terms represents double interactions, etc. Introducing this representation into Eq. (2.69) gives

$$\begin{aligned} \sum_{i=1}^{N} \nabla^2 \psi_{p,1}(\mathbf{r}) &+ \sum_{i \neq j}^{N} \sum^{N} \nabla^2 \psi_{p,2}(\mathbf{r}) \cdots \\ &+ 2 \sum_{i=1}^{N} \nabla \psi_c(\mathbf{r}) \cdot \nabla \psi_{p,1}(\mathbf{r}) + 2 \sum_{i \neq j}^{N} \sum^{N} \nabla \psi_c(\mathbf{r}) \cdot \nabla \psi_{p,2}(\mathbf{r}) + \cdots \\ &+ \left[\sum_{i=1}^{N} \nabla \psi_{p,1}(\mathbf{r}) \right]^2 + 2 \sum_{i=1}^{N} \nabla \psi_{p,1}(\mathbf{r}) \cdot \sum_{i \neq j}^{N} \sum^{N} \nabla \psi_{p,2}(\mathbf{r}) \\ &+ \left[\sum_{i \neq j}^{N} \sum^{N} \nabla \psi_{p,2}(\mathbf{r}) \right]^2 + \cdots = -k^2 \sum_{i=1}^{N} V_0(\mathbf{r}) \end{aligned} \tag{2.71}$$

where the full arguments of the complex phases have been suppressed but are obvious. Terms such as the first and third on the left as well as the one on the right describe single aerosol interactions; those such as the second and fourth describe double aerosol interactions. The sixth term represents triple aerosol interactions. Expanding the fifth term yields

$$\left[\sum_{i=1}^{N} \nabla \psi_{p,1}(\mathbf{r}) \right]^2 = \sum_{i=1}^{N} [\nabla \psi_{p,1}(\mathbf{r})]^2 + \sum_{i \neq j}^{N} \sum^{N} \nabla \psi_{p,1}(\mathbf{r}) \cdot \nabla \psi_{p,1}(\mathbf{r})$$

Thus, both single and double aerosol events are represented by this term. Similarly, the seventh term involves double and quadruple interactions. Equating the separate groups of single, double, etc., aerosol interaction terms of Eq. (2.71) gives a system of equations distributed throughout the multiplicity of interactions, viz.,

$$\sum_{i=1}^{N} \{\nabla^2\psi_{p,1}(\mathbf{r}) + 2\nabla\psi_c(\mathbf{r}) \cdot \nabla\psi_{p,1}(\mathbf{r}) + [\nabla\psi_{p,1}(\mathbf{r})]^2\} = -k^2 \sum_{i=1}^{N} V_0(\mathbf{r}) \qquad (2.72)$$

$$\sum^{N}\sum^{N}_{i\neq j} \{\nabla^2\psi_{p,2}(\mathbf{r}) + 2\nabla\psi_c(\mathbf{r}) \cdot \nabla^2\psi_{p,2}(\mathbf{r}) + [\nabla\psi_{p,2}(\mathbf{r})]^2\}$$

$$= -\sum^{N}\sum^{N}_{i\neq j} \nabla\psi_{p,1}(\mathbf{r}) \cdot \nabla\psi_{p,1}(\mathbf{r}) \qquad (2.73)$$

$$\vdots$$

Hence, within this formalism, the effect of the aerosols commences with the single, independent interactions with the N aerosols described by Eq. (2.72). Double and coupled interactions with pairs of aerosols by the perturbed complex phase field [previously determined by Eq. (2.72)] are represented by Eq. (2.73). So long as the concentration of the aerosols is such that the mean free path that separates them is much larger than the wavelength of propagation, the "cooperative" interaction events that are described by Eq. (2.73), and those that would follow it, can be neglected. As is well known,[20] the average distance d between a random distribution of particles characterized by the average number of particles per unit volume ϑ_0 (i.e., the concentration) is given by $d = 0.55396\vartheta^{-1/3}$. Thus, considering the largest concentration value of 1.22×10^{11} particles per cubic meter realized in the case of radiative fog as listed in Table 2.2, one has that $d = 112\ \mu\text{m}$, which exceeds optical wavelengths by a factor of 180. Hence, one can only be concerned with a level of interaction approximation that is correct up to single, independent events as given by Eq. (2.72).

Having shown that the propagation through an ensemble of aerosols is described by Eqs. (2.68) and (2.72) with the auxiliary relations of Eqs. (2.67) and (2.70), one can now relate the effective permittivity V_0 of an aerosol (which, as given in the formalism above, can be interpreted as the "scattering potential" from the point of view of the complex phase) to the fundamental properties that characterize the aerosol and are represented by ξ_i. To this end, consider the case where there are no fluctuations within the continuum field of the atmospheric permittivity, i.e., $\tilde{\varepsilon}_c(\mathbf{r}) = 0$, and, since one has admitted a single scattering theory, consider the problem of the interaction of the complex phase field with just one aerosol. Thus, Eq. (2.66) becomes

$$(\nabla^2 + k^2)E(\mathbf{r}) = -k^2V_0(\mathbf{r} - \mathbf{R}_i;\ \xi_i)E(\mathbf{r}) \qquad (2.74)$$

which is analogous to the problem considered in Sec. 1.3.1 of Chap. 1 and has as the solution for the scattered field $E_s(\mathbf{r})$

$$E_s(\mathbf{r}) = \frac{k^2}{4\pi}\int \frac{\exp(ik|\mathbf{r} - \mathbf{r}'|)}{|\mathbf{r} - \mathbf{r}'|} V_0(\mathbf{r}' - \mathbf{R}; \xi_i)E_s(\mathbf{r}')\, d\mathbf{r}' \qquad (2.75)$$

Since the characteristic size of the aerosols is taken to dominate over that of the wavelength, one can employ the paraxial approximation for the Green function, i.e.,

$$k|\mathbf{r} - \mathbf{r}'| \approx kr\left(1 - \hat{\mathbf{r}} \cdot \frac{\mathbf{r}'}{r}\right) = k(r - \hat{\mathbf{r}} \cdot \mathbf{r}') = kr - \mathbf{k} \cdot \mathbf{r}'$$

as well as the Born approximation, whereby one replaces the scattered field appearing in the integrand by the unperturbed field, taken here to be the plane wave

$$E_s(\mathbf{r}) = \exp(i\mathbf{k}_0 \cdot \mathbf{r}') \qquad |\mathbf{k}_0| = |\mathbf{k}|$$

Operating on Eq. (2.75) with these approximations and simplifying the resulting expression gives

$$E_s(\mathbf{r}) = \frac{k^2}{4\pi}\frac{\exp(ikr)}{r}\int \exp(-i\mathbf{K} \cdot \mathbf{r}')V_0(\mathbf{r}' - \mathbf{R}; \xi_i)\, d\mathbf{r}' \qquad (2.76)$$

where $\mathbf{K} = \mathbf{k} - \mathbf{k}_0$ is the scattered wave vector. Noting that this vector represents only a change in the direction but not the magnitude of the initial and scattered wave vectors, one has that

$$K = \sqrt{2k^2(1 - \cos\theta)} \approx k\theta$$

for small scattering angles θ. Thus, one can write Eq. (2.76) in the form

$$E_s(r, \theta) = \frac{\exp(ikr)}{r} f(\theta) \qquad (2.77)$$

where
$$f(\theta) \equiv \frac{k^2}{4\pi}\int \exp(-i\mathbf{K} \cdot \mathbf{r}')V_0(\mathbf{r}' - \mathbf{R}; \xi_i)\, d\mathbf{r}' \qquad (2.78)$$

is the scattering amplitude.

At this point, one must assign parameters for the aerosol property descriptor ξ_i. From the discussion in Sec. 2.2.3 of the present chapter, one can take the aerosols to possess a spherical geometry and be characterized by a random magnitude a for the radii, values of which are governed by Eq. (2.13). Since all values assumed by these random radii are taken to be large compared to the wavelength, and since, by their nature, the refractive index n_p of these aerosols deviates appreciably from the nominal value of n_c, it is a palatable assumption to represent the scattered electric field of a plane wave impinging upon an aerosol by the well-known approximation[21] that is obtained from Mie theory,

$$E_s(\mathbf{r}, \theta) \approx \frac{\exp(ikr)}{ikr}\left[\eta^2 \frac{J_1(\eta\theta)}{\eta\theta} + O\left(\frac{1}{\zeta}\right)\right] \tag{2.79}$$

where $\eta = ka$ and $\zeta = 2\eta|n_p - n_c|$. In the extreme case of min $(a) = 1\ \mu\text{m}$ and $n_p = 1.33$, one has at a nominal value of $\lambda = 0.63\ \mu\text{m}$ that $\eta = 10$ and $\zeta = 6.6$. Hence, the terms represented by the second term within the brackets of Eq. (2.79) are negligible relative to the first term in the cases considered here. Equating Eq. (2.79) with Eq. (2.77) and writing η in terms of K finally yields the relation

$$\int \exp(-i\mathbf{K}\cdot\mathbf{r}')V_0(\mathbf{r}'; a)\, d\mathbf{r}' = -i\left(\frac{4\pi a^2}{k}\right)\left[\frac{J_1(Ka)}{Ka}\right] \tag{2.80}$$

connecting the quantity V_0 with those of the aerosol, i.e., a. This Fourier transform relationship is independent of the particular value of n_p so long as the condition $\zeta \gg 1$ prevails.

It now remains to incorporate the relationship of Eq. (2.80), along with the statistics that govern $\mathbf{R}_i$ and a as given by Eqs. (2.7) to (2.10) and the statistical description of the continuum, viz., Eq. (2.61), into an overall description of the atmospheric permittivity (or refractive index) field that can be interfaced with the results of Chap. 1. This forms the subject of the next section.

2.5 A Unified Statistical Model of the Atmospheric Electromagnetic Propagation Environment

It is now a simple matter to obtain a complete statistical description of the atmospheric permittivity field by forming the characteristic functional from the foregoing developments. Substituting Eq. (2.64) into Eq. (1.85), and noting that the $\tilde{\varepsilon}_p(\mathbf{r})$ term will, by Eq. (2.80), contain complex terms, yields

$$\begin{aligned}\Phi_\varepsilon(\chi_1, \chi_2) &= \left\langle \exp\left\{i \int \tilde{\varepsilon}_c(\mathbf{r})[\chi_1(\mathbf{r}) + \chi_2(\mathbf{r})]\, d\mathbf{r}\right.\right.\\ &\qquad \left.\left. + i\int \tilde{\varepsilon}_p(\mathbf{r})\chi_1(\mathbf{r})\, d\mathbf{r} + i\int \tilde{\varepsilon}_p^*(\mathbf{r})\chi_2(\mathbf{r})\, d\mathbf{r}\right\}\right\rangle\\ &= \left\langle \exp i \int \tilde{\varepsilon}_c(\mathbf{r})\chi_3(\mathbf{r})\, d\mathbf{r}\right\rangle\\ &\qquad \times \left\langle \exp\left[i\int \tilde{\varepsilon}_p(\mathbf{r}\chi_1(\mathbf{r})\, d\mathbf{r} + i\int \tilde{\varepsilon}_p^*(\mathbf{r})\chi_2(\mathbf{r})\, d\mathbf{r}\right]\right\rangle \end{aligned} \tag{2.81}$$

where the second line results from the assumption that fluctuations of the continuum and aerosol components are uncorrelated and $\chi_3(\mathbf{r}) = \chi_1(\mathbf{r}) + \chi_2(\mathbf{r})$. (More rigorously, however, an aerosol particle displaces a volume of the continuum that is equal to the aerosol's random volume and therefore slightly alters the overall statistics of the continuum, thus inducing a correlation between these two random processes. This is consistent with the assumption that the aerosols are conservative passive additives since, in this case, the presence of the aerosols simply decreases the level, and not the mechanics, of the continuum fluctuations. This correlation will be a function of both the aerosol concentration and the structure parameter of the continuum field and, relative to their "autocorrelations" which are separately proportional to these parameters, can be neglected at least to first order.) Hence, one has for the characteristic functional of the permittivity field with uncorrelated continuum and aerosol components

$$\Phi_\varepsilon(\chi_1, \chi_2, \chi_3) = \Phi_{\varepsilon_c}(\chi_3)\Phi_{\varepsilon_p}(\chi_1, \chi_2) \tag{2.82}$$

where
$$\Phi_{\varepsilon_c}(\chi_3) \equiv \left\langle \exp\left[i \int \tilde{\varepsilon}_c(\mathbf{r})\chi_3(\mathbf{r})\, d\mathbf{r} \right]\right\rangle \tag{2.83}$$

is the characteristic functional for the continuum component and

$$\Phi_{\varepsilon_p}(\chi_1, \chi_2) \equiv \left\langle \exp\left[i \int \tilde{\varepsilon}_p(\mathbf{r})\chi_1(\mathbf{r})\, d\mathbf{r} + i \int \tilde{\varepsilon}_p^*(\mathbf{r})\chi_2(\mathbf{r})\, d\mathbf{r} \right]\right\rangle \tag{2.84}$$

is that for the aerosol component.

Since the statistics of the continuum component have been obtained in a previous section, there is no need to proceed further with its characteristic functional description beyond how it combines with the aerosol component as shown in Eq. (2.82). However, for completeness, a form of this particular functional will be briefly discussed. An assumption employed in Sec. 1.3.2 of Chap. 1 was that the values of the fluctuations of the permittivity field at a particular point in space (and, obviously, over a sufficiently long interval of time) are governed by gaussian statistics. Many times, such assumptions are necessary to obtain tractable expressions, viz., Eq. (1.27). The application of this assumption also alleviates the need for the types of plausibility arguments, such as that which follows Eq. (1.130) of Sec. 1.6, where, by not assuming specific statistics that prevail for the values of the permittivity field, one is able, through use of the central limit theorem, to treat the complex phase of a propagating wave as a gaussian random function. (However, such specific applications at the level of the complex phase are not as restrictive as those at the level of the permittivity.) In fact, it is through the use of the central limit theorem that one can motivate the use of gaussian statistics for the permittivity field. As was shown in Sec. 2.3.2, the spatial variation of the permittivity field subtends a continuously cascading chain of inhomogeneities within the inertial-convective subrange. The ever-present wind within the atmosphere translates this collection of inhomogeneities, with their attendant variety of permittivity values that deviate from the norm of zero, across a fixed point in the atmosphere. Such a large assembly of inhomogeneities will, by the central limit theorem, assume a gaussian distribution about the zero mean. Thus, with $\tilde{\varepsilon}_c(\mathbf{r})$ a gaussian random function where $\langle \tilde{\varepsilon}_c(\mathbf{r}) \rangle = 0$, the variable that appears as the argument of the characteristic functional of Eq. (2.83) (sometimes called the *second characteristic functional*), i.e.,

$$\zeta \equiv i \int \tilde{\varepsilon}_c(\mathbf{r}) \chi_3(\mathbf{r})\, d\mathbf{r} \qquad \langle \zeta \rangle = 0 \tag{2.85}$$

is also described by gaussian statistics. Hence, remembering Eq. (1.131) and substituting into this relation Eq. (2.85) yields

$$\left\langle \exp\left[i \int \tilde{\varepsilon}_c(\mathbf{r}) \chi_3(\mathbf{r})\, d\mathbf{r} \right] \right\rangle = \exp\left\{ \left\langle \frac{1}{2} \left[i \int \tilde{\varepsilon}_c(\mathbf{r}) \chi_3(\mathbf{r})\, d\mathbf{r} \right]^2 \right\rangle \right\}$$

Expanding the squared term in the argument on the right side and using Eq. (2.83) once again finally gives

$$\Phi_{\varepsilon_c}[\chi_3(\mathbf{r})] = \exp\left[-\frac{1}{2} \int \int \langle \tilde{\varepsilon}_c(\mathbf{r}_1) \tilde{\varepsilon}_c(\mathbf{r}_2) \rangle \chi_3(\mathbf{r}_1) \chi_3(\mathbf{r}_2)\, d\mathbf{r}_1\, d\mathbf{r}_2 \right] \tag{2.86}$$

Consider now the characteristic functional for the aerosol permittivity component. Unlike the case just addressed for the continuum component, the following development is critical for bringing the statistical description of the effect of aerosols to the same level as that already obtained for the continuum in Sec. 2.3.3. Substituting Eq. (2.65) into Eq. (2.84) gives for spherical aerosols of radius a

$$\Phi_{\varepsilon_p}(\chi_1, \chi_2) = \left\langle \exp\left[i \int \sum_{i=1}^{n} V_0(\mathbf{r} - \mathbf{R}_i; a) \chi_1(\mathbf{r})\, d\mathbf{r} \right.\right.$$
$$\left.\left. + \, i \int \sum_{i=1}^{n} V_0^*(\mathbf{r} - \mathbf{R}_i; a) \chi_2(\mathbf{r})\, d\mathbf{r} \right] \right\rangle \tag{2.87}$$

One must now evaluate the ensemble average indicated here, making use of the previously established statistics taken to model aerosol behavior in the atmosphere. This commences with averaging over the N aerosols that are taken to occupy the arbitrary volume V and which occur with a random radius described by the probability density $p(a)$. Remembering the statistical independence that allows the joint probability density of Eq. (2.8) to be written, one has for the characteristic functional $\Phi_{\varepsilon_p}(\chi_1, \chi_2; N)$ specific to N aerosols

$$\begin{aligned}\Phi_{\varepsilon_p}(\chi_1, \chi_2; N) &= \prod_{i=1}^{N} \int_0^\infty \int_{-\infty}^{\infty} \exp\left[i \int \sum_{i=1}^{N} V_0(\mathbf{r} - \mathbf{R}_i; a)\chi_1(\mathbf{r})\, d\mathbf{r} \right. \\ &\quad \left. + i \int \sum_{i=1}^{N} V_0^*(\mathbf{r} - \mathbf{R}_i; a)\chi_2(\mathbf{r})\, d\mathbf{r} \right] \frac{1}{V}\, d\mathbf{R}_i\, p(a)\, da \\ &= \left\{ \int_0^\infty \int_{-\infty}^{\infty} \exp\left[i \int V_0(\mathbf{r} - \mathbf{R}_0; a)\chi_1(\mathbf{r})\, d\mathbf{r} \right. \right. \\ &\quad \left. \left. + i \int V_0^*(\mathbf{r} - \mathbf{R}_0; a)\chi_2(\mathbf{r})\, d\mathbf{r} \right] \frac{1}{V}\, d\mathbf{R}_0\, p(a)\, da \right\}^N \end{aligned} \tag{2.88}$$

One must now average this specific characteristic functional for N aerosols over all possible values of N. Thus, using Eq. (2.9), one has

$$\Phi_{\varepsilon_p}(\chi_1, \chi_2) = \sum_{N=0}^{\infty} \frac{\langle N \rangle^N}{N!} \exp(-\langle N \rangle) \Phi_{\varepsilon_p}(\chi_1, \chi_2; N) \tag{2.89}$$

Using the identification

$$Z(V_0, \chi_1, \chi_2) \equiv \exp\left[i \int V_0(\mathbf{r} - \mathbf{R}_0; a)\chi_1(\mathbf{r})\, d\mathbf{r} + i \int V_0^*(\mathbf{r} - \mathbf{R}_0; a)\chi_2(\mathbf{r})\, d\mathbf{r} \right] \tag{2.90}$$

and substituting Eq. (2.88) into Eq. (2.89) gives

$$\begin{aligned}\Phi_{\varepsilon_p}(\chi_1, \chi_2) &= \exp(-\langle N \rangle) \sum_{N=0}^{\infty} \frac{1}{N!} \langle N \rangle^N \\ &\quad \times \left[\int_0^\infty \int_{-\infty}^{\infty} \frac{1}{V} Z(V_0, \chi_1, \chi_2)\, d\mathbf{R}_0\, p(a)\, da \right]^N \\ &= \exp\left[\int_0^\infty \int_{-\infty}^{\infty} \frac{\langle N \rangle}{V} Z(V_0, \chi_1, \chi_2)\, d\mathbf{R}_0\, p(a)\, da - \langle N \rangle \right] \end{aligned} \tag{2.91}$$

Noting that, because of the volume and aerosol number normalization adopted in Section 2.2.3,

$$\vartheta_0 \int_{-\infty}^{\infty} d\mathbf{R}_0 \int_0^{\infty} p(a)\, da = \langle N \rangle \qquad \vartheta_0 \equiv \frac{\langle N \rangle}{V}$$

and employing this relation to represent the second term within the exponential argument of Eq. (2.91) finally yields for the characteristic functional for the aerosols

$$\Phi_\varepsilon(\chi_1, \chi_2) = \exp\left\{ \vartheta_0 \int_0^{\infty} \int_{-\infty}^{\infty} [Z(V_0, \chi_1, \chi_2) - 1]\, d\mathbf{R}_0\, p(a)\, da \right\} \tag{2.92}$$

As was done for the continuum, one can now obtain the same assortment of overall statistics to characterize the contribution to the permittivity due to an assembly of atmospheric aerosols. In particular, using Eq. (D1.2.14) with Eqs. (2.90) and (2.92), one has for the first-order moment of the permittivity due to the presence of aerosols

$$\langle \tilde{\varepsilon}_p(\mathbf{r}) \rangle = \vartheta_0 \int_{-\infty}^{\infty} \int_0^{\infty} p(a) V_0(\mathbf{r} - \mathbf{R}; a)\, da\, d\mathbf{R} \tag{2.93}$$

One then takes the inverse Fourier transform of Eq. (2.80) and substitutes it into this relation and obtains

$$\langle \tilde{\varepsilon}_p(\mathbf{r}) \rangle = -i\vartheta_0 \left(\frac{1}{2\pi}\right)^3 \left(\frac{4\pi}{k}\right) \int_{-\infty}^{\infty} \int_{-\infty}^{\infty} \int_0^{\infty} p(a) a^2 \times \left[\frac{J_1(Ka)}{Ka}\right] \exp\,[i\mathbf{K} \cdot (\mathbf{r} - \mathbf{R})]\, d\mathbf{R}\, da\, d\mathbf{K} \tag{2.94}$$

Evaluating the $\mathbf{R}$ integration, i.e., integrating with respect to all possible aerosol positions, yields a δ function in the $\mathbf{K}$ variable times the factor $(2\pi)^3$, which makes the $\mathbf{K}$ integration straightforward and results in

$$\langle \tilde{\varepsilon}_p(\mathbf{r}) \rangle = -i\vartheta_0 \left(\frac{1}{2}\right) \left(\frac{4\pi}{k}\right) \int_0^{\infty} a^2 p(a)\, da \tag{2.95}$$

where the factor of ½ is the result of the combination of the first-order Bessel function and its argument evaluated at zero. Substituting Eq. (2.13) into Eq. (2.95) and evaluating the resulting integral finally gives

$$\langle \tilde{\varepsilon}_p(\mathbf{r}) \rangle = -i\vartheta_0 \left(\frac{2\pi}{k}\right) \left(\frac{\nu + 1}{\nu}\right) \langle a \rangle^2 \tag{2.96}$$

This relationship indicates that an assembly of aerosols that are each taken to only scatter and not attenuate the propagating electromagnetic radiation will collectively attenuate the wave field.

In a similar fashion, one can apply Eq. (D1.2.15) to Eqs. (2.90) and (2.92) to obtain the second moment of the permittivity field. Following the procedures outlined in Digression 1.2 to evaluate the required second functional derivative of Eq. (2.92) yields

$$\langle \tilde{\varepsilon}_p(\mathbf{r}_1)\tilde{\varepsilon}_p^*(\mathbf{r}_2)\rangle = \vartheta_0 \int_{-\infty}^{\infty}\int_0^{\infty} p(a)V_0(\mathbf{r}_1 - \mathbf{R}; a) \times \left[V_0^*(\mathbf{r}_2 - \mathbf{R}; a) + \vartheta_0 \int_{-\infty}^{\infty}\int_0^{\infty} p(a')V_0^*(\mathbf{r}_2 - \mathbf{R}'; a') \times da'\, d\mathbf{R}' \right] da\, d\mathbf{R} \quad (2.97)$$

The second term within the bracket is the result of the fact that the first-order moment is nonzero. However, it is also second-order in the concentration and aerosol size distribution and thus can be neglected in a first approximation. Especially when cases are considered where the "cooperative" effects can be neglected, as they are here, this is a palatable approximation. Hence, one has

$$\langle \tilde{\varepsilon}_p(\mathbf{r}_1)\tilde{\varepsilon}_p^*(\mathbf{r}_2)\rangle \approx \vartheta_0 \int_{-\infty}^{\infty}\int_0^{\infty} p(a)V_0(\mathbf{r}_1 - \mathbf{R}; a)V_0^*(\mathbf{r}_2 - \mathbf{R}; a)\, da\, d\mathbf{R} \quad (2.98)$$

Substituting the inverse of Eq. (2.80) into this relation and, as done above, simplifying by doing all the integrations that are possible yields

$$\langle \tilde{\varepsilon}_p(\mathbf{r}_1)\tilde{\varepsilon}_p^*(\mathbf{r}_2)\rangle = \vartheta_0 \left(\frac{2}{\pi}\right)\left(\frac{1}{k}\right)^2 \int_{-\infty}^{\infty}\int_0^{\infty} \left[\frac{J_1(Ka)}{Ka}\right]^2 \times a^4 p(a) \exp\left[i\mathbf{K}\cdot(\mathbf{r}_1 - \mathbf{r}_2)\right] da\, d\mathbf{K} \quad (2.99)$$

which shows that the aerosol permittivity statistics are homogeneous and isotropic (of course, this is to be expected, since each aerosol particle is modeled to be spherical and no other direction was specialized in the treatment), and also, upon remembering Eq. (D1.3.6), immediately allows one to write for the associated spectral density

$$\Phi_{\varepsilon_p}(|\mathbf{K}|) = \Phi_{\varepsilon_p}(K) = \vartheta_0 \left(\frac{2}{\pi}\right)\left(\frac{1}{k}\right)^2 \int_0^{\infty} \left[\frac{J_1(Ka)}{Ka}\right]^2 a^4 p(a)\, da \quad (2.100)$$

Using Eq. (2.13) in this relation and simplifying by changing variables gives

$$\Phi_{\varepsilon_p}(K) = C^2_{\varepsilon_p}\varphi_{\varepsilon_p}(K) \tag{2.101}$$

where

$$C^2_{\varepsilon_p} \equiv \vartheta_0 \left(\frac{2}{\pi}\right)\left(\frac{1}{k}\right)^2 \langle a\rangle \tag{2.102}$$

is the aerosol "structure parameter" and

$$\varphi_{\varepsilon_p}(K) \equiv \langle a\rangle \left(\frac{1}{K}\right)^2 \left(\frac{1}{\nu^2\Gamma(\nu)}\right) \int_0^\infty x^{\nu+1} J_1^2\left(K\frac{\langle a\rangle}{\nu}x\right) \exp(-x)\, dx \tag{2.103}$$

is the associated basic spectral density.

One can finally write the statistical modeling equations for the permittivity field of an atmosphere containing the conservative passive additives of temperature, water vapor, and aerosols and satisfying the prevailing assumptions of this chapter by combining the results of this section with those of Sec. 2.3.3. There is, however, one minor point that must still be established. Throughout the various discussions and developments of this chapter, the parameters of permittivity and refractive index have been used. Though these quantities endeavor to describe the same physical concept, they do assume different mathematical forms in the equations. In particular, from Eq. (2.1), which connects these parameters, one has that for small variations,

$$\tilde{\varepsilon}(\mathbf{r}) = \frac{\partial\varepsilon}{\partial n}\,\tilde{n}(\mathbf{r}) = 2n\tilde{n}(\mathbf{r}) \tag{2.104}$$

Thus, only the levels of the fluctuations in these two quantities are different. In terms of the relevant structure parameters, defined operationally via Eq. (2.59), for a random atmosphere with a nominal refractive index equal to one, Eq. (2.104) and a recollection of the definition of a structure function give for the structure parameters of permittivity and refractive index

$$C^2_\varepsilon = 4C^2_n \tag{2.105}$$

Hence, although the model equations that follow below will be stated in terms of the permittivity, they are easily connected to the refractive index description through Eqs. (2.1), (2.104), and (2.105).

In the case of the first-order moment of the permittivity, one has the decomposition

$$\langle\varepsilon(\mathbf{r})\rangle = \langle\varepsilon_{0c}(\mathbf{r})\rangle + \langle\varepsilon_{0p}(\mathbf{r})\rangle + \langle\tilde{\varepsilon}_c(\mathbf{r})\rangle + \langle\tilde{\varepsilon}_p(\mathbf{r})\rangle$$

where the first two terms on the right side represent the averages of the nominal values due to the continuum and the aerosols, and the last two terms are the averages of the fluctuations of these components. From what was said above, the "nominal atmosphere" is one in which

$$\langle \varepsilon_{0c}(\mathbf{r}) \rangle = 1 \qquad \langle \varepsilon_{0p}(\mathbf{r}) \rangle = 0$$

where the last condition arises from the requirement that the nominal atmosphere is one with no aerosols present. The fluctuations of the continuum component are then taken to have a zero mean, whereas the "fluctuations" due to the aerosol components (i.e., their presence) are given by Eq. (2.96). Hence, one has

$$\langle \varepsilon(\mathbf{r}) \rangle = 1 - i\vartheta_0 \left(\frac{2\pi}{k}\right)\left(\frac{\nu + 1}{\nu}\right) \langle a \rangle^2 \tag{2.106}$$

As for the second-order statistics, one has the spectral densities of Eqs. (2.61) and (2.100). Since the continuum and aerosol components are taken to be statistically independent and the homogeneous and isotropic statistics of the spectrum of Eq. (2.100) are a special case of the more general local statistics of the spectrum of Eq. (2.61), one has the locally homogeneous and isotropic spectrum

$$\begin{aligned} \Phi_\varepsilon(K) &= \Phi_{\varepsilon_c}(K) + \Phi_{\varepsilon_p}(K) \\ &= \frac{0.033 C_\varepsilon^2}{(K^2 + K_0^2)^{11/6}} \exp\left(\frac{-K^2}{K_m^2}\right) \\ &\quad + C_{\varepsilon_p}^2 \langle a \rangle \left(\frac{1}{K}\right)^2 \left(\frac{1}{\nu^2 \Gamma(\nu)}\right) \int_0^\infty x^{\nu+1} J_1^2 \left(K \frac{\langle a \rangle}{\nu} x\right) \\ &\quad \times \exp(-x)\, dx \end{aligned} \tag{2.107}$$

where, remembering Eq. (2.105),

$$C_\varepsilon^2 = 4 \left[\left(\frac{A_T}{\langle T \rangle}\right)^2 C_T^2 + 2 \left(\frac{A_T A_Q}{\langle T \rangle \langle Q \rangle}\right) C_{TQ} + \left(\frac{A_Q}{\langle Q \rangle}\right)^2 C_Q^2 \right] \tag{2.108}$$

from Eq. (2.60). Equations (2.106) to (2.108) compose the desired model for the random permittivity field of the atmosphere that, when used in conjunction with the results of Chap. 1, allows one to describe a plethora of statistical propagation problems at visible wavelengths.

It will prove convenient in the calculations that will follow in Chap. 4 to derive approximate analytical expressions for the aerosol spectral density $\Phi_{\varepsilon_p}(K)$ represented by the second term in Eq. (2.107) as obtained from Eqs. (2.100) to (2.103). In particular, the integral within this term is given in the form of an infinite series over a generalized hypergeometric function, viz.,[22]

$$\int_0^\infty x^{\nu+1} J_1^2\left(K\frac{\langle a\rangle}{\nu}x\right)\exp(-x)\,dx = \left(\frac{K\langle a\rangle}{2\nu}\right)^2 \sum_{m=0}^{\infty} \frac{\Gamma(\nu+3+2m)}{m!\Gamma(2+m)}\,{}_2F_1[-m, -(1+m); 2; 1] \times \left[-\left(\frac{K\langle a\rangle}{2\nu}\right)^2\right]^m \tag{2.109}$$

Consider now the case where $K\langle a\rangle/(2\nu) \ll 1$. From Table 2.2, one sees that for radiative fogs, this condition restricts the use of the spatial frequencies to $K < 2.4 \times 10^6\ \mathrm{m}^{-1}$. Similarly, from Table 2.3, one has for the MP rain distribution at a rain rate of 50 mm/h, $K < 4.2 \times 10^3\ \mathrm{m}^{-1}$. Hence, these ranges are well within the inertial subrange of the turbulence field which is governed by the spectral cutoff of $K < 5.9 \times 10^3\ \mathrm{m}^{-1}$ [see Eq. (2.57) with $l_0 = 1 \times 10^{-3}\ \mathrm{m}^{-1}$]. Therefore, when used in analyses with the turbulence background included, one can consider the aerosol component under the condition $K\langle a\rangle/(2\nu) \ll 1$, thus allowing

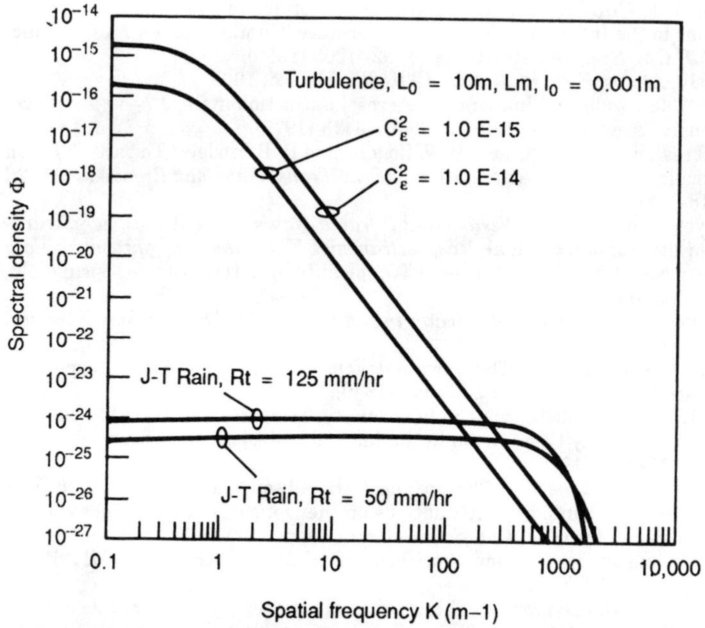

Figure 2.1 Plot of Eq. (2.107) using the approximation of Eq. (2.111) showing Kolmogorov and aerosol spectra components for typical parameter values.

Eq. (2.109) to be simplified considerably. Retaining only the first two terms in the series of Eq. (2.109) and substituting the resulting expression into Eq. (2.103), one has

$$\Phi_{\epsilon_p}(K) \approx \frac{\vartheta_0 \langle a \rangle^4}{2\pi k^2} \frac{(\nu + 2)(\nu + 1)}{\nu^3} \left[1 - (\nu + 4)(\nu + 3) \left(\frac{K\langle a \rangle}{2\nu} \right)^2 + \cdots \right] \tag{2.110}$$

Finally, for analytical purposes, one can consider the two terms within the brackets of Eq. (2.110) as the first of an exponential series, thus allowing one to write for the (approximate) aerosol spectral density

$$\Phi_{\epsilon_p}(K) \approx \frac{\vartheta_0 \langle a \rangle^4}{2\pi k^2} \frac{(\nu + 2)(\nu + 1)}{\nu^3} \exp \left[-(\nu + 4)(\nu + 3) \left(\frac{K\langle a \rangle}{2\nu} \right)^2 \right] \tag{2.111}$$

Figure 2.1 shows a comparison of the Kolmogorov and aerosol spectral densities.

References

1. R. J. Hill, S. F. Clifford, and R. S. Lawrence, "Refractive Index and Absorption Fluctuations in the Infrared Caused by Temperature, Humidity, and Pressure Fluctuations," *J. Opt. Soc. Am.* **70** (10), pp. 1192–1205 (1980).
2. R. M. Goody, *Atmospheric Radiation*. Claredon, Oxford, 1965.
3. B. Nilsson, "Meteorological Influence on Aerosol Extinction in the 0.2–40 μm Wavelength Range," *Appl. Opt.* **18** (20), pp. 3457–3473 (1979).
4. For a good review, see R. L. Olsen, D. V. Rogers, and D. B. Hodge, "The aR^b Relation in the Calculation of Rain Attenuation," *IEEE Trans. Antennas Propagat.* **AP-26** (2), pp. 318–329 (1978).
5. V. Ye. Zuyev, *Propagation of Visible and Infrared Waves in the Atmosphere,* translation from the Russian original, *Rasprostraneniye Vidimykh i Infrakrasnykh Voln v Atmosfere.* NASA TT F-707. National Technical Information Service, Springfield, Va., 1972. Chapter 5.
6. D. Deirmendjian, *Electromagnetic Scattering on Spherical Polydispersions.* Elsevier, New York, 1964.
7. R. Gunn and G. D. Kinzer, "The Terminal Velocity of Fall for Water Droplets in Stagnant Air," *J. Meteor.* **6**, pp. 243–248 (1949).
8. R. J. Hill, "Spectra of Fluctuations in Refractivity, Temperature, Humidity, and the Temperature-Humidity Cospectrum in the Inertial and Dissipation Ranges," *Radio Sci.* **13**, pp. 953–961 (1978).
9. C. A. Friehe, J. C. La Rue, F. H. Champagne, C. H. Gibson, and G. F. Dreyer, "Effects of Temperature and Humidity Fluctuations on the Optical Refractive Index in the Marine Boundary Layer," *J. Opt. Soc. Am.* **65** (12), pp. 1502–1511 (1975).
10. A. S. Monin and A. M. Yaglom, *Statistical Fluid Mechanics,* Vol. 1. MIT Press, Cambridge, Mass., 1973. Chap. 1.
11. O. Reynolds, "On the Dynamical Theory of Incompressible Viscous Fluids and the Determination of the Criterion," *Trans. Roy. Soc. London* **186**, pp. 123–161 (1894).
12. A. N. Kolmogorov, "Local Structure of Turbulence in an Incompressible Fluid at Very High Reynolds Numbers," *Dokl. Akad. Nauk SSSR* **30** (4), pp. 299–303 (1941).

13. A. S. Monin and A. M. Yaglom, *Statistical Fluid Mechanics,* Vol. 2. MIT Press, Cambridge, Mass., 1975. Chap. 7.
14. A. M. Obukhov, "Structure of the Temperature Field in a Turbulent Flow," *Izv. Akad. Nauk SSSR, Ser. Georg. i Geofiz* **13** (1), pp. 58–69 (1949).
15. N. K. Vinnichenko and J. A. Dutton, "Empirical Studies of Atmospheric Structure and Spectra in the Free Atmosphere," *Radio Sci.* **4** (12), pp. 1115–1126 (1969).
16. J. C. Wyngaard, Y. Izumi, and S. A. Collins, Jr., "Behavior of the Refractive Index Structure Parameter Near the Ground," *J. Opt. Soc. Am.* **61** (12), pp. 1646–1650 (1971).
17. J. Spencer, "Long-Term Statistics of Atmospheric Turbulence Near the Ground," Rome Air Development Center Technical Report, RADC-TR-78-182, August 1978.
18. I. V. Obukhov, "Effect of the Motion of an Optical Radiation Source on Heterodyne Reception in the Presence of Atmospheric Turbulence," *Radio Eng. Electron. Phys.* **28** (8), pp. 58–61 (1983).
19. V. P. Kukharets and L. R. Tsvang, "Structure Parameter of the Refractive Index in the Atmospheric Boundary Layer," *Atmos. Ocean. Phys.* **16** (2), pp. 73–77 (1980).
20. S. Chandrasekhar, "Stochastic Problems in Physics and Astronomy," *Rev. Mod. Phys.* **15** (1), pp. 1–89 (1943).
21. H. C. van de Hulst, *Light Scattering by Small Particles.* Wiley, New York, 1957.
22. I. S. Gradshteyn and I. M. Ryzhik, *Table of Integrals, Series, and Products.* Academic Press, New York, 1980. Eq. (6.626.1).

Chapter

3

The Formation and Characterization of Images Viewed through Random Media

3.1 Introduction

This chapter will address the basic theory that provides the statistical description of the formation of a diffraction image of an extended object as viewed through a random medium. In Sec. 3.2, the Green function solution of the stochastic parabolic equation, given in Sec. 1.5 of Chap. 1, is combined with the relations governing wave field transformation through a thin lens of finite aperture size to yield a relation that connects the electric field in the object plane (due to the radiation that scatters or is emitted by the object) with that which results in the image plane of the lens after propagating through the random atmosphere and the optics of the imaging system. Various concepts such as the lens law and equivalent image and object coordinates are established within the approximation of geometrical optics, and the Rayleigh criterion for resolution is discussed. The first-order moment of the random intensity distribution in the image plane of a lens is analyzed in Sec. 3.3 for the case of incoherent imaging. Here, the concepts of coherent versus incoherent imaging are discussed, as well as those of point spread functions for nonisoplanatic and isoplanatic propagation. This then allows the well-known Fourier optics description of the modulation transfer function approach to imaging which employs that of the lens and the associated mutual coherence function of spherical wave propagation. The temporal aspects of image formation are discussed in Sec. 3.4, where short-term and long-term imaging are discussed, as well as the mathematical formulation of the former in terms of statistical Zernike polynomials. Higher-order statistical moments of the random image plane intensity field are the subject of Sec. 3.5, and finally, in Sec. 3.6, a definition and assessment of object resolution are developed.

3.2 The Analysis of Diffraction Image Formation by a Lens of an Object in a Random Medium

Figure 3.1 shows the basic rudiments of a visible or infrared image formation system. The extended object being imaged is taken to exist in the two-dimensional *object plane* defined by the transverse coordinate $\boldsymbol{\rho}_0$ referred to the axis of the imaging system (as are all other transverse coordinates unless otherwise indicated); the longitudinal distance L which separates the object plane from the input aperture of the lens is such that all longitudinal dimensions of the object can be neglected and the object can be "projected" onto the object plane, allowing it to extend only in the transverse direction. The electric field $E_0(\boldsymbol{\rho}_0)$ that is responsible for the intensity distribution through which one sees an object could have as its origin the infrared emission due to heating (i.e., a self-luminous object or target) or the reflection of some remotely located and directed source, such as a spotlight or laser beam.

The initial field $E_0(\boldsymbol{\rho}_0)$ establishes a wave field $E(x, \boldsymbol{\rho})$ which propagates through the random medium that is taken to intervene between

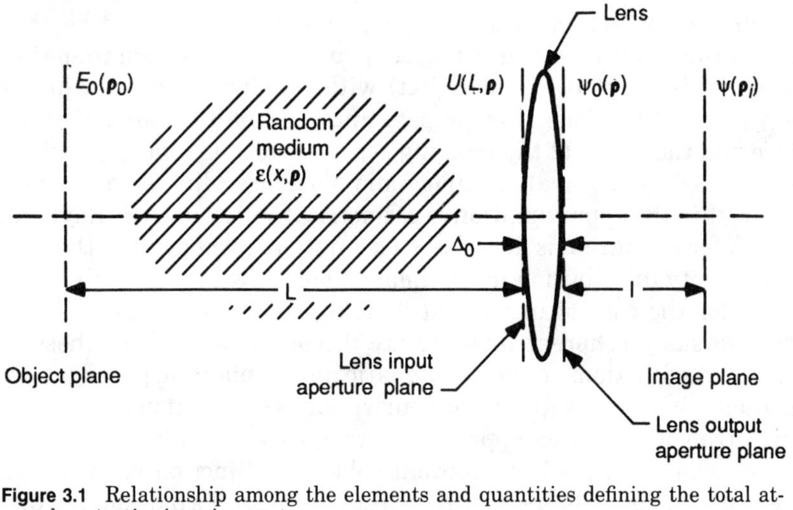

Figure 3.1 Relationship among the elements and quantities defining the total atmosphere/optics imaging system.

the object plane and the *input aperture plane* of the lens. The resulting field in the input aperture plane is connected with the field in the object plane by Eq. (1.93) with the use of Eq. (1.32), viz.,

$$E(L, \boldsymbol{\rho}) = \frac{ik}{2\pi L} \int_{-\infty}^{\infty} \exp\left[-i\frac{k(\boldsymbol{\rho} - \boldsymbol{\rho}_0)^2}{2L}\right] G_{\tilde{\varepsilon}}(L, \boldsymbol{\rho}; 0, \boldsymbol{\rho}_0) E_0(\boldsymbol{\rho}_0)\, d\boldsymbol{\rho}_0 \quad (3.1)$$

where the stochastic Green function $G_{\tilde{\varepsilon}}(L, \boldsymbol{\rho}; 0, \boldsymbol{\rho}_0)$ is the *spherical wave solution* of the various forms of equations derived in Chap. 1 that describe it, e.g., the parabolic equation of Eq. (1.94) or its Rytov transformation representation of Eq. (1.96). In the case where propagation takes place through a vacuum, or at least a homogeneous deterministic medium, one has $G_{\tilde{\varepsilon}} = 1$, and Eq. (3.1) becomes the well-known solution for diffraction propagation from object to lens obtained by application of the rigorous Fresnel-Kirchhoff diffraction theory.[1,2]

The input aperture plane field is then transformed, via the refractive capability of the lens, to the field $\Psi_0(L, \boldsymbol{\rho}')$ that is defined in the *output aperture plane* of the lens. Such field transformation properties of the lens, which act on the phase of the incoming wave field at the input so as to produce a convergent spherical wave at the output, are described by defining and deriving a transmission function $\tau(\boldsymbol{\rho}, \boldsymbol{\rho}')$ of the lens. In most cases, one can consider the lens in the "thin lens" approximation, where, using the verbiage of geometrical optics, a coordinate of an incoming ray in the input aperture plane is the same as that of the outgoing ray in the output aperture plane, i.e., $\boldsymbol{\rho} = \boldsymbol{\rho}'$. Then, by the well-known treatment[2] of such a circular thin lens with maximum thickness Δ_0 (which, by definition of a thin lens, must be much smaller than any other linear dimension entering the problem) and composed of material of refractive index n, one has that the fields $E(L, \boldsymbol{\rho})$ and $\Psi_0(L, \boldsymbol{\rho})$ are related by

$$\Psi_0(L, \boldsymbol{\rho}) = \tau(\boldsymbol{\rho}) E(L, \boldsymbol{\rho}) \quad (3.2)$$

where the transmission function is given by the phase transformation law

$$\tau(\boldsymbol{\rho}) = W(\boldsymbol{\rho}) \exp(-ikn\Delta_0) \exp\left(\frac{ik\rho^2}{2F}\right) \quad (3.3)$$

where the lens focal length F is related to the radius of curvature R_1 of the concave part of the lens on the output side and the radius of curvature R_2 of the convex part of the lens on the input side by the relation

$$\frac{1}{F} = (n - 1)\left(\frac{1}{R_1} - \frac{1}{R_2}\right) \quad (3.4)$$

and where $W(\boldsymbol{\rho})$ is the pupil function of the lens, which accounts for its transmission of a limited portion of the wave field due to its finite diameter D and is simply given by

$$W(\boldsymbol{\rho}) = \begin{cases} 1 & |\boldsymbol{\rho}| \leq \dfrac{D}{2} \\ 0 & |\boldsymbol{\rho}| > \dfrac{D}{2} \end{cases} \tag{3.5}$$

Finally, one must consider the transformation of the field distribution $\Psi_0(L, \boldsymbol{\rho})$ through the deterministic medium of the imaging system into that $\Psi(\boldsymbol{\rho}_i)$ which exists in the plane at which the diffraction image is formed, i.e., the *imaging plane*. The classical solution of this problem from the Maxwell equations takes the form of the well-known Helmholtz-Kirchhoff integral theorem and forms the basis of scalar diffraction theory; application of the Kirchhoff approximation then reduces it to the classical Fresnel-Kirchhoff diffraction relation, which, upon using the Fresnel approximation, yields

$$\Psi(\boldsymbol{\rho}_i) = \frac{ik}{2\pi l} \int_{-\infty}^{\infty} \Psi_0(\boldsymbol{\rho}) \exp\left[-i\,\frac{k(\boldsymbol{\rho}_i - \boldsymbol{\rho})^2}{2l}\right] d\boldsymbol{\rho} \tag{3.6}$$

One can also arrive at this expression by a judicious use of Eqs. (1.93) and (1.32) as employed in the derivation of Eq. (3.1) applied to a deterministic medium.

Making the appropriate substitutions between Eqs. (3.1) to (3.3) and Eq. (3.6) gives, upon rearranging terms,

$$\begin{aligned}\Psi(\rho_i) = A \int_{-\infty}^{\infty}\int_{-\infty}^{\infty} \exp\Bigg[&-\frac{ik}{2}\left(\frac{\rho_i^2}{l} + \frac{\rho_0^2}{L}\right) \\ &+ ik\left(\frac{\boldsymbol{\rho}_i}{l} + \frac{\boldsymbol{\rho}_0}{L}\right)\cdot\boldsymbol{\rho} - \frac{ik}{2}\left(\frac{1}{l} + \frac{1}{L} - \frac{1}{F}\right)\rho^2\Bigg] \\ &\times W(\boldsymbol{\rho}) G_{\hat{\varepsilon}}(L, \boldsymbol{\rho}; 0, \boldsymbol{\rho}_0) E_0(\boldsymbol{\rho}_0)\, d\boldsymbol{\rho}_0\, d\boldsymbol{\rho}\end{aligned} \tag{3.7}$$

where the A is a constant phase factor,

$$A \equiv -\left(\frac{k}{2\pi}\right)^2 \left(\frac{1}{Ll}\right) \exp\,(-ikn\Delta_0) \tag{3.8}$$

Equation (3.7) relates the field distribution in the imaging plane to that in the object plane. Because of the presence of the random medium through which the field must propagate, represented by the stochastic

Green function $G_{\tilde{\varepsilon}}$, the field distribution $\Psi(\boldsymbol{\rho}_i)$ is also a random function. Before analyzing Eq. (3.7) in its entirety, however, it is expedient to understand the essence of image formation by what is essentially the diffraction pattern $\Psi(\boldsymbol{\rho}_i)$ established in the image plane by the lens of the field distribution $E_0(\boldsymbol{\rho}_0)$ created by the object.

To this end, one first considers, for brevity of analysis, the application of Eq. (3.7) to the nonrandom propagation case where $G_{\tilde{\varepsilon}} = 1$. Second, the geometrical optics approximation will be employed, since it is only in this situation that one can classically define the intensity distribution $|\Psi(\boldsymbol{\rho}_i)|^2$ as the "image" of the object distribution $|E_0(\boldsymbol{\rho}_0)|^2$ by connecting these in a one-to-one mapping of an object point to an image point. Taking the condition

$$\frac{1}{l} + \frac{1}{L} - \frac{1}{F} = 0 \tag{3.9}$$

to prevail over the fundamental lengths that enter into the problem, one obtains from Eq. (3.7) in the nonrandom propagation case

$$\Psi(\boldsymbol{\rho}_i) = A\left(\frac{l}{k}\right)^2 \int_{-\infty}^{\infty}\int_{-\infty}^{\infty} \exp\left[-\frac{ik}{2l}(\rho_i^2 + M\rho_0^2) + i\mathbf{q}\cdot(\boldsymbol{\rho}_i + M\boldsymbol{\rho}_0)\right] \times W\left(l\frac{\mathbf{q}}{k}\right) E_0(\boldsymbol{\rho}_0)\, d\mathbf{q}\, d\boldsymbol{\rho}_0 \tag{3.10}$$

where the change of variables

$$\mathbf{q} \equiv \frac{k}{l}\boldsymbol{\rho} \qquad M \equiv \frac{l}{L}$$

was employed to facilitate going over to the geometrical optics approximation, where one has (relative to all other lengths that enter into the problem) $\lambda \to 0$, thus requiring $k \to \infty$. Noting that in this limit, $W(l(\mathbf{q}/k)) \to W(0) = 1$, Eq. (3.10) yields

$$\begin{aligned}\Psi(\boldsymbol{\rho}_i) &= A\left(\frac{l}{k}\right)^2 \int_{-\infty}^{\infty} \exp\left[-\frac{ik}{2l}(\rho_i^2 + M\rho_0^2)\right] \\ &\quad\times \int_{-\infty}^{\infty} \exp\left[i\mathbf{q}\cdot(\boldsymbol{\rho}_i + M\boldsymbol{\rho}_0)\right] d\mathbf{q}\, E_0(\boldsymbol{\rho}_0)\, d\boldsymbol{\rho}_0 \\ &= \left(\frac{L}{l}\right) \exp\left\{-ik\left[\frac{\rho_i^2(M-1)}{2lM} + n\Delta_0\right]\right\} E_0\left(-\frac{\boldsymbol{\rho}_i}{M}\right)\end{aligned} \tag{3.11}$$

Thus, the intensity distribution $I(\boldsymbol{\rho}_i)$ in the image plane is related to that in the object plane $I_0(\boldsymbol{\rho}_0)$ by

$$I(\boldsymbol{\rho}_i) = |\Psi(\boldsymbol{\rho}_i)|^2 = \left(\frac{L}{l}\right)^2 I_0\left(-\frac{\boldsymbol{\rho}_i}{M}\right) \tag{3.12}$$

indicating that the intensity at a point $\boldsymbol{\rho}_i$ in the image plane is connected via a one-to-one contraction mapping with the intensity at the point $\boldsymbol{\rho}_0 = -\boldsymbol{\rho}_i M^{-1}$ in the object plane; i.e., an exact reproduction of the image of the object can be obtained that is inverted and contracted by the factor $M = l/L$, sometimes referred to as the *magnification factor*. With respect to the image plane coordinate $\boldsymbol{\rho}_i$, the quantity $-\boldsymbol{\rho}_i M^{-1}$ is referred to as the *equivalent image coordinate* $\boldsymbol{\rho}_i'$. Analogously, with respect to the object plane coordinate $\boldsymbol{\rho}_0$, the quantity $-\boldsymbol{\rho}_0 M$ is referred to as the *equivalent object coordinate* $\boldsymbol{\rho}_0'$.

The derivation of the geometrical optics relation of Eq. (3.12) relies on two major points. First, the seemingly arbitrary condition of Eq. (3.9) was assumed. However, it is through this condition that one is able to obtain, within the approximation of geometrical optics, the reproduction in the image plane of the object intensity distribution, giving rise to what is classically defined as the image of the object. Equation (3.9) specifies the relation that must exist between the focal length F of the lens and the distances L and l for the object to be "focused" in the image plane, and for this reason is called the *lens law* of geometrical optics.

The second point is that Eq. (3.12) is derived on the assumption that the propagating radiation field is unrestricted, viz., the geometrical optics approximation allows one to neglect the effect of the pupil function $W(\boldsymbol{\rho})$ which, in the general case, forms a boundary condition on the radiation field upon entering the lens. In the more general case of Eq. (3.10) where the diffractive effects due to a finite pupil size are taken into account, each point of the object is represented by the Fraunhoffer diffraction pattern of the lens pupil. It is this effect that places fundamental restrictions on the resolution capability of an imaging system. This will now be briefly demonstrated.

From Eq. (3.10), one has for the deterministic intensity in the focal plane

$$\begin{aligned} I(\boldsymbol{\rho}_i) \equiv \Psi(\boldsymbol{\rho}_i)\Psi^*(\boldsymbol{\rho}_i) = |A|^2 \left(\frac{l}{k}\right)^4 \int_{-\infty}^{\infty}\int_{-\infty}^{\infty}\int_{-\infty}^{\infty} \exp\bigg[&-\frac{ik}{2l} M(\rho_0^2 - \rho_0'^2) \\ &+ iM(\mathbf{q}\cdot\boldsymbol{\rho}_0 - \mathbf{q}'\cdot\boldsymbol{\rho}_0') + i(\mathbf{q}-\mathbf{q}')\cdot\boldsymbol{\rho}_i\bigg] \\ \times W\left(\frac{l\mathbf{q}}{k}\right) W^*\left(\frac{l\mathbf{q}'}{k}\right) &E_0(\boldsymbol{\rho}_0)E_0^*(\boldsymbol{\rho}_0')\, d\mathbf{q}\, d\boldsymbol{\rho}_0\, d\boldsymbol{\rho}_0'\, d\mathbf{q}' \end{aligned} \tag{3.13}$$

Taking the source to be spatially incoherent, which defines the fact that there is no relation that exists between two distinct field points $\boldsymbol{\rho}_0$ and $\boldsymbol{\rho}_0'$ within the source field, and described by the model

$$E_0(\boldsymbol{\rho}_0)E_0^*(\boldsymbol{\rho}_0') = \frac{4\pi}{k^2} I_0(\boldsymbol{\rho}_0)\delta(\boldsymbol{\rho}_0 - \boldsymbol{\rho}_0') \tag{3.14}$$

where $I_0(\boldsymbol{\rho}_0)$ is the intensity emitted in the object plane at $\boldsymbol{\rho}_0$, substituting this relation into Eq. (3.13), and performing the required integration yields

$$I(\boldsymbol{\rho}_i) = |A|^2 \left(\frac{l}{k}\right)^4 \left(\frac{4\pi}{k^2}\right) \int_{-\infty}^{\infty} \int_{-\infty}^{\infty} \exp\left[-i\mathbf{q}_d \cdot (\boldsymbol{\rho}_i \cdot M\boldsymbol{\rho}_0)\right] \times I_0(\boldsymbol{\rho}_0) K\left(l\frac{\mathbf{q}_d}{k}\right) d\mathbf{q}_d\, d\boldsymbol{\rho}_0 \tag{3.15}$$

where

$$K\left(l\frac{\mathbf{q}_d}{k}\right) \equiv \int_{-\infty}^{\infty} W\left(l\frac{\mathbf{q}}{k}\right) W^*\left(l\frac{\mathbf{q}-\mathbf{q}_d}{k}\right) d\mathbf{q} \tag{3.16}$$

where $\mathbf{q}_d \equiv \mathbf{q} - \mathbf{q}'$. Equation (3.16) gives a quantity that is essentially the overlap area of two circles, each of radius lq/k, whose centers are separated by the distance lq_d/k; it is a parameter of some importance which will be discussed later in this chapter. Performing the indicated Fourier transform over the spatial coordinate $\mathbf{q}_d$ of the convolution defined by Eq. (3.16) yields, upon remembering Eq. (3.8),

$$I(\boldsymbol{\rho}_i) = \left(\frac{4\pi k^2}{L^2 l^2}\right) \int_{-\infty}^{\infty} \left|\hat{W}\left[\frac{k(\boldsymbol{\rho}_i + M\boldsymbol{\rho}_0)}{l}\right]\right|^2 I_0(\boldsymbol{\rho}_0)\, d\boldsymbol{\rho}_0 \tag{3.17}$$

where

$$\hat{W}(\boldsymbol{\kappa}_d) \equiv \left(\frac{1}{2\pi}\right)^2 \int_{-\infty}^{\infty} \exp(-i\boldsymbol{\rho} \cdot \boldsymbol{\kappa}_d) W^*(\boldsymbol{\rho})\, d\boldsymbol{\rho} \tag{3.18a}$$

is the Fourier transform of the pupil function. Using Eq. (3.5) and converting Eq. (3.18a) to plane polar coordinates, one obtains

$$\begin{aligned}
\hat{W}(\boldsymbol{\kappa}_d) &= \left(\frac{1}{2\pi}\right)^2 \int_0^{2\pi} \int_0^{D/2} \exp(-i\rho\kappa_d \cos\phi)\rho\, d\rho\, d\phi \\
&= \left(\frac{1}{2\pi}\right) \int_0^{D/2} J_0(\rho\kappa_d)\rho\, d\rho\, d\phi \\
&= \left(\frac{1}{2\pi}\right)\left(\frac{D}{2}\right)\left(\frac{1}{\kappa_d}\right) J_1\left(\frac{D\kappa_d}{2}\right)
\end{aligned} \tag{3.18b}$$

Substituting this result into Eq. (3.17) gives

$$I(\boldsymbol{\rho}_i) = \left(\frac{D^2}{4\pi L^2}\right) \int_{-\infty}^{\infty} J_1^2 \left(\frac{Dk|\boldsymbol{\rho}_i + M\boldsymbol{\rho}_0|}{2l}\right) \frac{I_0(\boldsymbol{\rho}_0)}{|\rho_i + M\rho_0|^2}\, d\boldsymbol{\rho}_0 \qquad (3.19)$$

which shows that the presence of the finite circular aperture diffracts the intensity such that each point $\boldsymbol{\rho}_0$ in the object plane is represented by the Fraunhoffer diffraction pattern of the lens pupil; the total image intensity at a fixed point $\boldsymbol{\rho}_i$ in the image plane is a composition of such diffraction patterns for all points of the intensity distribution within the object plane.

Remembering the well-known property that the function $J_1^2(x)$ is very sharply peaked at $x = 0$ and falls off rapidly as the value x increases to the value $x_{01} \equiv 3.832$, the first "zero" of the function, consider two points $\boldsymbol{\rho}_0$ and $\boldsymbol{\rho}_0'$ in the object plane viewed from a single point $\boldsymbol{\rho}_i$ in the image plane. Thus, one has two values for the arguments $\boldsymbol{\rho}_i + M\boldsymbol{\rho}_0$ and $\boldsymbol{\rho}_i + M\boldsymbol{\rho}_0'$ within the Bessel functions. Let the central peak of one of these functions coincide with the first zero of the other; disregarding values beyond the first zero of the first such function, it is this condition which assures the minimal effect of the image of the first point $\boldsymbol{\rho}_0$ on that of the second $\boldsymbol{\rho}_0'$, and thus can be used as a criterion for the ability to resolve these two separate points. This situation is represented by the following condition:

$$\frac{Dk}{2l}\left(|\boldsymbol{\rho}_i + M\boldsymbol{\rho}_0| - |\boldsymbol{\rho}_i + M\boldsymbol{\rho}_0'|\right) = x_{01} \qquad (3.20)$$

Defining $\Delta\boldsymbol{\rho}_0 \equiv \boldsymbol{\rho}_0 - \boldsymbol{\rho}_0'$ as the corresponding spatial displacement in the object plane, considering only collinear points, and admitting the condition $l \ll L$, which, using the lens law of Eq. (3.9), gives $l \approx F$, Eq. (3.20) can be rearranged to yield

$$\Delta\rho_0 = \frac{2Fx_{01}}{DkM} = 1.22\,\frac{F\lambda}{D}\left(\frac{1}{M}\right) \qquad (3.21)$$

which is the form of the classical Rayleigh criterion for the resolving capability for imaging scenarios limited by Fraunhoffer diffraction. Equation (3.21) gives the minimum displacement that two points can have in the object plane, and still be recognized as separate points, when viewed with a lens of finite diameter D. In the geometrical optics limit where $\lambda/D \to \infty$, one has $\Delta\boldsymbol{\rho}_0 \to 0$.

Given that one can expect the concepts of the lens law and the equivalent image and object coordinates motivated in the geometrical optics construction above to apply *a priori* to this more general diffractive case, one has, upon using Eq. (3.9) in Eq. (3.7),

$$\Psi(\boldsymbol{\rho}_i) = A \int_{-\infty}^{\infty} T(\boldsymbol{\rho}_i; \boldsymbol{\rho}_0) E_0(\boldsymbol{\rho}_0)\, d\boldsymbol{\rho}_0 \tag{3.22}$$

where

$$T(\boldsymbol{\rho}_i, \boldsymbol{\rho}_0) \equiv \int_{-\infty}^{\infty} \exp\left[-\frac{ik}{2}\left(\frac{\rho_i^2}{l} + \frac{\rho_0^2}{L}\right) + ik\left(\frac{\boldsymbol{\rho}_i}{l} + \frac{\boldsymbol{\rho}_0}{L}\right)\cdot \boldsymbol{\rho}\right] \times W(\boldsymbol{\rho}) G_{\tilde{\varepsilon}}(L, \boldsymbol{\rho}; 0, \boldsymbol{\rho}_0)\, d\boldsymbol{\rho} \tag{3.23}$$

is the composite transmission function which describes the stochastic wave propagation through both the focused imaging system and the atmosphere. One then sees that the Green function describing the random medium, through which the wave field that originates at the object must propagate, can also contribute to an even further reduction in image resolution. At the same time, it is evident from Eq. (3.14) that the pupil function $W(\boldsymbol{\rho})$ is a deterministic property of the imaging system, independent of the random wave field, which can be augmented with a definition much more complicated than that of Eq. (3.5) to counter the deleterious effects of $G_{\tilde{\varepsilon}}$. This, of course, forms the subject of adaptive optics, which is addressed in Chap. 5.

3.3 The Statistical Analysis of the Random Intensity Distribution in the Imaging Plane—The First-Order Moment

Because of the stochastic nature of Eq. (3.22), one can only characterize in a statistical manner quantities which are defined in terms of the diffraction field $\Psi(\boldsymbol{\rho}_i)$. In particular, one has from this relation the second-order spatial moment

$$B_{\Psi\Psi^*}(\boldsymbol{\rho}_{i_1}, \boldsymbol{\rho}_{i_2}) \equiv \langle \Psi(\boldsymbol{\rho}_{i_1}) \Psi^*(\boldsymbol{\rho}_{i_2}) \rangle \tag{3.24}$$

which reduces, at the point $\boldsymbol{\rho}_{i_1} = \boldsymbol{\rho}_{i_2} \equiv \boldsymbol{\rho}_i$, to the ensemble averaged intensity distribution of the diffraction image due to the object field $E_0(\boldsymbol{\rho}_0)$.

$$\begin{aligned} B_{\Psi\Psi^*}(\boldsymbol{\rho}_i, \boldsymbol{\rho}_i) &\equiv \langle |\Psi(\boldsymbol{\rho}_i)|^2 \rangle = \langle I(\boldsymbol{\rho}_i) \rangle \\ &= |A|^2 \int_{-\infty}^{\infty} \int_{-\infty}^{\infty} \langle T(\boldsymbol{\rho}_i; \boldsymbol{\rho}_{0_1}) T(\boldsymbol{\rho}_i^*; \boldsymbol{\rho}_{0_2}) \\ &\qquad \times E_{0_1}(\boldsymbol{\rho}_{0_1}) E_{0_2}^*(\boldsymbol{\rho}_{0_2}) \rangle\, d\boldsymbol{\rho}_{0_1}\, d\boldsymbol{\rho}_{0_2} \end{aligned} \tag{3.25}$$

Equation (3.25) has been written to account for the most general case where statistical correlation may prevail over $T(\boldsymbol{\rho}_i, \boldsymbol{\rho}_0)$ and $E_0(\boldsymbol{\rho}_0)$. To elucidate this possible connection, while, at the same time, bringing to the fore the implications of forming an ensemble average in such an imaging scenario, consider writing for the initial electric field emanating from the object plane

$$E_0(\boldsymbol{\rho}_0) = R(\boldsymbol{\rho}_0)E_I(\boldsymbol{\rho}_0) \tag{3.26}$$

where, in the case of a remotely illuminated target, $R(\boldsymbol{\rho}_0)$ is the complex reflection coefficient of the object surface and $E_I(\boldsymbol{\rho}_0)$ is the field of the illuminating radiation incident on the object surface. [In the case of a self-luminous target, which will be considered later, $R(\boldsymbol{\rho}_0)$ is the emissivity of the target and $E_I(\boldsymbol{\rho}_0)$ is a referenced field of emission.] The source of $E_I(\boldsymbol{\rho}_0)$ could be a laser beam that traverses some or all of the propagation path over which the object viewing is taking place. Hence, this quantity is also a random function that, if a portion of each of the two propagation paths is within the characteristic correlation distance from the other (within the atmosphere, that is, the outer scale of turbulence L_0), is statistically related to $T(\boldsymbol{\rho}_i, \boldsymbol{\rho}_0)$. The reflection coefficient may also be a random function of position on the target due to surface roughness of the object material. Thus, the situation arises where one is implicitly dealing with two distinct ensembles over which averaging takes place: one is the ensemble of the propagation medium statistics, and the other is that of the object reflection statistics.[3] Since the reflection properties of the object are independent of the propagation properties of the propagation medium, these two averages are statistically independent. Hence, substituting Eqs. (3.23) and (3.26) into the integrand of Eq. (3.25) yields

$$\begin{aligned}\langle TT^*E_0E_0^*\rangle &= \exp\left[-\frac{ik}{2L}(\rho_{0_1}^2 - \rho_{0_2}^2)\right] \\ &\quad \times \int_{-\infty}^{\infty}\int_{-\infty}^{\infty} \exp\left[+ik\left(\frac{\boldsymbol{\rho}_i}{l} + \frac{\boldsymbol{\rho}_{0_1}}{L}\right)\cdot\boldsymbol{\rho}' - ik\left(\frac{\boldsymbol{\rho}_i}{l} + \frac{\boldsymbol{\rho}_{0_2}}{L}\right)\cdot\boldsymbol{\rho}''\right] \\ &\quad \times W(\boldsymbol{\rho}')W^*(\boldsymbol{\rho}'') \times \langle G_{\tilde{\varepsilon}}(L, \boldsymbol{\rho}'; 0, \boldsymbol{\rho}_{0_1})G_{\tilde{\varepsilon}}^*(L, \boldsymbol{\rho}''; 0, \boldsymbol{\rho}_{0_2}) \times E_I(\boldsymbol{\rho}_{0_1})E_I^*(\boldsymbol{\rho}_{0_2})\rangle_p \\ &\quad \times \langle R(\boldsymbol{\rho}_{0_1})R^*(\boldsymbol{\rho}_{0_2})\rangle_o \, d\boldsymbol{\rho}' \, d\boldsymbol{\rho}''\end{aligned} \tag{3.27}$$

where the subscript o refers to averaging over the statistics of the object reflectivity, and the subscript p refers to averaging over the statistics of the propagation medium. Hence, the statistical characterization of the diffraction image formation as described by Eq. (3.25) encompasses the creation of these two ensemble averages. Equation (3.27) forms the starting point for the general analysis of partially coherent image propagation of an object of arbitrary surface texture in a random medium. The situation can simplify considerably when Eq. (3.27) is applied to more specific, yet realistic and common, imaging scenarios, as will now be done.

As with any statistical correlation, there is associated a characteristic correlation length, either temporal or spatial as the application may warrant. For example, in the case of the formation of the ensemble average over the object statistics,[3] one has the associated correlation length l_{obj}. Similarly, in the case of averaging over the propagation medium statistics, one has, as discussed in Chap. 1, a related correlation length l_0 which is usually associated with the inner scale of turbulence. There is also, as will be discussed later in this chapter, a temporal correlation time $\tau_0 \equiv D/V$ which is related to the characteristic velocity V of the medium as well as the spatial extent of the lens input aperture, in particular, the lens aperture diameter D. One also has a correlation length that lends itself to the statistical description of the spatial coherence of the incident wave field as it impinges on the target being illuminated;[3] this is the spatial coherence length l_{coh}. As mentioned earlier, the cases considered in this book are those dealing with incoherent sources and wave fields. This means that the coherence length is reduced to the wavelength of the field, i.e., $l_{\text{coh}} \sim \lambda$. Since, from the considerations in Chap. 1, the condition $l_0 \gg \lambda$ is taken to prevail, one has $l_{\text{coh}} \ll l_0$, thus allowing the ensemble average over the propagation medium as indicated in Eq. (3.27) to decouple, viz.,

$$\langle G_{\tilde{\varepsilon}} G_{\tilde{\varepsilon}}^* E_I E_I^* \rangle_p = \langle G_{\tilde{\varepsilon}} G_{\tilde{\varepsilon}}^* \rangle_p \langle E_I E_I^* \rangle_p$$

Using this fact and, analogous to Eq. (3.15), the incoherent source condition

$$\langle E_I(\boldsymbol{\rho}_{0_1}) E_I^*(\boldsymbol{\rho}_{0_2}) \rangle = \frac{4\pi}{k^2} I_I(\boldsymbol{\rho}_{0_1}) \delta(\boldsymbol{\rho}_{0_1} - \boldsymbol{\rho}_{0_2}) \tag{3.28}$$

in Eq. (3.27) yields, upon substitution into Eq. (3.25) and remembering the definition of Eq. (1.63) in the case for $n = m = 1$,

$$\langle I(\boldsymbol{\rho}_i) \rangle = \frac{k^2}{4\pi^3 L^2 l^2} \int_{-\infty}^{\infty} P(\boldsymbol{\rho}_i, \boldsymbol{\rho}_0) I_I(\boldsymbol{\rho}_0) \langle |R(\boldsymbol{\rho}_0)|^2 \rangle_o \, d\boldsymbol{\rho}_0 \tag{3.29}$$

where

$$P(\boldsymbol{\rho}_i, \boldsymbol{\rho}_0) \equiv \int_{-\infty}^{\infty} \int_{-\infty}^{\infty} \exp\left[ik \left(\frac{\boldsymbol{\rho}_i}{l} + \frac{\boldsymbol{\rho}_0}{L} \right) \cdot (\boldsymbol{\rho}' - \boldsymbol{\rho}'') \right] \times W(\boldsymbol{\rho}') W^*(\boldsymbol{\rho}'') \gamma_{11}(L, \boldsymbol{\rho}' - \boldsymbol{\rho}_0, \boldsymbol{\rho}'' - \boldsymbol{\rho}_0) \, d^2\rho' \, d^2\rho'' \tag{3.30}$$

is the *nonisoplanatic point spread function*[4,5] governing the formation of the image $\langle I(\rho_i)\rangle$ through the random medium and

$$\gamma_{11}(L, \boldsymbol{\rho}' - \boldsymbol{\rho}_0, \boldsymbol{\rho}'' - \boldsymbol{\rho}_0) \equiv \langle G_{\tilde{\varepsilon}}(L, \boldsymbol{\rho}'; 0, \boldsymbol{\rho}_0) G^*_{\tilde{\varepsilon}}(L, \boldsymbol{\rho}''; 0, \boldsymbol{\rho}_0)\rangle$$

is the second-order moment, defined by Eq. (1.63), of the random Green function describing the propagating wave field. One thus sees how consideration of the second-order moment γ_{11} of the stochastic electric field enters into the problem of imaging through a random medium. Equation (3.29) is written for the general case where the object is illuminated with (or emits) spatially nonuniform incoherent radiation of intensity $I_I(\boldsymbol{\rho}_0)$. This relation can be slightly simplified if one considers the relationship between the extent of the object size and the radiation field to be such that this intensity can be taken to be constant over the object plane, thus rendering it independent of position; whatever spatial variation in object intensity exists is relegated to the average of the square of the modulus of the object reflectivity. Hence, without loss of generality, Eq. (3.29) can be written as

$$\langle I(\boldsymbol{\rho}_i)\rangle = \frac{k^2 I_I}{4\pi^3 L^2 l^2} \int_{-\infty}^{\infty} P(\boldsymbol{\rho}_i, \boldsymbol{\rho}_0) \langle |R(\boldsymbol{\rho}_0)|^2\rangle_o \, d^2\rho_0 \qquad (3.31)$$

Defining the difference coordinate $\boldsymbol{\rho}_d \equiv \boldsymbol{\rho}' - \boldsymbol{\rho}''$ in the lens aperture plane allows Eq. (3.30) to be written

$$P(\boldsymbol{\rho}_i, \boldsymbol{\rho}_0) \equiv \int_{-\infty}^{\infty} \exp\left[\frac{ik}{l}\left(\boldsymbol{\rho}_i + \frac{l\boldsymbol{\rho}_0}{L}\right) \cdot \boldsymbol{\rho}_d\right] K_\gamma(\boldsymbol{\rho}_d, \boldsymbol{\rho}_0)\, d^2\rho_d$$

$$\equiv P(\boldsymbol{\rho}_i'', \boldsymbol{\rho}_0) \qquad (3.32)$$

where

$$K_\gamma(\boldsymbol{\rho}_d, \boldsymbol{\rho}_0) \equiv \int_{-\infty}^{\infty} \gamma_{11}(L, \boldsymbol{\rho}' - \boldsymbol{\rho}_0, \boldsymbol{\rho}' - \boldsymbol{\rho}_0 \cdot \boldsymbol{\rho}_d)$$

$$\times\, W(\boldsymbol{\rho}') W^*(\boldsymbol{\rho}' - \boldsymbol{\rho}_d)\, d^2\rho' \qquad (3.33)$$

and an auxiliary coordinate $\boldsymbol{\rho}_i''$ has been defined in the image plane that is a linear combination of the original image coordinate $\boldsymbol{\rho}_i$ and the inverse equivalent object coordinate, viz., $\boldsymbol{\rho}_i'' \equiv \boldsymbol{\rho}_i + (l/L)\boldsymbol{\rho}_0$. Equation (3.32) is the most general form of the point spread function and depends on the coordinates of the point sources that compose the imaging target, since different coordinates can correspond to different propagation paths with different local statistical properties. Such a case can occur in short-term (or short-exposure) imaging where a particular representation of what would otherwise be an ensemble is selected. Because of this circumstance, $P(\boldsymbol{\rho}_i'', \boldsymbol{\rho}_0)$ is also called the *instantaneous point spread function.* (Again, the fact that an imaging situation is *short term* is implicit in the time period over which the formation of the ensemble average related to γ_{11} occurs; until this is rigorously justified later, it is sufficient to consider a time interval smaller than the characteristic time $\tau_0 \equiv D/V$ to be "short term" for the particular situation.) Defining the Fourier transform relationships

$$\langle I(\boldsymbol{\rho}_i)\rangle = \int_{-\infty}^{\infty} \langle i(\boldsymbol{\kappa}_i)\rangle \exp\,(i\boldsymbol{\kappa}_i \cdot \boldsymbol{\rho}_i)\, d^2\kappa_i$$

and
$$\langle |R(\boldsymbol{\rho}_0)|^2\rangle_o = \int_{-\infty}^{\infty} \langle r(\boldsymbol{\kappa}_0)\rangle \exp\,(i\boldsymbol{\kappa}_0 \cdot \boldsymbol{\rho}_0)\, d^2\kappa_0$$

and applying these to Eq. (3.31) gives

$$\langle i(\boldsymbol{\kappa}_i)\rangle = \frac{k^2 I_I}{\pi L^2 l^2} \int_{-\infty}^{\infty} p_2\left(\boldsymbol{\kappa}_i,\, \boldsymbol{\kappa}_0 + \frac{l}{L}\,\boldsymbol{\kappa}_i\right) \langle r(\boldsymbol{\kappa}_0)\rangle\, d^2\kappa_0 \qquad (3.34)$$

where

$$p_2\left(\boldsymbol{\kappa}_i,\, \boldsymbol{\kappa}_0 + \frac{l}{L}\,\boldsymbol{\kappa}_i\right) \equiv \left(\frac{1}{2\pi}\right)^2 \int_{-\infty}^{\infty} p_1(\boldsymbol{\kappa}_i, \boldsymbol{\rho}_0) \times \exp\left\{i\left[\boldsymbol{\kappa}_0 + \left(\frac{l}{L}\right)\boldsymbol{\kappa}_i\right]\cdot \boldsymbol{\rho}_0\right\} d^2\rho_0 \qquad (3.35)$$

is the two-coordinate transform of $P(\boldsymbol{\rho}_i'', \boldsymbol{\rho}_0)$ in terms of the one-coordinate transform given by

$$p_1(\boldsymbol{\kappa}_i, \boldsymbol{\rho}_0) \equiv \left(\frac{1}{2\pi}\right)^2 \int_{-\infty}^{\infty} P(\boldsymbol{\rho}_i'', \boldsymbol{\rho}_0) \exp\,(-i\boldsymbol{\kappa}_i \cdot \boldsymbol{\rho}_i'')\, d^2\rho_i'' \qquad (3.36)$$

These relations show that each Fourier component at a fixed spatial frequency $\boldsymbol{\kappa}_i$ corresponds to the entire range of spatial frequencies $\boldsymbol{\kappa}_0$ of the object. Again, this is due to the nonisoplanaticity induced by the atmospheric propagation medium and the time scale over which the associated ensemble average is obtained, and is borne out in the dependence of the second-order moment γ_{11} on the coordinates in the object plane. Although this general situation seems to result in a rather poor reproduction of the object on the image plane, it has been shown[4] to be the basis for the phenomenon of superresolution, i.e., a reproduction of the object intensity distribution in the image plane well beyond the diffraction limit as given by Eq. (3.21). This superresolution effect owes its existence to the nonisoplanaticity of the random medium on the time scales corresponding to short-exposure imaging. This is an example of an adaptive optics method of image enhancement and will be discussed further in Chap. 5. However, for the remainder of the book, the more applicable case of isoplanatic imaging will be considered.

In the case where isoplanaticity obtains, i.e., when the characteristics of the random medium and the exposure conditions are such that the second-order moment γ_{11} is a statistically homogeneous function (as in the case of long-exposure imaging), viz.,

$$\gamma_{11}(L, \boldsymbol{\rho}' - \boldsymbol{\rho}_0, \boldsymbol{\rho}' - \boldsymbol{\rho}_0 - \boldsymbol{\rho}_d) = \gamma_{11}(L, \boldsymbol{\rho}_d)$$

Eqs. (3.34) to (3.36) simplify considerably. First, it should be noted that in this special circumstance, one can write, using Eq. (1.65),

$$C\gamma_{11}(L, \boldsymbol{\rho}_d) = \Gamma_{11}(L, \boldsymbol{\rho}_d, 0) \equiv \Gamma_{11}(L, \boldsymbol{\rho}_d) \tag{3.37}$$

where the coefficient C is equal to unity in the case of spherical wave propagation with unit source amplitude, as is the case here, since the actual amplitude and its variations along the target are convolved in the reflectivity factor $R(\boldsymbol{\rho}_0)$. Given the condition of Eq. (3.37), Eq. (3.32) also becomes explicitly independent of $\boldsymbol{\rho}_0$, thus giving

$$P(\boldsymbol{\rho}_i'', \boldsymbol{\rho}_0) = P(\boldsymbol{\rho}_i'') = \int_{-\infty}^{\infty} \exp\left(\frac{ik}{l}\boldsymbol{\rho}_i'' \cdot \boldsymbol{\rho}_d\right) K(\boldsymbol{\rho}_d) \times \Gamma_{11}(L, \boldsymbol{\rho}_d)\, d^2\rho_d \tag{3.38}$$

where $K(\boldsymbol{\rho}_d)$ is as defined by Eq. (3.16), with an appropriate change of variable; and as introduced in Eq. (3.32), $\boldsymbol{\rho}_i'' \equiv \boldsymbol{\rho}_i + \boldsymbol{\rho}_0'$, where the modified object coordinate $\boldsymbol{\rho}_0' \equiv -(l/L)\boldsymbol{\rho}_0$ is employed. This circumstance now allows the $\boldsymbol{\rho}_0$ integration in Eq. (3.35) to be easily performed, yielding

$$p_2\left(\boldsymbol{\kappa}_i, \boldsymbol{\kappa}_0 + \frac{l}{L}\boldsymbol{\kappa}_i\right) = p_1(\boldsymbol{\kappa}_i)\delta\left(\boldsymbol{\kappa}_0 + \frac{l}{L}\boldsymbol{\kappa}_i\right) \tag{3.39}$$

Substituting Eq. (3.39) into Eq. (3.34) allows the latter to reduce to

$$\langle i(\boldsymbol{\kappa}_i)\rangle = \left(\frac{k^2 I_I}{\pi L^2 l^2}\right) p_1(\boldsymbol{\kappa}_i)\left\langle r\left(-\frac{l}{L}\boldsymbol{\kappa}_i\right)\right\rangle \tag{3.40}$$

Finally, substituting Eq. (3.38) into Eq. (3.36), performing the required integration, and using this result in Eq. (3.40) gives

$$\langle i(\boldsymbol{\kappa}_i)\rangle = \left(\frac{I_I}{\pi L^2}\right) K\left(\frac{l}{k}\boldsymbol{\kappa}_i\right) \Gamma_{11}\left(L, \frac{l}{k}\boldsymbol{\kappa}_i\right)\left\langle r\left(-\frac{l}{L}\boldsymbol{\kappa}_i\right)\right\rangle \tag{3.41}$$

This fundamental and important result of incoherent, isoplanatic imaging gives rise to the concept of "transfer functions" in optics. Analogous to the treatment of well-known "linear-systems" theory, Eq. (3.41) describes the transfer of a frequency-dependent amplitude of the spatial Fourier transform of the object $\langle r[-(l/L)\boldsymbol{\kappa}_i]\rangle$ cascaded with the second-order moment of the associated random electric field of the electromagnetic wave propagating through the random atmosphere, $\Gamma_{11}[L, (l/k)\boldsymbol{\kappa}_i]$, and the convolution of the aperture function that defines the optics through which the image is acquired, $K[(l/k)\boldsymbol{\kappa}_i]$, which results in the frequency-dependent amplitude which corresponds to the intensity distribution in the image plane $\langle i(\boldsymbol{\kappa}_i)\rangle$. For this reason, the second-order moment Γ_{11} of the random electric field associated with image transmission of the object plane intensity distribution is called the *modulation transfer function* (MTF) of the propagation medium, K is the MTF of the imaging optics, and $\langle i\rangle$ is the MTF of the "total" (i.e., atmosphere plus imaging system) optical system. [Many times, one finds reference made to the *optical transfer function* (OTF) rather than the MTF. These quantities are related by the fact that the OTF is the general, complex-valued parameter and the MTF is the modulus of the OTF; hence, in the many cases where the OTF is a real-valued quantity, the OTF and the MTF are synonymous. Complicating the verbiage even further, the OTF or MTF of the propagation medium is, by its statistical formulation, also called the *mutual coherence function* (MCF). In the remainder of the book, the MTF will be employed in the cases of quantities of nonstatistical origin, such as K, but the MCF will be used in the case of the statistical originating quantity Γ_{11}.] The functional argument $-(l/L)\boldsymbol{\kappa}_i$ is the spatial frequency $\boldsymbol{\kappa}_i$ of the image modified by the magnification and inversion factor $-(l/L)$ due to the lens; analogous to the result of Eq. (3.12), this corresponds to the spatial frequency of the object, $\boldsymbol{\kappa}_0$. It is related to the linear spatial frequency $\mathbf{f}_0$ associated with the transform of the intensity profile of the object in the object plane (measured in units of cycles or line pairs per unit length) by the relation $2\pi\mathbf{f}_0 = \boldsymbol{\kappa}_0$. The argument $(l/k)\boldsymbol{\kappa}_i$ appearing in both Γ_{11} and K replaces the difference coordinate $\boldsymbol{\rho}_d$, thus allowing the identification $2\pi\mathbf{f}_i = \boldsymbol{\kappa}_i = k\boldsymbol{\rho}_d/l$, where $\mathbf{f}_i$ is the linear spatial frequency of the transform of the intensity profile of the image in the image plane.

It must be remembered that the derivation of Eq. (3.41), and its contrast to the more general case of Eq. (3.34), made use of the assumption that Γ_{11} is the long-term MCF of the random medium. It is this quantity that is derived from the application of the methods given in Chap. 1 to the results of Chap. 2. In particular, the statistical model of the atmosphere developed in Chap. 2 necessarily entails the use of averaging over *all* ensembles of the turbidity field, thus relegating any subsequent derivation from these results (which will be the subject of Chap. 4) as *long term*. It now remains to discern the explicit statistical difference between the short-term and long-term MCF and, at the same time, recover a short-term description of the MCF from a result that holds only in the long term.

3.4 Short-Term versus Long-Term Imaging through a Random Medium—Statistical Zernike Polynomial Expansions

The fact that the physical processes that distinguish between a *long-term* and a *short-term* image are concatenated in, respectively, the use of a *partial* versus a *full* ensemble of all the configurations that the random wave field can assume was first noted by Fried.[6] To understand this process in its entirety, one must again consider the complex phase representation of the stochastic Green function; viz., using Eqs. (1.98) and (1.108), one has

$$G_{\tilde{\varepsilon}}(\mathbf{r}, \mathbf{r}') = \exp\left[\chi(\mathbf{r}, \mathbf{r}') + iS_1(\mathbf{r}, \mathbf{r}')\right] \tag{3.42}$$

in terms of the random log-amplitude $\chi(\mathbf{r}, \mathbf{r}')$ and phase $S_1(\mathbf{r}, \mathbf{r}')$. The various configurations that the propagating wave field can assume while an ensemble is being formed are thus represented by variations in $\chi(\mathbf{r}, \mathbf{r}')$ and $S_1(\mathbf{r}, \mathbf{r}')$; the former results in intensity fluctuations within the field, and the latter describes distortions of the phase front of the field.

Before a rigorous analysis is conducted, it is fruitful to digress with the following qualitative considerations. Although intensity fluctuations within the field from which an image is formed have effects that are analogous to a random lens apodization, it is the phase structure of the field that most severely affects image formation. It thus becomes of interest to just quantitatively describe the structure of each of the member configurations of the random phase distortion of the propagating wave that composes the ensemble; in particular, it is desired to statistically describe the spatial structure of the distorted wave front in terms of the component aberrations, such as tilt, defocus, astigmatism, coma, etc., that have been classically used to model phase distortions.[7] It is intuitively obvious that the higher-order aberrations, such as astigmatism and coma, are associated with higher spatial frequency content than the lower-order aberrations, which translates to higher corresponding temporal frequencies when one considers a moving medium, i.e., the atmosphere with a velocity V transverse to the direction of propagation. The exposure time over which the image is formed essentially acts as a low-pass filter to these high temporal frequencies. The larger the exposure time, the smaller the temporal frequencies that are within the image. According to these considerations, the lowest-order tilt component of the phase front will therefore establish a point of demarcation between images that contain short-term versus long-term effects.

3.4.1 Short-term imaging in a moving random medium

To demonstrate the temporal filtering aspect of exposure time in obtaining an image through a random turbulent medium possessing an average constant transverse velocity V across the propagation path, as well as account for the spatial averaging effect of a finite-size aperture,[8] consider the instantaneous flux $P(t)$ that enters the lens input aperture at time t,

$$P(t) = \int_{-\infty}^{\infty} W(\boldsymbol{\rho}) I(\boldsymbol{\rho}, t)\, d\boldsymbol{\rho} \tag{3.43}$$

where $I(\boldsymbol{\rho}, t)$ is the instantaneous intensity at a point in the aperture. Taking τ_{ex} to be an arbitrary exposure time, it is desired to obtain the temporal aspects of the total flux during this exposure time. To this end, one can write, similar to Eq. (3.43),

$$P(t + \tau_{ex}) = \int_{-\infty}^{\infty} W(\boldsymbol{\rho}) I(\boldsymbol{\rho}, t + \tau_{ex})\, d\boldsymbol{\rho} \tag{3.44}$$

apply the frozen field hypothesis [cf. Eq. (D1.1.18] to the instantaneous random intensity field to obtain the relationship

$$I(\boldsymbol{\rho}, t + \tau_{ex}) = I(\boldsymbol{\rho} - \mathbf{v}\tau_{ex}, t) \tag{3.45}$$

and finally form the temporal correlation

$$R(\tau_{ex}) \equiv \langle P(t) P(t + \tau_{ex}) \rangle = \int_{-\infty}^{\infty} \int_{-\infty}^{\infty} W(\boldsymbol{\rho}) W(\boldsymbol{\rho}') \times B_I(\boldsymbol{\rho} - \boldsymbol{\rho}' + \mathbf{v}\tau_{ex})\, d\boldsymbol{\rho}\, d\boldsymbol{\rho}' \tag{3.46}$$

where

$$B_I(\boldsymbol{\rho} - \boldsymbol{\rho}' + \mathbf{v}\tau_{ex}) = B_I(\boldsymbol{\rho}, \boldsymbol{\rho}' - \mathbf{v}\tau_{ex}) \equiv \langle I(\boldsymbol{\rho}, t) I(\boldsymbol{\rho} - \mathbf{v}\tau_{ex}, t) \rangle \tag{3.47}$$

is the correlation of the intensity fluctuations, the statistics of which are assumed to be homogeneous. [A more rigorous approach to this would entail the use of Eq. (3.1), resulting in the association of the fourth-order moment of the random wave field with the correlation of intensities within the lens aperture; symmetry considerations of the parabolic equation for the fourth-order moment in a homogeneous and isotropic situation would then lead to the same results as will be obtained in this more straightforward approach.] Defining the spatial Fourier transform

$$B_I(\boldsymbol{\rho}) = \int_{-\infty}^{\infty} F_I(\boldsymbol{\kappa}) \exp(i\boldsymbol{\kappa} \cdot \boldsymbol{\rho})\, d\boldsymbol{\kappa}$$

and applying it to Eq. (3.46) gives, after rearranging factors,

$$R(\tau_{\text{ex}}) = (2\pi)^4 \int_{-\infty}^{\infty} |\hat{W}(\boldsymbol{\kappa})|^2 F_I(\boldsymbol{\kappa}) \exp(i\boldsymbol{\kappa} \cdot \mathbf{v}\tau)\, d\boldsymbol{\kappa} \tag{3.48}$$

where $\hat{W}(\boldsymbol{\kappa})$ is defined in Eq. (3.18*a*) and explicitly given by Eq. (3.18*b*). Applying to Eq. (3.48) the temporal Fourier transform given by

$$w(\omega) = \frac{1}{2\pi} \int_{-\infty}^{\infty} R(\tau) \exp(i\omega\tau)\, d\tau \tag{3.49}$$

and using Eq. (3.18*b*) yields

$$w(\omega) = (2\pi R^2)^2 \int_0^{2\pi} \int_0^{\infty} \frac{J_1^2(\kappa R)}{(\kappa R)^2} F_I(\kappa) \times \delta(\kappa v \cos\phi - \omega)\kappa\, d\kappa\, d\phi$$

Finally, performing the angular integration over the function within the δ function gives

$$w(\omega) = (2\pi R^2)^2 \int_{\omega/v}^{\infty} \frac{J_1^2(\kappa R)}{(\kappa R)^2} \frac{2F_I(\kappa)}{\sqrt{(\kappa v)^2 - \omega^2}}\, \kappa\, d\kappa \tag{3.50}$$

Changing variables via the prescription

$$\kappa' = \sqrt{\kappa^2 - \left(\frac{\omega}{v}\right)^2}$$

allows Eq. (3.50) to be written

$$w(\omega) = 2(2\pi R^2)^2 \left(\frac{1}{v}\right) \int_0^{\infty} \left[\frac{J_1(R\sqrt{\kappa'^2 + (\omega/v)^2})}{R\sqrt{\kappa'^2 + (\omega/v)^2}}\right]^2 \times F_I(\sqrt{\kappa'^2 + (\omega/v)^2})\, d\boldsymbol{\kappa}' \tag{3.51}$$

This relationship shows that the first factor within the integrand of Eq. (3.51) acts as a temporal-spatial filter on the transition from the spatial spectrum of intensity fluctuations at the lens aperture to the temporal spectrum that governs the phenomena of the attendant exposure time. In particular, when the exposure time τ_{ex} is such that the temporal frequency and transverse wind velocity satisfy the condition $R\omega/v \gg 1$, where $\omega \leq 2\pi/\tau_{ex}$, one has that this filter factor is small for all values of κ', thus rendering negligible in $w(\omega)$ the effect of all turbulent inhomogeneities described by the second factor F_I. If, however, τ_{ex} is such that $R\omega/v \ll 1$, there is a range of κ' values where $\kappa' \ll 1/R$, thus allowing the filter factor to retain values on the order of unity and therefore enabling the stochastic effects of F_I within this range to govern the spectrum $w(\omega)$. Hence, so long as the condition $2\pi R/V \gg \tau_{ex}$ obtains, one obtains a temporal spectrum of the image that is not entirely degraded by the effects of turbulent inhomogeneities within the random medium. Only when $2\pi R/V \leq \tau_{ex}$ does the spectrum become perturbed with the random inhomogeneities, but only those associated with the spatial frequencies $\kappa' \ll 1/R$. This demonstrates not only the benefits of short-term imaging but also the phenomenon of spatial aperture averaging.

Now that short-term imaging and its attendant constraints have been sufficiently described, these considerations will now be given a firm analytical basis through the use of Zernike polynomial expansions of the random phase front.

3.4.2 Zernike polynomial expansion of a random phase front at the input aperture of a lens

The aberration description of an arbitrary phase front is well known and is easily expedited using a system of orthogonal polynomials typically called Zernike polynomials.[7] Such a system of polynomials is defined on a unit circle used to represent the input aperture of a circular lens. Thus, letting $\Phi_j(\boldsymbol{\rho})$ represent the phase surface at points $\boldsymbol{\rho}$ across the lens aperture described by the jth aberration, one defines the polynomial expansion

$$S(\boldsymbol{\rho}) = \sum_{j=1}^{\infty} \Phi_j(\boldsymbol{\rho}) \qquad \Phi_j(\boldsymbol{\rho}) = a_j Z_j\left(\frac{\boldsymbol{\rho}}{R}\right) \tag{3.52}$$

where a_j is the jth expansion coefficient and Z_j is the jth Zernike polynomial, the system of which is defined by, using a plane polar coordinate system the origin of which is placed on the optical axis of the lens aperture,[9]

$$Z_j(\mathbf{r}) \equiv \begin{cases} \sqrt{n+1}R_n^m(r)\sqrt{2}\cos(m\theta) & j \text{ even, } m \neq 0 \\ \sqrt{n+1}R_n^m(r)\sqrt{2}\sin(m\theta) & j \text{ odd, } m \neq 0 \\ \sqrt{n+1}R_n^0(r) & \text{any } j,\ m = 0 \end{cases} \tag{3.53}$$

with

$$R_n^m(r) = \sum_{s=0}^{(n-m)/2} \frac{(-1)^s (n - s)!r^{n-2s}}{s![(n + m)/2 - s]![(n - m)/2 - s]!} \tag{3.54}$$

using the normalized radial vector $\mathbf{r} \equiv \boldsymbol{\rho}/R$. The quantities n, called the *radial degree,* and m, called the *azimuthal frequency,* must also satisfy the condition that $n - |m|$ is an even quantity with $m \leq n$. The index j is the *mode number* of the expansion as well as of the particular Zernike polynomial. The angular functions within Eq. (3.53) form a two-dimensional rotation group, and the radial functions are formed from Jacobi polynomials. Table 3.1 lists the commonly used polynomials showing the corresponding radial degree, azimuthal frequency, and mode numbers as well as the names of the classically associated aberrations.

The polynomials Z_j satisfy the orthogonality property

$$\int_{-\infty}^{\infty} Z_j\left(\frac{\boldsymbol{\rho}}{R}\right) Z_k\left(\frac{\boldsymbol{\rho}}{R}\right) W(\boldsymbol{\rho})\, d^2\rho = \pi R^2 \delta_{jk} \tag{3.55}$$

It is convenient to redefine the aperture function $W(\boldsymbol{\rho})$ to accommodate the normalized coordinates that are required by the Zernike polynomials, viz.,

$$\int_{-\infty}^{\infty} Z_j(\mathbf{r}) Z_k(\mathbf{r}) W_N(\mathbf{r})\, d^2r = \delta_{jk} \tag{3.56}$$

where the normalized aperture function is defined as

$$W_N(\mathbf{r}) \equiv \begin{cases} \dfrac{1}{\pi} & |\mathbf{r}| \leq 1 \\ 0 & |\mathbf{r}| > 1 \end{cases} \tag{3.57}$$

Applying to Eq. (3.52) the property exhibited by Eq. (3.56) yields the expression for the expansion coefficients

$$\begin{aligned} a_j &= \int_{-\infty}^{\infty} W_N(\mathbf{r}) S(R\mathbf{r}) Z_j(\mathbf{r})\, d^2r \\ &= \left(\frac{1}{R^2}\right) \int_{-\infty}^{\infty} W_N\left(\frac{\boldsymbol{\rho}}{R}\right) S(\boldsymbol{\rho}) Z_j\left(\frac{\boldsymbol{\rho}}{R}\right) d^2\rho \end{aligned} \tag{3.58}$$

As the phase $S_1(\boldsymbol{\rho})$ is a random function across the lens aperture, so too are the coefficients given by Eq. (3.58). Thus, in the case of stochastic propagation, these coefficients can only be described statistically and are related to the phase statistics through Eq. (3.58).

TABLE 3.1 Zernike Polynomials as a Function of Azimuthal Frequency *m* and Radial Degree *n*

Radial degree	Azimuthal frequency *m*					
n	0	1	2	3	4	5
0	$Z_1 = 1$ Constant					
1		$Z_2 = 2r\cos\theta$ $Z_3 = 2r\sin\theta$ Tilts (lateral position)				
2	$Z_4 = \sqrt{3}(2r^2 - 1)$ Defocus (longitudinal position)		$Z_5 = \sqrt{6}r^2\sin 2\theta$ $Z_6 = \sqrt{6}r^2\cos 2\theta$ Astigmatism (third order)			
3		$Z_7 = \sqrt{8}$ $\times (3r^3 - 2r)\sin\theta$ $Z_8 = \sqrt{8}$ $\times (3r^3 - 2r)\cos\theta$ Coma (third order)		$Z_9 = \sqrt{8}r^3\sin 3\theta$ $Z_{10} = \sqrt{8}r^3\cos 3\theta$		
4	$Z_{11} = \sqrt{5}(6r^4 - 6r^2 + 1)$ Third-order spherical		$Z_{12} = \sqrt{10}$ $\times (4r^4 - 3r^2)\cos 2\theta$ $Z_{13} = \sqrt{10}$ $\times (4r^4 - 3r^2)\sin 2\theta$		$Z_{14} = \sqrt{10}r^4\cos 4\theta$ $Z_{15} = \sqrt{10}r^4\sin 4\theta$	
5		$Z_{16} = \sqrt{12}$ $\times (10r^5 - 12r^3 + 3r)\cos\theta$ $Z_{17} = \sqrt{12}$ $\times (10r^5 - 12r^3 + 3r)\sin\theta$		$Z_{18} = \sqrt{12}$ $\times (5r^5 - 4r^3)\cos 3\theta$ $Z_{19} = \sqrt{12}$ $\times (5r^5 - 4r^3)\sin 3\theta$		$Z_{20} = \sqrt{12}r^5\cos 5\theta$ $Z_{21} = \sqrt{12}r^5\sin 5\theta$
6	$Z_{22} = \sqrt{7}$ $\times (20r^6 - 30r^4 + 12r^2 - 1)$ Fifth-order spherical		Z_{23} Z_{24}		Z_{25} Z_{26}	

Table 3.1 shows that the description of the lateral tilts of the phase front that, as mentioned above, distinguishes between a short- and a long-term exposure is achieved by employing modes up to third order in Eq. (3.52). Hence, if one desires to describe the mutual coherence function of the wave field in a short-exposure situation, one must commence with Eq. (3.42) with this component of the random phase removed, i.e.,

$$G_{\tilde{\varepsilon}}(L, \boldsymbol{\rho}; 0, \boldsymbol{\rho}_0) \equiv G_{\tilde{\varepsilon}}(\boldsymbol{\rho}) = \exp\left\{\chi(\boldsymbol{\rho}) + i\left[S_1(\boldsymbol{\rho}) - \sum_{j=1}^{3} \Phi_j(\boldsymbol{\rho})\right]\right\} \qquad (3.59)$$

The development of the "tilt corrected," i.e., short-term (ST), second-order moment from this expression follows along the same lines as would be followed using Eq. (3.42). In particular, employing only the assumption that the fluctuating quantities are governed by isotropic gaussian statistics, one has from Eq. (3.59), upon using the same calculational procedure employed to derive Eq. (1.132),

$$\begin{aligned}\langle G_{\tilde{\varepsilon}}(\boldsymbol{\rho})G^*_{\tilde{\varepsilon}}(\boldsymbol{\rho}')\rangle_{ST} &= \exp\{\tfrac{1}{2}\langle[\chi(\boldsymbol{\rho}) + \chi(\boldsymbol{\rho}') - 2\langle\chi\rangle]^2 - [S_1(\boldsymbol{\rho}) - S_1(\boldsymbol{\rho}')]^2\rangle + 2\langle\chi\rangle\} \\ &\quad \times \exp\left\{\left\langle [S_1(\boldsymbol{\rho}) - S_1(\boldsymbol{\rho}')]\left[\sum_{j=1}^{3} \Phi_j(\boldsymbol{\rho}) - \sum_{j=1}^{3} \Phi_j(\boldsymbol{\rho}')\right]\right.\right. \\ &\qquad \left.\left. - \frac{1}{2}\left[\sum_{j=1}^{3} \Phi_j(\boldsymbol{\rho}) - \sum_{j=1}^{3} \Phi_j(\boldsymbol{\rho}')\right]^2\right\rangle\right\} \end{aligned} \qquad (3.60)$$

[Unlike the subsequent expression derived from Eq. (1.132), i.e., Eq. (1.134), no assumption is used here to rid of some of the log-amplitude terms. Since the assumption of isotropic gaussian statistics is ultimately used in the application of any of the statistical approaches given in Chap. 1, Eq. (3.60) is applicable with any of these methods.] Noting that the first exponential within Eq. (3.60) is that associated with the long-term isoplanatic MCF, i.e., the MCF Γ_{11} that has been considered hitherto and is the result of the modeling given in Chaps. 1 and 2, one can write for the short-term MCF $\gamma_{11_{ST}}$

$$\begin{aligned}\gamma_{11_{ST}}(\boldsymbol{\rho}, \boldsymbol{\rho}') &\equiv \langle G_{\tilde{\varepsilon}}(\boldsymbol{\rho})G^*_{\tilde{\varepsilon}}(\boldsymbol{\rho}')\rangle_{ST} \\ &= \Gamma_{11}(\boldsymbol{\rho} - \boldsymbol{\rho}')\Delta_C(\boldsymbol{\rho}, \boldsymbol{\rho}') \end{aligned} \qquad (3.61)$$

where the MCF phase tilt correction factor $\Delta_C(\boldsymbol{\rho}, \boldsymbol{\rho}')$ is given by the second exponential function in Eq. (3.60).

To analytically evaluate this correction factor, one commences with the infinite series expansion for the phase $S_1(\mathbf{r})$, and carrying out the operations indicated in the ensemble average within the second exponential function of Eq. (3.60) yields

$$\Delta_C(\boldsymbol{\rho}, \boldsymbol{\rho}') \equiv \exp(\langle ... \rangle)$$

$$\langle ... \rangle = \sum_{k=1}^{3} \left[\sum_{j=1}^{\infty} \langle \Phi_j(\boldsymbol{\rho}) \Phi_k(\boldsymbol{\rho}) \rangle - \frac{1}{2} \sum_{j=1}^{3} \langle \Phi_j(\boldsymbol{\rho}) \Phi_k(\boldsymbol{\rho}) \rangle \right]$$
$$- \sum_{k=1}^{3} \left[\sum_{j=1}^{\infty} \langle \Phi_j(\boldsymbol{\rho}) \Phi_k(\boldsymbol{\rho}') \rangle - \frac{1}{2} \sum_{j=1}^{3} \langle \Phi_j(\boldsymbol{\rho}) \Phi_k(\boldsymbol{\rho}') \rangle \right]$$
$$- \text{(two similar terms with subscripts } j \text{ and } k \text{ transcribed on polynomials } \Phi) \quad (3.62)$$

At this point, one notes from Table 3.1 that statistical spatial correlation can potentially exist between Zernike polynomials of like azimuthal frequency but different radial degree, viz., the two Zernike polynomials Z_2 and Z_3, both of radial degree $n = 1$ and azimuthal frequency $m = 1$, and the polynomials Z_7 and Z_8, as well as Z_{16} and Z_{17}, of the same azimuthal frequency but, respectively, of radial degree $n = 3$ and $n = 5$. The further removed from radial degree $n = 1$, the smaller will be the correlation. One must pay heed to at least one such correlative connection if one is not to have an expression that is rather restrictive in its application.[10] Hence, the level of approximation adopted here will require one to retain only those polynomials of radial degree $n = 3$ in Eq. (3.62). This requires that terms within the infinite series in Eq. (3.62) be retained up to the mode number $j = 8$. Employing this truncation of the series, using the second expression in Eq. (3.52), and simplifying yields

$$\langle ... \rangle = \langle a_2^2 \rangle \left[\frac{1}{2} Z_2^2\left(\frac{\boldsymbol{\rho}}{R}\right) + \frac{1}{2} Z_2^2\left(\frac{\boldsymbol{\rho}'}{R}\right) - Z_2\left(\frac{\boldsymbol{\rho}}{R}\right) Z_2\left(\frac{\boldsymbol{\rho}'}{R}\right) \right]$$
$$+ \langle a_2 a_8 \rangle \left[Z_2\left(\frac{\boldsymbol{\rho}}{R}\right) Z_8\left(\frac{\boldsymbol{\rho}}{R}\right) + Z_2\left(\frac{\boldsymbol{\rho}'}{R}\right) Z_8\left(\frac{\boldsymbol{\rho}'}{R}\right) - Z_2\left(\frac{\boldsymbol{\rho}}{R}\right) Z_8\left(\frac{\boldsymbol{\rho}'}{R}\right) - Z_2\left(\frac{\boldsymbol{\rho}'}{R}\right) Z_8\left(\frac{\boldsymbol{\rho}}{R}\right) \right]$$
$$+ \text{(two similar terms with subscripts 1 and 2 exchanged with 3 and 4 on } a \text{ and } Z \text{ factors)} \quad (3.63)$$

Finally, using the relevant functional forms given in Table 3.1 for the Zernike polynomials required in Eq. (3.63) and noting that, because of the rotational symmetry of the problem, $\langle a_2^2 \rangle = \langle a_3^2 \rangle$ and $\langle a_2 a_8 \rangle =$

$\langle a_3 a_7\rangle$, one obtains for the MCF phase tilt correction factor with aberrational correlations up to mode number $j = 8$ taken into account

$$\Delta_C(\boldsymbol{\rho}, \boldsymbol{\rho}') = \exp\left\{2\langle a_2^2\rangle \frac{(\rho - \rho')^2}{R^2} + \sqrt{8}\langle a_2 a_8\rangle \times \left[6\,\frac{(\rho^4 + \rho'^4)}{R^4} - 6\,\frac{\boldsymbol{\rho}\cdot\boldsymbol{\rho}'(\rho^2 + \rho'^2)}{R^4} - 4\,\frac{(\rho - \rho')^2}{R^2}\right]\right\} \tag{3.64}$$

The averages over the products of the expansion coefficients are easily expressed in terms of those of the phase using Eq. (3.58), viz., forming the ensemble average of the jth coefficient a_k with the kth coefficient a_k yields

$$\langle a_j a_k\rangle = \int_{-\infty}^{\infty}\int_{-\infty}^{\infty} W_N(\mathbf{r})W_N(\mathbf{r}')B_S(\boldsymbol{\rho}, \boldsymbol{\rho}') \times Z_j(\mathbf{r})Z_k(\mathbf{r}')\, d^2r\, d^2r' \tag{3.65}$$

where, as in Chap. 1,

$$B_S(\boldsymbol{\rho}, \boldsymbol{\rho}') \equiv \langle S(\boldsymbol{\rho})S(\boldsymbol{\rho}')\rangle$$

is the phase correlation function. It should be noted that this approach, requiring the calculation of the phase correlation function, necessarily requires the use of the Rytov method in the evaluation of such statistics. However, as was pointed out in Sec. 1.7.1, the accuracy of the phase statistics is not affected by the limitations of the first-order Rytov analysis and can be expected to hold in turbulence conditions more severe than the first-order solution for the log-amplitude statistics. Hence, one can employ the Rytov expression for the phase correlation, viz., Eq. (1.127), in conditions of strong turbulence. One must only include higher-order Zernike terms, e.g., defocus and astigmatism, to compensate for effects prevalent in stronger turbulence conditions.

Hence, the use of Eqs. (3.61), (3.64), and (3.65) makes possible the "reversal" of the long-term averaging that is implicit in the modeling of the atmospheric turbulence field and the attendant electromagnetic fields that propagate, thus allowing a description of the short-term phenomena from which the long-term statistics derive.

The use of Zernike polynomials is not unique to the orthogonal expansion of a random phase field of a propagating wave. In particular, the statistical correlation that exists between the expansion coefficients of the various mode numbers of the Zernike polynomials of the same azimuthal frequency, which needed to be accounted for in the development of Eq. (3.64), suggests the use of another set of orthogonal polynomials which are characterized by statistically independent expansion coefficients. Such a set of polynomial basis functions has been shown to be given by the solution of the Karhunen-Loeve homogeneous integral equation.[11] Comparison (in terms of the minimization of the squared residual phase error across a lens aperture) of the Karhunen-Loeve expansion with the Zernike expansion shows[12] that the Zernike polynomials are optimum for low-order phase aberration representations, such as those that enter Eq. (3.64).

3.4.3 The short-exposure imaging transfer function

A slight complication arises from the fact that the short-term MCF is nonisoplanatic, as demonstrated by the existence of the first two terms within the brackets of Eq. (3.64). Hence, the corrected MCF of Eq. (3.61) cannot be directly applied to Eq. (3.41), which is an isoplanatic result. One must return to the general nonisoplanatic relations of Eqs. (3.34) to (3.36) for appropriate treatment.

In particular, one can obtain an approximate yet analytically amenable expression for the short-term tilt-corrected analog of Eq. (3.41) by assuming that the nonisoplanatic MCF appearing in Eq. (3.33) can be taken to be independent of the object coordinate $\boldsymbol{\rho}_0$. In this case, Eqs. (3.32) and (3.33) become

$$P(\boldsymbol{\rho}_i'') \equiv \int_{-\infty}^{\infty} \exp\left(\frac{ik}{l}\,\boldsymbol{\rho}_i'' \cdot \boldsymbol{\rho}_d\right) K_\gamma(\boldsymbol{\rho}_d)\, d^2\rho_d \qquad (3.66)$$

where

$$K_\gamma(\boldsymbol{\rho}_d) \equiv \int_{-\infty}^{\infty} \gamma_{11}(L, \boldsymbol{\rho}', \boldsymbol{\rho}' - \boldsymbol{\rho}_d) \times W(\boldsymbol{\rho}')W^*(\boldsymbol{\rho}' - \boldsymbol{\rho}_d)\, d^2\rho' \qquad (3.67)$$

Equation (3.66) can then be substituted into Eq. (3.36) to yield the relationship

$$\begin{aligned} p_1(\boldsymbol{\kappa}_i) &= \left(\frac{1}{2\pi}\right)^2 \int_{-\infty}^{\infty}\int_{-\infty}^{\infty} \exp\left(\frac{ik}{l}\,\boldsymbol{\rho}_d \cdot \boldsymbol{\rho}_i'' - i\boldsymbol{\kappa}_i \cdot \boldsymbol{\rho}_i''\right) \\ &\quad \times K_\gamma(\boldsymbol{\rho}_d)\, d\rho_d^2\, d\rho_i''^2 \\ &= \int_{-\infty}^{\infty} \delta\left(\frac{k}{l}\,\boldsymbol{\rho}_d - \boldsymbol{\kappa}_i\right) K_\gamma(\boldsymbol{\rho}_d)\, d\rho_d^2 \\ &= \left(\frac{l}{k}\right)^2 K_\gamma\left(\frac{l}{k}\,\boldsymbol{\kappa}_i\right) \end{aligned} \qquad (3.68)$$

As Eq. (3.39) still holds, Eq. (3.34) gives, upon using Eqs. (3.61) and (3.67),

$$\langle i(\boldsymbol{\kappa}_i)\rangle = \left(\frac{I_I}{\pi L^2}\right) K_C\left(\frac{l}{k}\boldsymbol{\kappa}_i\right)\Gamma_{11}\left(L, \frac{l}{k}\boldsymbol{\kappa}_i\right)\left\langle r\left(-\frac{l}{L}\boldsymbol{\kappa}_i\right)\right\rangle \tag{3.69}$$

where K_C is the "corrected" transfer function of the imaging system which reflects the removal of the phase tilt of the incoming image due to the short time duration over which the image is obtained,

$$K_C(L, \boldsymbol{\rho}_d) \equiv \int_{-\infty}^{\infty} \Delta_C(\boldsymbol{\rho}', \boldsymbol{\rho}' - \boldsymbol{\rho}_d) W(\boldsymbol{\rho}') W^*(\boldsymbol{\rho}' - \boldsymbol{\rho}_d)\, d\rho_d^2 \tag{3.70}$$

Hence, the corrections afforded by the short-term exposure operation correct the overall transfer function of the atmosphere/optical system not through the MCF Γ_{11} but through the MTF of the optical system. It is instructive to note that, as established earlier in Eq. (3.33), imaging through a nonisoplanatic propagation medium involves an involution of the position coordinates in the lens aperture that does not allow for a separation of components into those solely due to the propagation medium and those due to the optics used to form the image. Furthermore, since dealing with a short-exposure image is the simplest form of an adaptive imaging scheme whereby the tilting component of the incoming phase front is removed, this example shows how an adaptive scheme invariably requires a modification of the "structure" of the function that describes the processing capability of the optical system [i.e., $W(\boldsymbol{\rho}')W^*(\boldsymbol{\rho}'')$]. This will be a major subject of discussion in Chap. 5. To combine the results of Eqs. (3.41) and (3.69), one can form the general image transfer function relationship

$$\langle i(\boldsymbol{\kappa}_i)\rangle = \left(\frac{I_I}{\pi L^2}\right) t\left(L, \frac{l}{k}\boldsymbol{\kappa}_i\right)\left\langle r\left(-\frac{l}{L}\boldsymbol{\kappa}_i\right)\right\rangle \tag{3.71}$$

where the combined atmosphere/optics transfer function is given by

$$t\left(L, \frac{l}{k}\boldsymbol{\kappa}_i\right) = K\left(\frac{l}{k}\boldsymbol{\kappa}_i\right)\Gamma_{11}\left(L, \frac{l}{k}\boldsymbol{\kappa}_i\right) \tag{3.72}$$

for long-exposure imaging (i.e., where the image is secured over a time period $\tau_{\text{ex}} \geq 2\pi R/V$, as discussed earlier) and, for the opposite case of short-exposure imaging,

$$t\left(L, \frac{l}{k}\boldsymbol{\kappa}_i\right) = K_C\left(\frac{l}{k}\boldsymbol{\kappa}_i\right)\Gamma_{11}\left(L, \frac{l}{k}\boldsymbol{\kappa}_i\right) \tag{3.73}$$

3.5 The Statistical Analysis of the Random Intensity Distribution in the Imaging Plane—Higher-Order Moments

Since the intensity within the focal plane of the lens is a random quantity, higher-order moments, other than the first order described above, can be established. In particular, one can define the variance of the intensity fluctuations, i.e., $\langle I^2(\boldsymbol{\rho}_i)\rangle$, as well as the related normalized variance of these fluctuations,[3] viz.,

$$\sigma^2 \equiv \frac{\langle I^2(\boldsymbol{\rho}_i)\rangle_{o,p} - \langle I(\boldsymbol{\rho}_i)\rangle_{o,p}^2}{\langle I(\boldsymbol{\rho}_i)\rangle_{o,p}^2} \tag{3.74}$$

where, as dealt with in Sec. 3.3, the averaging is taken over the object as well as the propagation medium statistics. The square root of this quantity is also known as the *speckle contrast.* In order to obtain an expression for the normalized variance of intensity fluctuations, one must return to Eqs. (3.7), (3.9), and (3.26) and form the average of the square of the random intensity, $\langle I^2(\boldsymbol{\rho}_i)\rangle_{o,p}$, from which one finds the need to evaluate the fourth-order moments

$$\gamma_{22}(L, \boldsymbol{\rho} - \boldsymbol{\rho}_0, \boldsymbol{\rho}' - \boldsymbol{\rho}_0', \boldsymbol{\rho}'' - \boldsymbol{\rho}_0'', \boldsymbol{\rho}''' - \boldsymbol{\rho}_0''')$$

and

$$\langle R(\boldsymbol{\rho}_0)R^*(\boldsymbol{\rho}_0')R(\boldsymbol{\rho}_0'')R^*(\boldsymbol{\rho}_0''')\rangle_o$$

Such considerations can lead to potentially unwieldy expressions which can only be treated in approximation.[3,13] There is, however, a large class of imaging situations where one need not be concerned with such higher-order statistical perturbations. The attribute that distinguishes this class is the relative magnitude of the characteristic scale of the object (necessarily, an extended object) being imaged compared to that of the instantaneous point spread function of the combined imaging system and the random medium.

To demonstrate this,[14] consider once again the general nonisoplanatic situation described by Eqs. (3.19) and (3.30); since this is connected with imaging in the short term, one is assured of incorporating fluctuations of the short-term average intensity in the analysis to follow. In the case where

$$\left(\frac{\rho_i}{l} + \frac{\rho_0}{L}\right)(2R) \gg \max(\rho_0) \tag{3.75}$$

one has that the exponential function in Eq. (3.30) contributes substantially to the integral over the object coordinate $\boldsymbol{\rho}_0$, while, within the same integration region, the function $\langle |R(\boldsymbol{\rho}_0)|^2 \rangle_o$ remains relatively constant. One can then approximately write

$$\langle I(\boldsymbol{\rho}_i)\rangle = \left(\frac{k^2}{4\pi^3 L^2 l^2}\right) I_l(\boldsymbol{\rho}_0)\langle |R(\boldsymbol{\rho}_0)|^2\rangle$$
$$\times \int_{-\infty}^{\infty}\int_{-\infty}^{\infty}\int_{-\infty}^{\infty} \exp\left[ik\left(\frac{\boldsymbol{\rho}_i}{l}\right)\cdot(\boldsymbol{\rho}' - \boldsymbol{\rho}'')\right]$$
$$\times W(\boldsymbol{\rho}')W^*(\boldsymbol{\rho}'')\gamma_{11}(L, \boldsymbol{\rho}' - \boldsymbol{\rho}_0, \boldsymbol{\rho}'' - \boldsymbol{\rho}_0)\, d^2\rho'\, d^2\rho''\, d^2\rho_0 \quad (3.76)$$

where the integration over $\boldsymbol{\rho}_0$ now involves only the second-order moment γ_{11}. At this point, one must now consider what such an integral over this moment with respect to the object coordinate implies. To this end, one must return to Eq. (1.56), applied between the source point $(x, \boldsymbol{\rho}_0)$ and the observation point $(L, \boldsymbol{\rho}')$, and its complex conjugate, applied between the same source point and the observation point $(L, \boldsymbol{\rho}'')$, viz.,

$$\nabla^2_{\rho_0} G(L, \boldsymbol{\rho}'; x, \boldsymbol{\rho}_0) - 2ik\frac{\partial G(L, \boldsymbol{\rho}'; x, \boldsymbol{\rho}_0)}{\partial x}$$
$$+ k^2\tilde{\varepsilon}(x, \boldsymbol{\rho}_0)G(L, \boldsymbol{\rho}'; x, \boldsymbol{\rho}_0) = 0 \quad (3.77)$$

and $$\nabla^2_{\rho_0} G^*(L, \boldsymbol{\rho}''; x, \boldsymbol{\rho}_0) + 2ik\frac{\partial G^*(L, \boldsymbol{\rho}''; x, \boldsymbol{\rho}_0)}{\partial x}$$
$$+ k^2\tilde{\varepsilon}(x, \boldsymbol{\rho}_0)G^*(L, \boldsymbol{\rho}''; x, \boldsymbol{\rho}_0) = 0 \quad (3.78)$$

Following the usual procedure of multiplying the first equation by G^*, multiplying the second equation by G, subtracting the second result from the first, integrating the composite expression over the volume bounded by $0 \le x \le L$, $-\infty < \rho_0 < \infty$, and noting that the "surface terms" give a zero result at $\pm\infty$, one obtains

$$\int_{-\infty}^{\infty} [G(L, \boldsymbol{\rho}'; L, \boldsymbol{\rho}_0)G^*(L, \boldsymbol{\rho}''; L, \boldsymbol{\rho}_0) - G(L, \boldsymbol{\rho}'; 0, \boldsymbol{\rho}_0)G^*(L, \boldsymbol{\rho}''; 0, \boldsymbol{\rho}_0)]\, d^2\rho_0$$

Finally, employing the boundary condition given by Eq. (1.57) yields an orthogonality property of the parabolic equation Green functions, viz.,

$$\delta(\boldsymbol{\rho}' - \boldsymbol{\rho}'') = \int_{-\infty}^{\infty} G(L, \boldsymbol{\rho}'; 0, \boldsymbol{\rho}_0)G^*(L, \boldsymbol{\rho}''; 0, \boldsymbol{\rho}_0)\, d^2\rho_0 \quad (3.79)$$

Ensemble averaging this relation and using it in Eq. (3.76) gives

$$\langle I(\boldsymbol{\rho}_i)\rangle = \left(\frac{k^2}{4\pi^3 L^2 l^2}\right) I_I(\boldsymbol{\rho}_0)\langle|R(\boldsymbol{\rho}_0)|^2\rangle \int_{-\infty}^{\infty} |W(\boldsymbol{\rho}')|^2 \, d^2\rho' \tag{3.80}$$

Equation (3.80) indicates that the short-term ensemble average of the intensity distribution is a constant, thus relegating to zero the higher-order moments, in particular the normalized intensity variance of Eq. (3.74). Hence, so long as the dimensions of the imaging problem satisfy Eq. (3.75), no variation of the average intensity of the image will be induced by the random medium. Only when the emissivity properties of the object are endowed with a temporal variation will there be a nonzero intensity variance.

In cases that do not satisfy Eq. (3.75), e.g., earth-based astronomical observations or objects with small angular size accompanied by sharp boundaries, one must resort to more complicated analyses involving the fourth-order moments of the propagating wave field. Here, the random fluctuations of the propagating medium can come into play by "steering" the image of the object into different regions of isoplanaticity on the lens input aperture. However, as indicated in a series of discussions in the literature,[15,16] care must be taken in the evaluation of the fourth-order field moment, especially when one uses the extended Huygens-Fresnel principle and its associated phase approximation as was shown in Sec. 1.7.2 of Chap. 1.

In the cases considered in this book, image intensity fluctuations will not be addressed; it is implicitly assumed that the condition of Eq. (3.75) prevails, as is the case with the imaging of *extended* objects. The rudiments of the analysis required when considering intensity variations can be found elsewhere.[13]

3.6 Image Resolution and Its Assessment

As has been previously noted in addressing the quantitative measurement of the *resolution* (i.e., he discrimination quality of two closely spaced points in the object plane) of an object through the random atmospheric permittivity field,[6] there are several ways by which one can define what are essentially indirect measures of this important performance parameter of an optical system.[17] One such definition, suggested by Fried,[6] employs the result of the integral over all the spatial frequencies of the total optical system MTF as a measure of the resolution R,

$$R \equiv \int_{-\infty}^{\infty} \langle i(\boldsymbol{\kappa}_i)\rangle \, d^2\kappa_i \tag{3.81}$$

One can interpret this definition as a measure of the total spatial frequency bandwidth of the imaging "channel." If this relation is normalized with the resolution R_{max} that would be obtained if the lens diameter is taken to be arbitrarily large and the atmospheric component is removed from Eq. (3.41), the resulting ratio R/R_{max} can be related to the classical Strehl ratio, which is the maximum intensity in the image plane of an object viewed trough a random medium normalized by the maximum intensity of the same object viewed without the atmospheric component.

Another definition that can be used[12,17] is that of the normalized measure

$$R \equiv \frac{\int_{-\infty}^{\infty} \langle i(\boldsymbol{\kappa}_i)\rangle^2 \, d^2\kappa_i}{\int_{-\infty}^{\infty} \langle i_0(\boldsymbol{\kappa}_i)\rangle^2 \, d^2\kappa_i} \tag{3.82}$$

where $\langle i_0(\boldsymbol{\kappa}_i)\rangle$ is the MTF of the total optical system with the MCF of the atmospheric component $\Gamma_{11} = 1$, i.e., where the atmospheric propagation path is deterministic and, without loss of generality, can be taken to be absent. With this definition, R is a direct measure of the "information content" of the image. Thus, its value can be much more directly related to the resolution ability (i.e., the discriminating ability) of the total optical system than that given by Eq. (3.81).

These definitions, as well as any other arbitrary ones that can be considered, are individual measures of imaging performance and cannot be expected to give the same numerical result when applied to a given imaging scenario. Although each one has its merits and drawbacks,[17] it must be remembered that aside from obtaining a relative performance index of various imaging systems, which is what Eqs. (3.81) and (3.82) actually give, what one ultimately desires is what is actually implied by the concept of resolution; that is, one wants *the* measure of the smallest distance by which two points in the object plane can be separated and still be discernible in the imaging plane after propagation through the atmosphere and imaging optics (as was introduced at the end of Sec. 3.2). What is more, it is preferable, especially for design purposes, to relate this minimum distance to other fundamental quantities of the optical system, e.g., lens aperture diameter, focal length, etc., and those of the propagation medium, i.e., propagation distance, turbulence strength, etc.

In an analysis of the problem by Ishimaru,[18] a direct comparison of two limiting forms of the point spread function $P(\boldsymbol{\rho}_i'')$ of the total imaging system, instead of integral relationships such as Eqs. (3.81) or

(3.82), was employed to obtain analytical relationships between the quantities of the propagation medium and those of the imaging system that must be satisfied in order to assure acquisition of a target image. Although such relationships are important for imaging system design, the treatment in Ref. 18 does not allow a direct indication of what can be expected for the total system resolution.

One can, however, modify the approach of Ref. 18 to obtain a direct measure of object resolution while, at the same time, capturing the essence of the performance measure of Eq. (3.82). Consider the image plane intensity distribution in the case of isoplanatic imaging described by the substitution of Eq. (3.38) into Eq. (3.31),

$$\langle I(\boldsymbol{\rho}_i)\rangle = \frac{k^2}{4\pi^3 l^4}\int_{-\infty}^{\infty}\int_{-\infty}^{\infty} \exp\left[\frac{ik}{l}\,\boldsymbol{\rho}_d\cdot(\boldsymbol{\rho}_i - \boldsymbol{\rho}_0')\right] K(\boldsymbol{\rho}_d) \times \Gamma_{11}(L, \boldsymbol{\rho}_d)\langle|R(\boldsymbol{\rho}_0')|^2\rangle\, d^2\rho_d\, d^2\rho_0' \tag{3.83}$$

where the integration over the object coordinate has been replaced with that over the modified object coordinate. One now considers the MCF $\Gamma_{11}(L, \boldsymbol{\rho}_d)$ of the spherical wave propagating from the object plane to the lens aperture plane; assuming that the limit

$$\lim_{\rho_d\to\infty} \Gamma_{11}(L, \boldsymbol{\rho}_d) \equiv \Gamma_c(L) \tag{3.84}$$

exists, one can define the *coherent MCF* $\Gamma_c(L)$. This is that part of the total MCF that is due solely to the scattering of the wave field that remains directed along the propagation path. By virtue of the fact that the difference coordinate is taken to be arbitrarily large, $\Gamma_c(L)$ is devoid of any statistical process taking place on smaller scales. One can then define the *incoherent MCF* $\Gamma_i(L, \boldsymbol{\rho}_d)$,

$$\Gamma_i(L, \boldsymbol{\rho}_d) \equiv \Gamma_{11}(L, \boldsymbol{\rho}_d) - \Gamma_c(L) \tag{3.85}$$

and obtain that part of the total MCF that is completely determined by the scattering of the wave field, originally within the radius $\boldsymbol{\rho}_d$ in the transverse plane, out of the propagation path. Note that by Eq. (3.84),

$$\lim_{\rho_d\to\infty} \Gamma_i(L, \boldsymbol{\rho}_d) = 0$$

indicating that there is a decreasing amount of electromagnetic energy to be scattered out of the propagation direction as the transverse distance increases.

At this point, a small digression is in order concerning the origin of these concepts. The definition of Eq. (3.84) and the significance of the coherent and incoherent MCF are fundamentally derived by considering the stochastic imaging process described by Eqs. (3.71) and (3.72). In particular, for arbitrarily large values of the spatial frequency $\boldsymbol{\kappa}_i$, the portion of the image spectrum that remains unaffected by the random medium is the coherent or unscattered spectrum

$$\langle i(\boldsymbol{\kappa}_i)\rangle_c = \left(\frac{I_l}{\pi L^2}\right) K\left(\frac{l}{k}\boldsymbol{\kappa}_i\right) \Gamma_c(L) \left\langle r\left(-\frac{l}{L}\boldsymbol{\kappa}_i\right)\right\rangle$$

where one is now concerned with the limit

$$\lim_{\kappa_i \to \infty} \Gamma_{11}\left(L, \frac{l}{k}\boldsymbol{\kappa}_i\right) \equiv \Gamma_c(L)$$

which gives rise to the analogous expression in coordinate space of Eq. (3.84). Similarly, one has the incoherent or scattered spectrum

$$\langle i(\boldsymbol{\kappa}_i)\rangle_i = \left(\frac{I_l}{\pi L^2}\right) K\left(\frac{l}{k}\boldsymbol{\kappa}_i\right) \Gamma_i\left(L, \frac{l}{k}\boldsymbol{\kappa}_i\right) \left\langle r\left(-\frac{l}{L}\boldsymbol{\kappa}_i\right)\right\rangle$$

thereby allowing one to write for the total image spectrum

$$\langle i(\boldsymbol{\kappa}_i)\rangle = \langle i(\boldsymbol{\kappa}_i)\rangle_i + \langle i(\boldsymbol{\kappa}_i)\rangle_c$$

Strictly speaking, it is only the coherent term that carries all the available information concerning the object; the incoherent term only contributes to the attendant "noise" in the acquisition of the image.

Solving Eq. (3.85) for the total MCF and substituting the result into Eq. (3.83) allows the identification of the corresponding coherent and incoherent point spread functions, i.e.,

$$\langle I(\boldsymbol{\rho}_i)\rangle = \frac{k^2}{4\pi^3 l^4} \int_{-\infty}^{\infty} [P_i(\boldsymbol{\rho}_i - \boldsymbol{\rho}_0') + P_c(\boldsymbol{\rho}_i - \boldsymbol{\rho}_0')]\langle |R(\boldsymbol{\rho}_0')|^2\rangle \, d^2\rho_0' \tag{3.86}$$

where

$$P_i(\boldsymbol{\rho}_i - \boldsymbol{\rho}_0') \equiv \int_{-\infty}^{\infty} \exp\left[\frac{ik}{l}\boldsymbol{\rho}_d \cdot (\boldsymbol{\rho}_i - \boldsymbol{\rho}_0')\right] K(\boldsymbol{\rho}_d)\Gamma_i(L, \boldsymbol{\rho}_d)\, d^2\rho_d \tag{3.87}$$

is the incoherent point spread function and

$$P_c(\boldsymbol{\rho}_i - \boldsymbol{\rho}_0') \equiv \Gamma_c(L) \int_{-\infty}^{\infty} \exp\left[\frac{ik}{l}\boldsymbol{\rho}_d \cdot (\boldsymbol{\rho}_i - \boldsymbol{\rho}_0')\right] K(\boldsymbol{\rho}_d)\, d^2\rho_d \tag{3.88}$$

is the coherent point spread function. Since the incoherent MCF entails the electromagnetic radiation from the object that scatters out of the propagation direction and thus does not come directly from the object, the first term within the brackets of Eq. (3.86) describes the "background" illumination that appears as a halo around the object and degrades the contrast of its image. The second term within the brackets describes the radiation that emanates directly from the object and thus becomes a quantity of interest in the ability to secure an image of the object. Hence, the ability to resolve the image involves an interplay between the coherent and incoherent scattering components and the degradation of image contrast that is induced.

The pertinent quantity that characterizes the quality of an image in such a scenario, while at the same time allowing a quantitative estimate of the resolution capability as described above, is the *spatial spectral frequency/contrast (SF/C) ratio,* the definition of which is derived as follows. Let the reflectivity of the target be described as

$$\langle |R(\boldsymbol{\rho}_0')|^2\rangle = R_{\text{avg}}[1 + R_0 \cos(\boldsymbol{\kappa}_0 \cdot \boldsymbol{\rho}_0')] \tag{3.89}$$

where R_{avg} is the average reflectivity of the object and R_0 is the maximum reflectivity of the cosinusoidal variation with spatial frequency $\boldsymbol{\kappa}_0$ across the object. (Here, it will be assumed that the image of the target completely fills the field of view of the imaging system, thus allowing this analysis to consider the object as unbounded; however, in the general case of an unbounded object, the analysis must be modified from that which is given here.) The initial *contrast* C_0 of such a reflectivity distribution is defined by

$$C_0 \equiv \frac{\langle |R(\boldsymbol{\rho}_0')|^2\rangle_{\max} - \langle |R(\boldsymbol{\rho}_0')|^2\rangle_{\min}}{\langle |R(\boldsymbol{\rho}_0')|^2\rangle_{\max} + \langle |R(\boldsymbol{\rho}_0')|^2\rangle_{\min}}$$

$$= R_0 \tag{3.90}$$

One can similarly define the resulting image contrast C_i as

$$C_i \equiv \frac{\langle I(\boldsymbol{\rho}_i)\rangle_{\max} - \langle I(\boldsymbol{\rho}_i)\rangle_{\min}}{\langle I(\boldsymbol{\rho}_i)\rangle_{\max} + \langle I(\boldsymbol{\rho}_i)\rangle_{\min}} \tag{3.91}$$

and use Eqs. (3.86) to (3.89) to relate it to C_0, thus assessing the degradation by the intervening random medium. The ratio of Eq. (3.91) to Eq. (3.90), in terms of the spatial frequency of the object, is the SF/C ratio.

To this end, substituting Eqs. (3.87) and (3.88) into Eq. (3.86), using the worst-case approximation $\Gamma_i(L, \boldsymbol{\rho}_d) \approx \Gamma_i(L, 0)\delta(\boldsymbol{\rho}_d)$, and employing Eq. (3.89) yields

$$\langle I(\boldsymbol{\rho}_i)\rangle = \frac{R_{\text{avg}}}{\pi l^2}\Bigg[K(0)\Gamma_i(L, 0) + R_0 K(0)\Gamma_i(L, 0)\delta(\boldsymbol{\kappa}_0) + K(0)\Gamma_c(L) + R_0 K\left(\frac{l}{k}\boldsymbol{\kappa}_0\right)\Gamma_c(L)\cos(\boldsymbol{\rho}_i \cdot \boldsymbol{\kappa}_0)\Bigg] \tag{3.92}$$

The first term of this equation describes the intensity scattered out of and back into the propagation direction and thus comprises the backscatter intensity induced by that which is associated with the transmission of the image, i.e., the process described by the third and fourth terms involving the coherent intensity. The second term, however, is the incoherent intensity associated with the image structure at zero spatial frequency; its form arises from the worst-case approximation used for the incoherent MCF. Hence, rearranging these terms by separating the first term from the others and regrouping yields for $\boldsymbol{\kappa}_0 \neq 0$

$$\langle I(\boldsymbol{\rho}_i)\rangle = \frac{1}{\pi l^2}\{\langle I_F(L)\rangle[1 + R_F \cos(\boldsymbol{\rho}_i \cdot \boldsymbol{\kappa}_0)] + \langle I_{FB}(L)\rangle\} \tag{3.93}$$

where

$$\langle I_F(L)\rangle \equiv R_{\text{avg}} K(0)\Gamma_c(L) \tag{3.94}$$

is the average intensity across the image plane of the lens, determined only by the action of the propagation medium by way of the coherent MCF,

$$R_F \equiv R_F(L, \boldsymbol{\kappa}_0) = \frac{K[(l/k)\boldsymbol{\kappa}_0]}{K(0)} R_0 \tag{3.95}$$

is the final contrast of the object image determined only by the action of the optical system, and

$$\langle I_{FB}(L)\rangle \equiv R_{\text{avg}} K(0)\Gamma_i(L, 0) \tag{3.96}$$

is the average backscatter intensity solely determined by the incoherent intensity.

Equation (3.93) shows that the original intensity distribution of the object, as given by Eq. (3.89), is reproduced on the focal plane of the lens with appropriate modifications in the average intensity and contrast, as well as the appearance of an intensity term that accounts for the backscatter radiation. This latter term is deleterious to image reproduction, as is easily shown by applying the definition of Eq. (3.91) for the image contrast. In particular, using Eq. (3.93), one has

$$C_i = \frac{R_F(L, \boldsymbol{\kappa}_0)}{1 + I_{FB}(L)/I_F(L)} = \frac{R_F(L, \boldsymbol{\kappa}_0)}{1 + \Gamma_i(L, 0)/\Gamma_c(L)} \tag{3.97}$$

Furthermore, making use of Eq. (3.90), one has for the SF/C ratio of image contrast to that of the initial object distribution,

$$C_R = C_R(\boldsymbol{\kappa}_0) \equiv \frac{C_i(\boldsymbol{\kappa}_0)}{C_0}$$

$$= \frac{R_F(L, \boldsymbol{\kappa}_0)/R_0}{1 + \Gamma_i(L, 0)/\Gamma_c(L)} \tag{3.98}$$

Finally, rearranging factors and making use of the fact that $\Gamma_{11}(L, 0) = 1$, one has for the SF/C ratio the simple but lucid result

$$C_R = \Gamma_c(L) \frac{K[(l/k)\boldsymbol{\kappa}_0]}{K(0)} \tag{3.99}$$

which, if a minimum contrast ratio is specified, can be easily solved for the corresponding spatial frequency.

It should be noted that the analysis given above is an approximate "worst-case" analysis in that the incoherent intensity was taken to be a δ function of position across the lens rather than some general functional relation. This is why, for example, an equation such as Eq. (3.99) ensues where the spatial frequency response is totally due to the imaging optics, and the level is due solely to the coherent MCF. However, these approximate relations can be useful in the design and performance analysis of optical imaging situations. For example, in addition to Eq. (3.99), the second term of Eq. (3.93) is necessary if one is to account for the background illumination contribution to the mean-square shot current of a photomultiplier imaging system.

However, it is the very subjective estimation of the minimum contrast at which one can distinguish structural features of a target that severely complicates the problem and, in many cases, makes the problem specific to the particular imaging configuration, e.g., the human eye, image intensifier, etc. Thus, even this rigorous approach to the resolution problem is transcended by other factors, some of which can be of physiological origin, that determine the ability to secure an image.

In the case of the human eye, a quasi-empirical resolution requirement[19,20] has been established where the maximum resolution realized for any propagation situation is given by the spatial frequency $\kappa_{i,\max}$ (in this case, this spatial frequency is defined in the image plane of the lens) that corresponds to a minimum value of the atmosphere/optics transfer function, viz., $t(L, (l/k)\boldsymbol{\kappa}_{i,\max}) \approx 0.02$, which is a function of the minimum image contrast that the eye can tolerate and still be able to distinguish light from dark line pairs. However, this spatial frequency must be augmented with the magnification factor $M = l/L$; in particular, using the identification $\boldsymbol{\kappa}_o = -M\boldsymbol{\kappa}_i$, one has for the corresponding maximum spatial frequency $\boldsymbol{\kappa}_{o,\max}$ in the object plane $t(L, (L/k)\boldsymbol{\kappa}_{o,\max}) \approx 0.02$, where inversion through the lens has also been included. In terms of the maximum spatial frequency referred to the object plane, one can then employ empirical minimum resolution requirements that incorporate the above-mentioned elusive physiological effects. Using the linear spatial frequency in the object plane, with units of line pairs per meter, $f_{o,\max} = \kappa_{o,\max}/2\pi$ that are able to be resolved on the object, one has the various levels of object recognition given in Table 3.2. In what is to follow below and in Chap. 4, the atmosphere/optics transfer function will be referred to the spatial frequency κ_i in the image plane. The conversion to that in the object plane for application of the information in Table 3.2 then follows according to the above discussion.

To obtain a corresponding minimum value for the atmospheric MCF, one must, in general, consider the lens MTF for the specific cases considered. Remembering the definition of the lens MTF given by Eq. (3.16) as the overlap area of two circles of radius R, corresponding to that of the lens, separated by the distance $\boldsymbol{\rho}_d$, one has, upon using a straightforward trigonometric construction [one can also take the inverse Fourier transform of the square of Eq. (3.18*b*)],

$$K(\boldsymbol{\rho}_d) = \int_{-\infty}^{\infty} W(\boldsymbol{\rho})W^*(\boldsymbol{\rho} - \boldsymbol{\rho}_d)\, d^2\rho_d$$

$$= \frac{1}{2}(2R)^2 \left\{ \cos^{-1}\left(\frac{\rho_d}{2R}\right) - \frac{\rho_d}{2R}\left[1 - \left(\frac{\rho_d}{2R}\right)^2\right]^{1/2} \right\} \qquad (3.100)$$

which holds for $\rho_d \le 2R$; otherwise, $K(\boldsymbol{\rho}_d) = 0$. Hence, the maximum spatial resolution imposed solely by the diffraction limitations of the lens is given by $\rho_{d,\max} = 2R = l\kappa_{\max}/k$. For the extreme case of $R = 2.54$ cm, $l = 10$ cm, and $\lambda = 0.63\ \mu$m, this corresponds to $\kappa_{\max} = 5 \times 10^6\ \text{m}^{-1}$ or $f_{\max} = 8.0 \times 10^5$ line pairs/m. (This latter quantity is called

TABLE 3.2 Minimum Resolution Requirements

Recognition level	$f_{0,\max}$ Line pairs per minimum target dimension, m
Detection	1.0 ± 0.25
Orientation	1.4 ± 0.35
Recognition	4.0 ± 0.8
Identification	6.4 ± 1.5

the cutoff frequency of the lens, as will be discussed in the next chapter.) However, in the event that any atmospheric MCF has a spatial frequency $\kappa_i \ll \kappa_{\max}$ such that $\Gamma_{11}(\mathrm{L}, (l/k)\kappa_i) = 0.02$, one can effectively neglect the lens contribution in a resolution assessment. This is particularly the case in situations with strong turbulence or in the presence of aerosols.

In what is to follow in the remaining chapters, no specific resolution requirement will be employed, although the use of the quasi-empirical condition just mentioned will be useful in the comparison of the impact of various propagation situations on imaging.

References

1. M. V. Klein and T. E. Furtak, *Optics,* 2nd ed. Wiley, New York, 1986.
2. J. W. Goodman, *Statistical Optics*. Wiley, New York, 1984.
3. R. L. Fante, "Imaging of an Object Behind a Random Phase Screen Using Light of Arbitrary Coherence," *J. Opt. Soc. Am.* **A2** (12), pp. 2318–2328 (1985).
4. M. I. Charnotskii, V. A. Myakinin, and V. U. Zavorotnyi, "Observation of Superresolution in Nonisoplanatic Imaging Through Turbulence," *J. Opt. Soc. Am.* **A7** (8), pp. 1345–1350 (1990).
5. F. Roddier, "The Effects of Atmospheric Turbulence in Optical Astronomy," in *Progress in Optics,* E. Wolf, ed. North-Holland, Amsterdam, 1981. Vol. 19, pp. 281–376.
6. D. L. Fried, "Optical Resolution through a Randomly Inhomogeneous Medium for Very Long and Very Short Exposures," *J. Opt. Soc. Am.* **56** (10), pp. 1372–1379 (1966).
7. M. Born and E. Wolf, *Principles of Optics*. Pergamon, New York, 1965. Sec. 9.2.
8. V. I. Tatarskii, *The Effects of the Turbulent Atmosphere on Wave Propagation*. U.S. Dept. of Commerce, TT-68-50464, Springfield, Va., 1971. Chap. 4.
9. R. J. Noll, "Zernike Polynomials and Atmospheric Turbulence," *J. Opt. Soc. Am.* **66** (3), pp. 207–211 (1976).
10. J. Y. Wang, "Optical Resolution through a Turbulent Medium with Adaptive Phase Compensations," *J. Opt. Soc. Am.* **67** (3), pp. 383–390 (1977).
11. D. L. Fried, "Probability of Getting a Lucky Short-Exposure Image through Turbulence," *J. Opt. Soc. Am.* **68** (12), pp. 1651–1658 (1978).
12. J. Y. Wang and J. K. Markey, "Modal Compensation of Atmospheric Turbulence Phase Distortion," *J. Opt. Soc. Am.* **68** (1), pp. 78–87 (1978).
13. V. U. Zavorotnyi, "Image Intensity Fluctuations of an Incoherent Source Observed through a Turbulent Medium," *Radiophys. Quantum Electron.* **28** (12), pp. 972–977 (1985).
14. V. U. Zavorotnyi, "Origin of Intensity Fluctuations in the Image of an Incoherent Object Observed through a Turbulent Medium," *Opt. Spectrosc.* (*USSR*) **65** (4), pp. 575–576 (1988).
15. V. U. Zavorotnyi, "Imaging of an Object Behind a Random Phase Screen Using Light of Arbitrary Coherence: Comment," *J. Opt. Soc. Am.* **A5** (2), pp. 263–264 (1988).
16. R. L. Fante, "Imaging of an Object Behind a Random Phase Screen Using Light of Arbitrary Coherence: Reply to Comment," *J. Opt. Soc. Am.* **A5** (2), p. 365 (1988).
17. E. H. Linfoot, "Transmission Factors and Optical Design," *J. Opt. Soc. Am.* **46** (9), pp. 740–752 (1956).
18. A. Ishmaru, "Limitation on Image Resolution Imposed by a Random Medium," *Appl. Opt.* **17** (3), pp. 348–352 (1978).
19. R. F. Lutomirski, "Atmospheric Degradation of Electro-Optical System Performance," *Appl. Opt.* **17** (24), pp. 3915–3921 (1978).
20. *RCA Electro-Optics Handbook,* R. E. Simon, ed. RCA, Harrison, N.J., 1974.

Chapter

4

The Analysis of Image Propagation in the Atmosphere

4.1 Introduction

It is the purpose of this chapter to demonstrate the methods developed hitherto and apply them to the assessment of long- and short-exposure image propagation. This effort commences with the use of the Rytov method in Sec. 4.2. Expressions for the log-amplitude and phase structure functions for wave propagation through turbulence are derived, and the opportunity is taken to qualitatively discuss the various propagation mechanisms at this level. In particular, a "random lens" model of the turbulent atmosphere is given to correspond to the rigorous results. The corresponding wave structure function as well as the related long-term and short-term MCF are then derived. Finally, the wave structure function and MCF are obtained for the turbid atmospheric propagation case. Graphical results of the atmosphere/optics transfer function for both the long-term and short-term MCF cases are given and discussed for a wide range of propagation scenarios. Section 4.3 then addresses the application of the MCF derived directly from the parabolic wave equation. Once the modified equation, as specified in Sec. 1.5, is reduced to the canonical form of the parabolic MCF equation, a general solution is obtained, and, as in the case of the Rytov method earlier in the chapter, it is applied to both the turbulent and turbid atmosphere cases. This requires the explicit use of the characteristic functionals that describe the statistics of the turbulent and turbid atmospheric permittivity field as derived in Chap. 2. Finally, the relationship between the parabolic equation for the MCF and the classical radiative transfer equation in the approximation of small scattering angle is established. This connection is then used to justify the use of the contrast criterion for image quality presented in Chap. 3 by showing that it is essentially analogous to the well-known spatial spectral contrast used in phenomenological radiative transfer theory.

4.2 Application of the Rytov Method

Of the various techniques for the analysis of stochastic electromagnetic wave propagation discussed in Chap. 1, the Rytov method allows an elucidation of many details of the propagation mechanisms that are only implicit in the other treatments. For this reason, the derivation of the atmospheric spherical wave MCF needed for the assessment of atmospheric imaging will commence here with the calculation of the log-amplitude and phase structure functions that will give rise to the associated wave structure function from which the MCF is found. The statistical description of the propagation process at the level of the log-amplitude and phase will give insight into a qualitative model of the underlying process that perturbs the propagation of an image through the atmosphere.

It is thus desired here to obtain the MDF to within the first Rytov approximation, given by Eq. (1.134), for a spherical wave of unit intensity, viz.,

$$\Gamma_{11}(L; \boldsymbol{\rho}_d) = \exp\left[-\tfrac{1}{2}D_\phi(L, \boldsymbol{\rho}_d)\right] \tag{4.1}$$

where $D_\phi(L, \boldsymbol{\rho}_d)$ is the associated spherical wave structure function, given by Eq. (1.135), which is the sum of the spherical wave log-amplitude and phase structure functions, $D_\chi(L, \boldsymbol{\rho}_d)$ and $D_S(L, \boldsymbol{\rho}_d)$, respectively, prescribed by Eq. (1.129),

$$D_{\substack{\chi \\ S}}(L, \boldsymbol{\rho}_d) = \frac{k^2}{4}\int_0^L \int_{-\infty}^{\infty} \operatorname{Re}\{[|H(L, x, \boldsymbol{\kappa})|^2 \mp H^2(L, x, \boldsymbol{\kappa})]$$

$$\times [1 - \exp(i\boldsymbol{\kappa}\cdot\mathbf{Q})]\} F_\varepsilon(x, \boldsymbol{\kappa})\, d\boldsymbol{\kappa}\, dx \tag{4.2}$$

where

$$\mathbf{Q} \equiv \gamma(x)(\boldsymbol{\rho}_d) \tag{4.3}$$

and

$$H(L, x, \boldsymbol{\kappa}) \equiv \exp\left[-\frac{i\kappa^2}{2k}\gamma(x)(L - x)\right] \tag{4.4}$$

with $\gamma(x) = x/L$ for spherical wave propagation. At this point, the two-dimensional spectral density $F_\varepsilon(x, \boldsymbol{\kappa})$ of the δ-correlated random permittivity field must be related to the spectral density of the turbulence and aerosol fields as derived in Chap. 2. To this end, one substitutes Eq. (D1.3.16) into Eq. (1.122),

$$F_\varepsilon(x, \boldsymbol{\kappa}) = \left(\frac{1}{2\pi}\right) \int_{-\infty}^{\infty} \int_{-\infty}^{\infty} A(x, \boldsymbol{\rho}_d)$$

$$\times \exp\,(i\boldsymbol{\kappa}' \cdot \boldsymbol{\rho}_d - i\boldsymbol{\kappa} \cdot \boldsymbol{\rho}_d)\Phi_\varepsilon(x, \boldsymbol{\kappa}')\, d\boldsymbol{\kappa}'\, d\rho_d$$

$$= 2\pi\Phi_\varepsilon(x, \boldsymbol{\kappa}) \tag{4.5}$$

where the explicit dependency on the x coordinate is due to the general case of position-dependent values for the refractive index and aerosol structure parameters. Since the analytical models of the atmospheric permittivity statistics are such that $\Phi_\varepsilon(x, \boldsymbol{\kappa})$ is taken to be isotropic in $\boldsymbol{\kappa}$, i.e., $\Phi_\varepsilon(x, \boldsymbol{\kappa}) = \Phi_\varepsilon(x, |\kappa|) = \Phi_\varepsilon(x, \kappa)$, one can write, upon substituting Eqs. (4.3) to (4.5) into Eq. (4.2) and converting the κ integration to one over plane polar coordinates,

$$D_{\underset{S}{\chi}}(L, \rho_d) = \frac{(2\pi k)^2}{4} \int_0^L \int_0^\infty \mathrm{Re}\left(\left\{1 \mp \exp\left[\frac{-i\kappa^2 x(L-x)}{kL}\right]\right\}\right.$$

$$\left.\times \left[1 - J_0\left(\frac{\kappa\rho_d x}{L}\right)\right]\right) \Phi_\varepsilon(x, \kappa)\kappa\, d\kappa\, dx \tag{4.6}$$

where, as usual, the Bessel function $J_0(\ldots)$ is the result of the angular integration of the series expansion of $\exp\,(i\boldsymbol{\kappa} \cdot \mathbf{Q})$. Finally, one has from the development of Chap. 2 that the spectral density $\Phi_\varepsilon(x, \kappa)$ is the sum of the continuum and aerosol spectral densities, viz.,

$$\Phi_\varepsilon(x, \kappa) = \Phi_{\varepsilon_c}(x, \kappa) + \Phi_{\varepsilon_p}(x, \kappa) \tag{4.7}$$

where

$$\Phi_{\varepsilon_c}(x, \kappa) = \frac{0.033 C_\varepsilon^2(x)}{(\kappa^2 + K_0^2)^{11/6}} \exp\left(\frac{-\kappa^2}{K_m^2}\right) \tag{4.8}$$

for the continuum [Eq. (2.61)] and

$$\Phi_{\varepsilon_p}(x, \kappa) \approx \frac{\vartheta(x)\langle a\rangle^4}{2\pi k^2} \frac{(\nu+2)(\nu+1)}{\nu^3}$$

$$\times \exp\left[-(\nu+4)(\nu+3)\left(\frac{\kappa\langle a\rangle}{2\nu}\right)^2\right] \tag{4.9}$$

for the (approximate) aerosol component [Eq. (2.111)]. Here the position dependency of the refractive index structure parameter $C_\varepsilon^2(x)$ and the aerosol concentration $\vartheta(x)$ are explicitly shown. Finally, to make contact with the accepted usage, the turbulence structure parameter will be stated in terms of the refractive index rather than the permittivity, viz., Eq. (2.105) will be employed. Thus, substitution of Eq. (4.7) into

Eq. (4.6) will yield separate log-amplitude and phase structure functions for the continuum (i.e., turbulence) and aerosol (i.e., turbid) components. Likewise, there will then exist two corresponding wave structure functions. This situation, of course, is due to the statistical independence of these two components. In what is to follow, these will be individually considered.

4.2.1 Log-amplitude and phase structure functions for atmospheric turbulence

Substituting Eq. (4.8) into Eq. (4.6) yields

$$D_{\substack{\chi \\ S}}(L, \rho_d) = 1.3028k^2 \int_0^L \int_0^\infty \operatorname{Re}\left(\left\{1 \mp \exp\left[\frac{-i\kappa^2 x(L-x)}{kL}\right]\right\}\right.$$
$$\times \left.\left[1 - J_0\left(\frac{\kappa\rho_d x}{L}\right)\right]\right) (\kappa^2 + K_0^2)^{-11/6}$$
$$\times \exp\left(\frac{-\kappa^2}{K_m^2}\right) C_n^2(x)\kappa \, d\kappa \, dx \quad (4.10)$$

Consider now the integrand in the limit as $\kappa \to 0$ but with an arbitrary value for K_0. Here, one has that

$$1 \mp \exp\left(\frac{-i\kappa^2 x(L-x)}{kL}\right) \longrightarrow \begin{cases} \dfrac{i\kappa^2 x(L-x)}{kL} \\ 2 - \dfrac{i\kappa^2 x(L-x)}{kL} \end{cases}$$

and

$$1 - J_0\left(\frac{\kappa\rho_d x}{L}\right) \longrightarrow \frac{1}{2}\left(\frac{\kappa\rho_d x}{L}\right)^2$$

One immediately sees that in the case of the log-amplitude, the integrand is identically zero, and in the case of the phase, the integrand approaches zero as $\kappa^3 K_0^{-11/3}$ if $K_0 \neq 0$ and as $\kappa^{-2/3}$ if $K_0 = 0$ (i.e., $L_0 \to \infty$). Hence, in the latter case, the integral will still converge to zero and one can neglect the contribution of the outer inertial scale in Eq. (4.10). This, of course, is an affectation of the nature of the structure function and its characteristic ability to "filter out" large-scale processes. Having established this, one now distributes the quantity within the second set of brackets of Eq. (4.10) over the two terms in the first set of brackets and obtains

$$D_{\substack{\chi \\ S}}(L, \rho_d) = 1.3028k^2 \int_0^L \operatorname{Re}\{[I_1(x, \rho_d) \mp I_2(x, \rho_d)]\} C_n^2(x) \, dx \quad (4.11)$$

where

$$I_n(x, \rho_d) \equiv \int_0^\infty \left[1 - J_0\left(\frac{\kappa\rho_d x}{L}\right)\right] \kappa^{-11/3} \exp\left(-\frac{\kappa^2}{D_n^2}\right) \kappa\, d\kappa \quad (4.12)$$

$$\text{with} \quad D_n \equiv \begin{cases} K_m & n = 1 \\ [ix(L - x)/(kL) + 1/K_m^2]^{-1/2} & n = 2 \end{cases} \quad (4.13)$$

The evaluation of the general expression of Eq. (4.12) commences with replacing the quantity within the brackets with its series representation.[1] Distributing the integral over the summation of this convergent series and using the change of variables $z \equiv A_n\kappa^2$ yields

$$I_n(x, \rho_d) = \sum_{p=1}^{\infty} \frac{(-1)^p}{(p!)^2} \left(\frac{\rho_d x}{2L}\right)^{2p} \frac{D_n^{2p-5/3}}{2} \int_0^\infty z^{(p-5/6)-1} \exp(-z)\, dz$$

Noting that the indicated integral is the definition of the gamma function $\Gamma(p - 5/6)$, one has, after rearranging factors,

$$I_n(x, \rho_d) = -\frac{1}{2} D_n^{-5/3} \sum_{p=1}^{\infty} \frac{1}{(p!)^2} \left(-\frac{\rho_d^2 x^2 D_n^2}{4L^2}\right)^p \Gamma(p - 5/6)$$

Finally, comparing this with the series of the confluent hypergeometric function ${}_1F_1(a, b; z)$ gives the solution

$$I_n(x, \rho_d) = \frac{1}{2} D_n^{-5/3} \Gamma(-5/6) \left[1 - {}_1F_1\left(-5/6, 1; -\frac{\rho_d^2 x^2 D_n^2}{4L^2}\right)\right] \quad (4.14)$$

Thus, substituting Eq. (4.14) into Eq. (4.11) and remembering the definition established by Eq. (4.13) gives for the general solution for the spherical wave log-amplitude and phase structure functions [note: $\Gamma(-5/6) = -6.67957$]

$$D_{\chi \atop S}(L, \rho_d)$$

$$= 4.351 k^2 L \int_0^1 \left\{ K_m^{-5/3} \left[{}_1F_1\left(-5/6, 1; -\frac{\rho_d^2 \eta^2 K_m^2}{4}\right) - 1 \right] \right.$$

$$\mp \operatorname{Re} \left[\left[\frac{i\eta L(1-\eta)}{k} + \frac{1}{K_m^2} \right]^{5/6} \right.$$

$$\left. \left. \times \left({}_1F_1\left\{ -5/6, 1; -\frac{\rho_d^2 \eta^2}{4} \left[\frac{i\eta L(1-\eta)}{k} + \frac{1}{K_m^2} \right]^{-1} \right\} - 1 \right) \right] \right\}$$

$$\times C_n^2(\eta)\, d\eta \quad (4.15)$$

where the change of variables $\eta \equiv x/L$ has been evoked. For most optical imaging problems met with in the applications, this seemingly unwieldy expression can take on a more palatable form. In particular, consider the argument of the first hypergeometric function in Eq. (4.15) and suppose that it satisfies the condition

$$\frac{\rho_d^2 \eta^2 K_m^2}{4} \gg 1 \tag{4.16}$$

throughout most of the integration region in η; remembering the definition of K_m, this condition corresponds to the situation where $\rho_d > l_0$, i.e., the transverse dimension over which the wave statistics are taken is larger than the smallest dimension (the inner scale) of the turbulence field. One can then consider the asymptotic expansion of this function

$$\begin{aligned} {}_1F_1\left(-5/6, 1; -\frac{\rho_d^2 \eta^2 K_m^2}{4}\right) &\approx \frac{\Gamma(1)}{\Gamma(11/6)}\left(\frac{\rho_d^2 \eta^2 K_m^2}{4}\right)^{5/6} \\ &= 0.33485 \rho_d^{5/3} \eta^{5/3} K_m^{5/3} \end{aligned} \tag{4.17}$$

Similarly, one can consider the argument of the second such function in Eq. (4.15); a simplification can be realized if one has, again over most of the η integration region, that

$$\left|\frac{i\eta L(1 - \eta)}{k}\right| \gg \frac{1}{K_m^2} \tag{4.18}$$

which reduces to the condition $l_0 < \sqrt{L\lambda}$, involving the Fresnel zone of the propagation scenario, which is realized in all optical imaging applications. (At this point, it may be instructive to refer back to Sec. 1.4.1 of Chap. 1 concerning the definitions and relations of these fundamental propagation characteristics.) Applying these conditions to Eq. (4.15) yields

$$\begin{aligned} D_{\underset{S}{\chi}}(L, \rho_d) &= 4.351 k^2 L \int_0^1 \Bigg[K_m^{-5/3}(0.33485 \rho_d^{5/3} \eta^{5/3} K_m^{5/3} - 1) \\ &\quad \mp \eta^{5/6}(1 - \eta)^{5/6} L^{5/6} k^{-5/6} \\ &\quad \times \mathrm{Re}\left((i)^{5/6}\left\{ {}_1F_1\left[-5/6, 1; i\frac{\rho_d^2 \eta k}{4L(1 - \eta)}\right] - 1\right\}\right)\Bigg] \\ &\quad \times C_n^2(\eta)\, d\eta \\ &\qquad l_0 < \rho_d \qquad l_0 < \sqrt{L\lambda} \end{aligned} \tag{4.19}$$

which describes the log-amplitude and phase structure functions for optical wavelength propagation in the turbulent atmosphere.

Proceeding in the same manner, one can specialize the argument of the ${}_1F_1$ function in Eq. (4.19) for conditions that correspond to well-known limits encountered in optical propagation problems.[1] Consider the case where

$$\left| i \frac{\rho_d^2 \eta k}{4L(1 - \eta)} \right| \ll 1 \tag{4.20}$$

from which one finds the corresponding condition $\rho_d < \sqrt{L\lambda}$. The condition of Eq. (4.20) renders negligible the effects due to diffraction. In this case, one finds that the second term of Eq. (4.19) is negligible with respect to the first because of the factors that multiply the factor Re{...}. Hence, in this case, Eq. (4.19) further reduces to

$$D_{\underset{S}{\chi}}(L, \rho_d) = 1.457k^2L\rho_d^{5/3} \int_0^1 \eta^{5/3}C_n^2(\eta)\, d\eta$$

$$- 4.351k^2LK_m^{-5/3} \int_0^1 C_n^2(\eta)\, d\eta$$

$$l_0 < \rho_d < \sqrt{L\lambda} \tag{4.21}$$

which gives, in the homogeneous case of $C_n^2(\eta) = C_n^2$,

$$D_{\underset{S}{\chi}}(L, \rho_d) = 0.546k^2L\rho_d^{5/3}C_n^2 - 4.351k^2LK_m^{-5/3}C_n^2 \tag{4.22}$$

Thus, within the scale range given by $l_0 < \rho_d < \sqrt{L\lambda}$, the statistics of both the log-amplitude and phase behave similarly. The results of Eq. (4.22) are sometimes considered as the *far field solution* due to the fact that they apply for propagation distances $L > \rho_d^2/\lambda$. The case opposite to that given in Eq. (4.20) corresponds to the propagation condition $\rho_d > \sqrt{L\lambda}$; here, one employs the asymptotic expansion given by

$${}_1F_1(-5/6, 1; x) \xrightarrow[|x| \gg 1]{} \frac{\Gamma(1)}{\Gamma(11/6)} (-x)^{5/6} \left[1 - \left(\frac{5}{6}\right)^2 \left(\frac{1}{x}\right) + \cdots \right]$$

[There is another term to this expansion that involves the product $\exp(x)x^{-11/6}$; however, since in the case considered here, x is purely imaginary, $|\exp(x)| = 1$. Thus, this term is $\sim x^{-11/6}$, thus allowing it to be neglected.] In this instance, Eq. (4.19) becomes

$$D_{\chi,S}(L, \rho_d)$$

$$= 4.351k^2L \int_0^1 \Bigg[K_m^{-5/3}(0.33485\rho_d^{5/3}\eta^{5/3}K_m^{5/3} - 1)$$

$$\mp 0.33485\rho_d^{5/3}\eta^{5/3} \pm \cos\left(\frac{5\pi}{12}\right) \eta^{5/6}(1 - \eta)^{5/6}L^{5/6}k^{-5/6} \Bigg]$$

$$\times C_n^2(\eta)\, d\eta$$

$$= 1.457k^2L\rho_d^{5/3} \int_0^1 \eta^{5/3}C_n^2(\eta)\, d\eta \mp 1.457k^2L\rho_d^{5/3}$$

$$\times \int_0^1 \eta^{5/3}C_n^2(\eta)\, d\eta \pm 1.126k^{7/6}L^{11/6}$$

$$\times \int_0^1 \eta^{5/6}(1 - \eta)^{5/6}C_n^2(\eta)\, d\eta - 4.351k^2LK_m^2 \int_0^1 C_n^2(\eta)\, d\eta$$

$$l_0 < \sqrt{L\lambda} < \rho_d \qquad (4.23)$$

Once again, in the case of the homogeneous deposition of $C_n^2(\eta)$ along the propagation path, Eq. (4.23) simplifies to

$$D_{\chi,S}(L, \rho_d) = 0.546k^2L\rho_d^{5/3}C_n^2 \mp 0.546k^2L\rho_d^{5/3}C_n^2$$

$$\pm 0.248k^{7/6}L^{11/6}C_n^2 - 4.351k^2LK_m^{-5/3}C_n^2 \qquad (4.24)$$

where it is noted that the third term in Eq. (4.23) involves the integral evaluation

$$\int_0^1 \eta^{5/6}(1 - \eta)^{5/6}\, d\eta = B(11/6, 11/6) = 0.2205$$

where $B(\ldots)$ is the beta function. Because of the region of applicability of Eq. (4.24), i.e., $L < \rho_d^2/\lambda$, it is termed the *near field solution*.

These relationships show how the spherical wave field statistics D_χ and D_S evolve as ρ_d covers the various scale ranges defined by the parameters that are significant to the problem, i.e., l_0 and $\sqrt{L\lambda}$, and reveal numerous insights into the propagation process that ultimately leads to image degradation through the wave structure function and the MCF given by Eq. (4.1). Before the analysis in this direction is continued, such insights will now be given. At the outset, it should be noted in Eq. (4.24) that D_χ tends to a constant value as ρ_d increases beyond the Fresnel zone, whereas D_S continues to increase as $\rho_d^{5/3}$. This latter fact is due to the neglect of another characteristic quantity of the problem, viz., the outer scale of turbulence and its associated spatial frequency K_0 in the development following Eq. (4.10). Retaining it would only lead to mathematical complexities in the ensuing analysis. Although one can reintroduce the outer scale in the case where $\rho_d > \sqrt{L\lambda}$, and at the same time neglect the inner scale contribution, and show that D_S also tends to a constant value, it is not a necessity for the purposes of the present discussion. However, if one were to derive the corresponding MCF, which will be done later, and attempt to calculate the coherent part of the MCF, as defined by Eq. (3.76), the result would be zero, indicating that all of the energy associated with the wave propagation is scattered out of the propagation direction. This, of course, is due to the absence of the upper bound on the sizes of the turbulent inhomogeneities. In the treatment later in this chapter of the propagation problem via the parabolic wave equation, an approximation procedure will be employed that will allow the inclusion of the outer scale contribution.

To elucidate the statistical process that governs the log-amplitude throughout the region $l_0 < \rho_d < \sqrt{L\lambda}$ and into the saturation regime at $\rho_d > \sqrt{L\lambda}$, it is instructive to consider, not the structure function, but the correlation function of this quantity. As we pointed out in Digression 1.4, correlation functions contain more information about the random field than do the structure functions. However, with the information already obtained above, one need not explicitly evaluate the log-amplitude correlation functions, given by Eq. (1.127), to obtain such correlation expressions. In particular, one can return to Eq. (1.128) and, after rearrangement, obtain

$$B_\chi(L, \rho_d) = \frac{2\sigma_\chi^2(L) - D_\chi(L, \rho_d)}{2} \tag{4.25}$$

Thus, in addition to the structure function, one needs an expression for the variance $\sigma_\chi^2(L)$. This, too, can be obtained from what exists in the above development. To accomplish this, one can use a very convenient property of structure functions. To obtain the variance, one returns once again to Eq. (1.128) and considers the limiting form

$$\lim_{\rho_d \to \infty} D_\chi(L, \rho_d) = 2\sigma_\chi^2(L) - 2 \lim_{\rho_d \to \infty} B_\chi(L, \rho_d) \tag{4.26}$$

The limit of the second term on the left side of this relation can be shown to be zero by employing Eq. (D1.3.12) and evoking the Riemann-Lebesgue lemma for Fourier coefficients, viz.,

$$\lim_{\rho_d \to \infty} B(x, \rho) = \lim_{\rho_d \to \infty} \int_{-\infty}^{\infty} \exp(i\kappa \cdot \rho) F(x, \kappa)\, d\kappa = 0$$

so long as $F(x, \kappa)$ is absolutely integrable, which it is. Hence, one has from Eqs. (4.26) and (4.24) that

$$\sigma_\chi^2(L) = 0.124 k^{7/6} L^{11/6} C_n^2 - 2.176 k^2 L K_m^{-5/3} C_n^2 \qquad (4.27)$$

for the variance of the χ fluctuations of a spherical wave. [Note that by using Eq. (4.27) in Eq. (1.133), one now has an analytical expression for the constraint governing the application of the first-order Rytov approximation to the propagation quantities k, L, and C_n^2 for the spherical wave case.] One can now finally obtain an expression for the spatial correlation function by using the general expression of Eqs. (4.19) and (4.27) in Eq. (4.25). To further aid in the analysis, one can define a normalized correlation function $b_\chi(L, \rho_d)$:

$$b_\chi(L, \rho_d) \equiv \frac{B_\chi(L, \rho_d)}{\sigma_\chi^2(L)} \qquad (4.28)$$

A plot of Eq. (4.28) is shown in Fig. 4.1. Here, $b_\chi(L, \rho_d)$ is plotted versus the transverse coordinate ρ_d normalized with respect to the Fresnel zone length $\sqrt{L\lambda}$. The salient feature of this functional behavior is the negative tail of the correlation in the region $0.7 < \rho_d/\sqrt{L\lambda} < 1.4$ and the fact that the midpoint of this region occurs

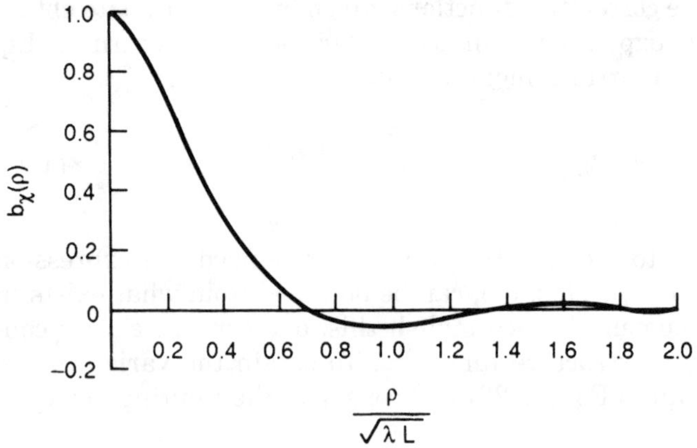

Figure 4.1 Graph of normalized log-amplitude correlation function vs. transverse distance.

approximately at the transverse distance $\rho_d = \sqrt{L\lambda}$. The interpretation of the negative region of $b_\chi(L, \rho_d)$ is that, for any two points in the transverse plane that are separated by approximately one Fresnel zone, the log-amplitude (and hence the intensity) at one point will be greater than the average (the intensity will be brighter than the average intensity) and the log-amplitude at the other point will be less than the average (the intensity will be darker than the average intensity). This behavior is indicative of diffraction phenomena induced by the turbulent inhomogeneities along the propagation path.

The Fresnel zone length provides the point of demarcation after which diffractive effects become important. The physical meaning of this quantity becomes obvious when one considers the following geometrical construction. Envision a point on the longitudinal axis of the propagation path at a distance d from the reception plane. One now considers at what distance ρ in the reception plane the distance to the source point increases by an amount $\lambda/2$, the condition for maximum phase variation. One thus has for this situation, by the simple application of the Pythagorean theorem,

$$d^2 + \rho^2 = \left(d + \frac{\lambda}{2}\right)^2$$

from which one obtains, upon noting that $d \gg \lambda/2$, the relation $\rho = \sqrt{d\lambda}$. Hence, for each distance along the propagation path, there exists a Fresnel zone. However, as can be seen from the equations for the structure functions in the case of the inhomogeneous deposition of $C_\varepsilon^2(\eta)$ along the path [e.g., Eqs. (4.21) and (4.23)], the turbulent fluctuations closest to the source are more heavily weighted than those closest to the reception plane; in particular, the weighting is $\sim\eta^{5/3}$. Thus, one can expect the Fresnel zone that corresponds to $d = L$ to have the greatest impact on the propagation dynamics.

As for the phase fluctuations in general, a simple qualitative model can be employed to describe the statistical behavior derived above. Using the parlance of geometrical optics, consider two rays that emanate from a phase front of a spherical wave having its origin at a distance L from the reception plane upon which the rays are at a transverse spacing ρ_d. Thus, from a simple geometrical construction, one has that at any intermediate x distance along the propagation path, the transverse spacing of the rays is given by

$$\rho(x) = \left(\frac{x}{L}\right) \rho_d \tag{4.29}$$

Assume that $l_0 < \rho(x) < \sqrt{L\lambda}$. (This, of course, does not hold at positions for which $x < Ll_0/\rho_d$, but, as was the case in the derivation of the equations given above, ramifications of this are deemed negligible.) For every position x along the path, one has, from considerations for the common volume of the turbulent medium that can, within the inertial subrange of refractive index fluctuations where [cf. Eq. (2.59), i.e., the 2/3 law]

$$\langle[\Delta n(x)]^2\rangle \sim D_n(\rho) = C_n^2\rho^{2/3}(x) \tag{4.30}$$

simultaneously affect both diverging rays, a corresponding ray segment length $\rho(x)$, of which there are a total number $N(x) = L/\rho(x)$ for each intermediate position. One then finally notes that the phase difference of the wave after traversing such a pair of ray segments is given by $\Delta S(\rho) = k\rho\,\Delta n$. The square of the total phase variation, averaged over the propagation statistics, due to all the ray segments along the entire propagation path as determined at the intermediate point x is thus given by

$$\langle[\Delta S(\rho)]^2\rangle = k^2\rho^2(x)\langle[\Delta n(x)]^2\rangle N(x)$$

Hence, one has, upon using Eqs. (4.29) and (4.30) and the definition for $N(x)$, and averaging the resulting expression over the entire propagation path length,

$$\langle[\Delta S(\rho)]^2\rangle_L = \frac{1}{L}\int_0^L \langle[\Delta S(x,\rho)]^2\rangle\,dx$$

$$= \frac{3}{8}\,k^2L\rho_d^{5/3}C_n^2 \tag{4.31}$$

To within the constant coefficient, this agrees with the major term obtained in Eqs. (4.22) and (4.23).

Similarly, one can qualitatively model the log-amplitude statistics. Here, the atmosphere is modeled as a set of plano-convex lenses that have random focal lengths that are determined by the random refractivity field. Returning to Eq. (3.4) of Chap. 3, one has for the focal length F of a lens in the plano-convex case, i.e., with $R_2 \to \infty$,

$$\frac{1}{F} = (n - 1)\frac{1}{R_2}$$

Letting the radius of curvature of the lens be given by $R_1 \sim l$, where, since a geometrical optics approach is being used, l must be constrained by the relationship

$$l \geq \sqrt{\lambda L} \tag{4.32}$$

and letting the "effective lens" refractive index variation be given by $(n - 1) \equiv \Delta n(\mathbf{r})$ for a lens at a position $\mathbf{r}$ within the atmosphere, the statistics of which are determined by the 2/3 law of Eq. (4.30), one can write for the mean square average of the focal lengths of such random lenses

$$\langle F^2 \rangle \sim \frac{\langle R_1^2 \rangle}{\langle [\Delta n(\Delta r)]^2 \rangle} = \frac{l^2}{C_n^2 (\Delta r)^{2/3}} \tag{4.33}$$

where $\Delta r \equiv |\Delta \mathbf{r}|$. Finally, assuming that the lenses are separated from each other by the same characteristic distance l, one has from Eq. (4.33)

$$\langle F^2 \rangle \sim \frac{l^{4/3}}{C_n^2} \tag{4.34}$$

of which there are approximately N such lenses throughout the length L of the propagation path, where

$$N = \frac{L}{l} \tag{4.35}$$

One now needs two other results from Chap. 3, viz., the lens law of Eq. (3.9) and the geometrical optics result of Eq. (3.12). Using the latter expression, one has for the variation of intensity behind a lens as compared to its absence in a field with incident intensity I_0

$$\delta I \equiv I - I_0 = \left[\left(\frac{L}{l} \right)^2 - 1 \right] I_0 \tag{4.36}$$

where it must be remembered that l in this expression has a meaning different from that used in this chapter. Applying the lens law in the case where $L \gg F$, one has that $l = F$ in Eq. (4.36), and one can obtain

$$\delta I = \left[\left(\frac{L}{F} \right)^2 - 1 \right] I_0 \approx \left(\frac{L}{F} \right)^2 I_0$$

To make contact with the log-amplitude, it is noted that $\delta A \sim \sqrt{\delta I}$ and also $A_0 \sim \sqrt{I_0}$, thus giving the relation

$$\delta A = \left(\frac{L}{F} \right) A_0$$

which is used with the definition of Eq. (1.109) to yield

$$\chi = \ln \left(\frac{A}{A_0} \right) = \ln \left(\frac{A_0 + \delta A}{A_0} \right) \approx \frac{\delta A}{A_0}$$

$$= \left(\frac{L}{F} \right) \tag{4.37}$$

The mean square amplitude variation due to the ensemble of the N atmospheric lenses described by Eq. (4.34) is now easily obtained. Using the equality in Eq. (4.32) yields

$$\langle \chi^2 \rangle = \frac{L^2}{\langle F^2 \rangle} N = \frac{L^2}{l^{4/3}} C_n^2 N$$

$$= C_n^2 L^3 l^{-7/3} \sim C_n^2 L^{11/6} k^{7/6} \tag{4.38}$$

which is the form of the major term obtained in Eq. (4.24).

Although the methods used in the construction of these qualitative models lack mathematical rigor throughout their development, they do aid in obtaining a physical interpretation of the phenomena. In fact, heuristic models of this type have been advanced to describe some propagation phenomena that have hitherto eluded rigorous mathematical description.[2]

4.2.2 The wave structure function and the long-exposure mutual coherence function for atmospheric turbulence

Having obtained expressions for the log-amplitude and phase structure functions for a spherical wave, it is now a straightforward matter to find the spherical wave MCF. As indicated by Eq. (4.1), one only needs to derive the corresponding wave structure function, viz., using Eqs. (1.135) and (4.19),

$$D_\phi(L, \rho_d) \equiv D_\chi(L, \rho_d) + D_S(L, \rho_d)$$

$$= 2.914 k^2 L \rho_d^{5/3} \int_0^1 \eta^{5/3} C_n^2(\eta)\, d\eta$$

$$- 8.702 k^2 L K_m^{-5/3} \int_0^1 C_n^2(\eta)\, d\eta \tag{4.39}$$

in the case of an inhomogeneous atmosphere and

$$D_\phi(L, \rho_d) = 1.093 k^2 L \rho_d^{5/3} C_n^2 - 8.702 k^2 L K_m^{-5/3} C_n^2 \tag{4.40}$$

for a homogeneous atmosphere. Both these expressions hold for the general case $l_0 < \rho_d$, $l_0 < \sqrt{L\lambda}$. To facilitate any further analysis, one can neglect the contribution of the inner scale of turbulence, letting $l_0 \to 0$, whereby $K_m \to \infty$. Thus, using Eqs. (4.39) and (4.40) in this approximation, Eq. (4.1) yields for the corresponding MCFs after rearrangement of factors

$$\Gamma_{11}(L; \rho_d) = \exp\left[-\left(\frac{\rho_d}{r_0}\right)^{5/3}\right] \tag{4.41}$$

where

$$r_0 \equiv \begin{cases} \left[1.457k^2L \int_0^1 \eta^{5/3}C_n^2(\eta)\, d\eta\right]^{-3/5} & \text{inhomogeneous atmosphere} \\ (0.546k^2LC_n^2)^{-3/5} & \text{homogeneous atmosphere} \end{cases} \tag{4.42}$$

is a fundamental length, first introduced by Fried,[3] that indicates the *coherence length* in the transverse plane of a spherical wave propagating over a distance L in the atmosphere. That is, the coherence length r_0 is the distance ρ_d in the transverse plane at which the MCF decreases by a factor of exp (-1) from its nominal value at $\rho_d = 0$. Table 4.1 shows values for the coherence length at the wavelength $\lambda = 0.63\ \mu$m and for various values of L and C_n^2. (The values for C_n^2 are representative of those listed in Table 2.4.) It is important to note that there are two cases in Table 4.1 that are formally not applicable (N/A) to the first-order Rytov approximation, as they violate the constraint $\sigma_\chi^2(L) = 0.124k^{7/6}L^{11/6}C_n^2 \ll 1$.

Figure 4.2 displays the MCF of Eq. (4.41) for a homogeneous atmosphere in the two cases of $r_0 = 2.29 \times 10^{-2}$ m and $r_0 = 9.10 \times 10^{-2}$ m. This plot demonstrates the typical behavior of the MCF between two points in a plane transverse to the direction of propagation as the distance of separation of these points increases. It is this fundamental phenomenon that is the source of image degradation in the atmosphere.

A more meaningful and universal form describing the impact of these results on imaging follows from the consideration of the long-exposure image transfer function relation of Eq. (3.72). Substituting into Eq. (3.72) the results of Eq. (3.89) and Eqs. (4.41) and (4.42) for the homogeneous atmosphere coherence length and using the connection

TABLE 4.1 Typical Values for the Coherence Length r_0

C_n^2 \ L	1 km	5 km	10 km
1×10^{-14} m$^{-2/3}$	2.29×10^{-2} m	N/A	N/A
1×10^{-15} m$^{-2/3}$	9.10×10^{-2} m	3.46×10^{-2} m	2.29×10^{-2} m

$\kappa_i = 2\pi \mathbf{f}_i$ between the angular spatial frequency κ_i and the linear spatial frequency $\mathbf{f}_i$ yields, upon normalization,

$$t_N(L, f) \equiv \frac{t(L, f)}{t(L, 0)}$$
$$= \frac{2}{\pi}\left\{\cos^{-1}\left(\frac{f}{f_c}\right) - \left(\frac{f}{f_c}\right)\left[1 - \left(\frac{f}{f_c}\right)^2\right]^{1/2}\right\}$$
$$\times \exp\left[-\left(\frac{2R}{r_0}\right)^{5/3}\left(\frac{f}{f_c}\right)^{5/3}\right] \quad (4.43)$$

where the *aperture cutoff frequency* f_c is defined by

$$f_c \equiv \frac{2R}{\lambda l} \quad (4.44)$$

in accordance with the fact that, as specified by Eq. (3.92), $K(l\kappa/k) = K(l\lambda\mathbf{f}) = 0$ for frequencies f such that $l\lambda f > 2R$. This formation of the problem involves a dimensionless parameter that incorporates two fundamental quantities of the atmosphere/optics imaging system, viz., $2R/r_0$, which provides a comparison of the diameter of the entrance pupil of the imaging system lens to the atmospheric coherence length. Figure 4.3*a* shows a plot of the normalized transfer function $t_N(L, f)$ versus the normalized spatial frequency f/f_c for various representative

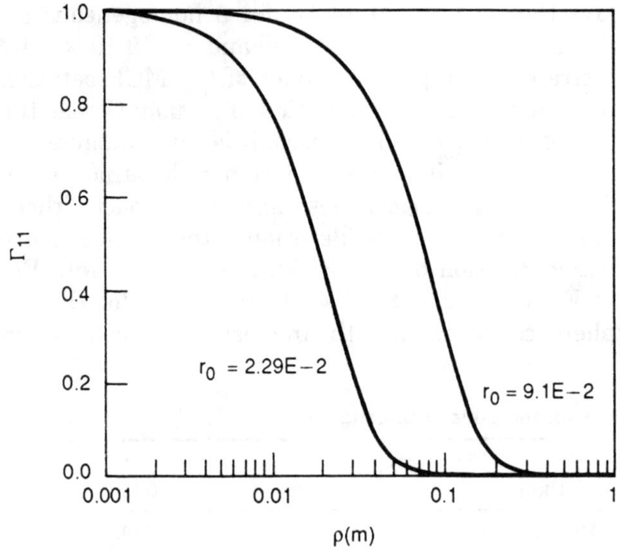

Figure 4.2 Plot of spherical wave MCF vs. transverse separation.

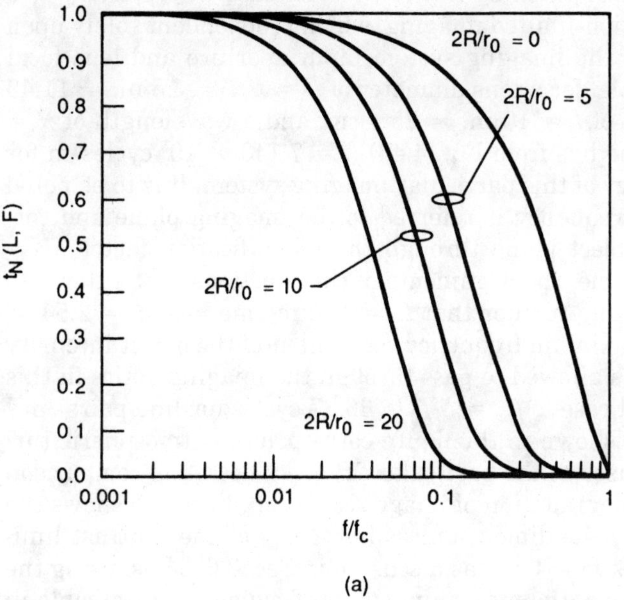

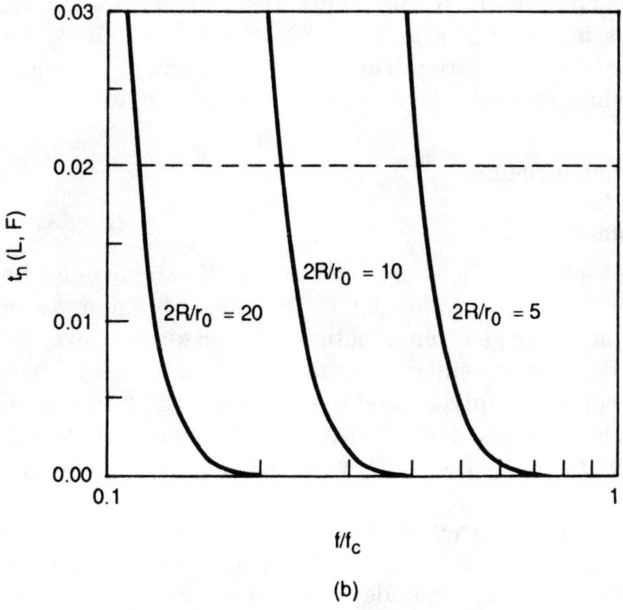

Figure 4.3 Plot of Eq. (4.43) for various $2R/r_0$ values.

values of $2R/r_0$ as dictated by Eq. (4.43). The case where $2R/r_0 = 0$ represents diffraction-limited imaging, which is dependent solely upon the dimensions of the imaging system input aperture and lens focal length. For example, for a lens diameter of $D = 2R = 4.5$ in. $= 11.43$ cm, a focal length of $l = 10$ in. $= 25.4$ cm, and a wavelength of $\lambda = 0.63 \times 10^{-6}$ m, one has from Eq. (4.44) $f_c = 7.143 \times 10^5$ cycles/m for the cutoff frequency of this particular imaging system. It is to be noted that this spatial frequency is referred to the imaging plane and corresponds to the object plane through the magnification factor $M = l/L$. [See, for example, the arguments of the functions that enter into Eq. (3.41).] Assuming further that $L = 10$ km, one has $M = 2.54 \times 10^{-5}$. Thus, the maximum frequency component of the object intensity distribution that is allowed to pass through the imaging optics in this diffraction-limited case is $f_{c,i} = Mf_c = 36.27$ cycles (or line pairs)/m.

The other cases shown in the figure correspond to atmospheric turbulence-limited cases at the particular $2R/r_0$ values. As a comparison of the turbulence degradation of image resolution, Fig. 4.3*b* shows the values assumed by the dimensionless frequency at the contrast limitation level of $t_N(L, f) = 0.02$, as discussed in Sec. 3.6. Thus, using the example given above, the maximum spatial frequencies resolvable in the object plane for $2R/r_0 = 5$, 10, and 20 are, respectively, 7.43, 3.97, and 2.18 line pairs/m. According to Table 3.2, these cases allow, respectively, the identification, recognition, and orientation of an object whose maximum characteristic size is on the order of a meter.

4.2.3 The short-exposure mutual coherence function for atmospheric turbulence

An analytical expression for the short-exposure MCF can now also be derived. As shown in Sec. 3.4.2 of Chap. 3, the effect of a short-term sample of the propagating electromagnetic field necessarily involves an interaction with the functional description of the imaging optics at a level where one cannot simply append the long-term MCF to that of the short term. Thus, one must deal directly with the short-term image transfer function as given by Eq. (3.73) with the subsidiary relations of Eqs. (3.70), (3.64), and (3.65).

Considering the form required by Eq. (3.70) and, attendant with the difference coordinate $\boldsymbol{\rho}_d$, defining the "centroid" coordinate $\boldsymbol{\rho}_c \equiv (2\boldsymbol{\rho}' - \boldsymbol{\rho}_d)/2$, Eq. (3.64) can be expanded and written as

$$\Delta_C(\boldsymbol{\rho}', \boldsymbol{\rho}' - \boldsymbol{\rho}_d) = \exp\left(2\langle a_2^2\rangle \frac{\rho_d^2}{R^2} - 4\sqrt{8}\langle a_2 a_8\rangle \frac{\rho_d^2}{R^2}\right)$$
$$\times \exp\left\{6\sqrt{8}\langle a_2 a_8\rangle \left[2(\boldsymbol{\rho}_c \cdot \boldsymbol{\rho}_d)^2 + \rho_c^2\rho_d^2 + \frac{\rho_d^4}{4}\right]\right\} \quad (4.45)$$

where the second exponential term involves the coordinate $\boldsymbol{\rho}_c$. Substituting this expression into Eq. (3.70) now involves an integration over the product

$$W\left(\boldsymbol{\rho}_c + \frac{\boldsymbol{\rho}_d}{2}\right) W^*\left(\boldsymbol{\rho}_c - \frac{\boldsymbol{\rho}_d}{2}\right)$$

which immediately results in analytical difficulties. To circumvent such problems, it proves expedient to describe the circular lens aperture not by the realistic model of Eq. (3.57) but by the analytically amenable model that takes the transmission function of the aperture as a gaussian function of position within the aperture, viz.,

$$W(\boldsymbol{\rho}') = \exp\left(-\frac{\rho'^2}{2R^2}\right) \tag{4.46}$$

This gaussian aperture function is defined in such a way that

$$K(\rho_d) = \int_{-\infty}^{\infty} W\left(\boldsymbol{\rho}_c + \frac{\boldsymbol{\rho}_d}{2}\right) E^*\left(\boldsymbol{\rho}_c - \frac{\boldsymbol{\rho}_d}{2}\right) d^2\rho_d$$

$$= \pi R^2 \exp\left(-\frac{\rho_d^2}{4R^2}\right) \tag{4.47}$$

and therefore $K(0) = \pi R^2$ as before. Thus, in this approximation, the long-term transfer function analogous to Eq. (4.43) becomes

$$t_N(L, f) \approx \exp\left[-\left(\frac{f}{f_c}\right)^2\right] \exp\left[-\left(\frac{2R}{r_0}\right)^{5/3}\left(\frac{f}{f_c}\right)^{5/3}\right] \tag{4.48}$$

Figure 4.4 shows the comparison of Eqs. (4.43) and (4.48) in the case where $2R/r_0 = 10$. Employing Eqs. (4.45) and (4.46) in Eq. (3.70) yields, upon converting the integration into one over plane polar coordinates,

$$K_C(\rho_d) = \pi R^2 \left(1 - 50.912\langle a_2 a_8\rangle \frac{\rho_d^2}{R^2}\right)^{-1}$$

$$\times \exp\left[\left(2\langle a_2^2\rangle - 11.314\langle a_2 a_8\rangle - \frac{1}{4}\right)\frac{\rho_d^2}{R^2} + 4.243\langle a_2 a_8\rangle \frac{\rho_d^4}{R^4}\right] \tag{4.49}$$

It is now necessary to obtain explicit expressions for the Zernike coefficient correlations $\langle a_j a_k\rangle$ given by Eq. (3.65) and related to the phase statistics of the propagating wave. To this end, it is easiest to deal with the Fourier transform of Eq. (3.65). Thus, using the known[4,5] transform relations governing the Zernike polynomials, viz.,

$$Z_k\left(\frac{\boldsymbol{\rho}}{R}\right) W_N\left(\frac{\boldsymbol{\rho}}{R}\right) = \int_{-\infty}^{\infty} Q_j(\boldsymbol{\kappa}) \exp\left(\frac{i\boldsymbol{\kappa}\cdot\boldsymbol{\rho}}{R}\right) d^2\kappa \tag{4.50}$$

where

$$Q_j(\boldsymbol{\kappa}) = \sqrt{n+1}\,\frac{J_{n+1}(\kappa)}{\pi\kappa} \times \begin{cases} (-1)^{(n-m)/2} - i^m \cos(m\theta)/(\pi\sqrt{2}) & j \text{ even}, m \neq 0 \\ (-1)^{(n-m)/2} - i^m \sin(m\theta)/(\pi\sqrt{2}) & j \text{ odd}, m \neq 0 \\ (-1)^{n/2} & \text{any } j, m = 0 \end{cases} \qquad (4.51)$$

where the meaning of the indices is given following Eq. (3.54), one has, upon transforming Eq. (3.65) in the usual way and remembering the definition for the normalized coordinate $\mathbf{r} \equiv \boldsymbol{\rho}/R$,

$$\langle a_j a_k \rangle = \left(\frac{2\pi}{R}\right)^2 \int_{-\infty}^{\infty} Q_j(\boldsymbol{\kappa}) Q_k^*(\boldsymbol{\kappa}) \beta_S \left(\frac{\boldsymbol{\kappa}}{R}\right) d^2\kappa \qquad (4.52)$$

where

$$\beta_s\left(\frac{\boldsymbol{\kappa}}{R}\right) \equiv \int_{-\infty}^{\infty} B_S(\boldsymbol{\rho}_d) \exp\left(\frac{i\boldsymbol{\kappa}\cdot\boldsymbol{\rho}_d}{R}\right) d^2\rho_d \qquad (4.53)$$

is the Fourier transform of the phase correlation. (It must be noted that because of the use of normalized coordinates, the spatial frequency

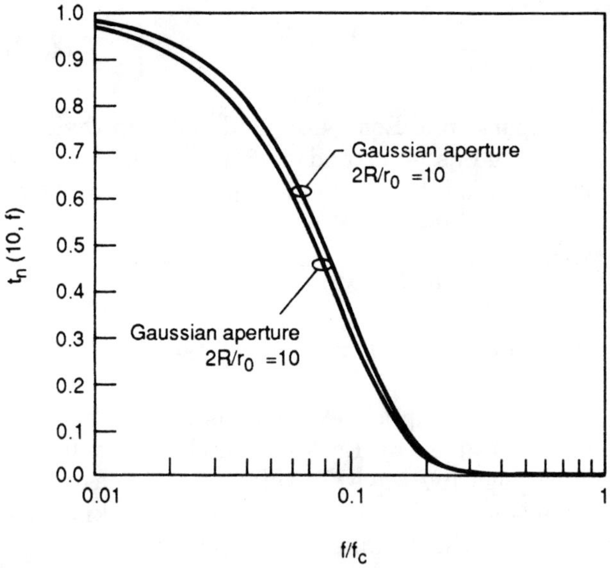

Figure 4.4 Comparison of exact and approximate transfer functions.

vector $\boldsymbol{\kappa}$ is dimensionless.) Since an expression has already been obtained for the related structure function of the random phase field [e.g., Eq. (4.24) for the near field case], one can follow the same procedure used to derive Eq. (D1.4.10) from Eq. (D1.4.9) so as to relate the transform $\beta_s(\kappa)$ of the correlation $B_S(\rho_d)$ to that of the structure function $D_S(\rho_d)$. Hence, one has

$$\beta\left(\frac{\kappa}{R}\right) = \left(\frac{1}{2\kappa^2}\right)\int_0^\infty J_0\left(\frac{\kappa\rho_d}{R}\right) \times \left\{\frac{1}{\rho}\frac{\partial}{\partial\rho}\,\rho\,\frac{\partial}{\partial\rho}\,[D_S(\rho_d)]\right\}\rho_d\,d\rho_d$$

$$= \left(\frac{\pi R^2}{\kappa^2}\right)(2)\left(\frac{1}{r_0}\right)^{5/3}\left(\frac{5}{3}\right)^2\int_0^\infty J_0\left(\frac{\kappa\rho_d}{R}\right)\rho_d^{2/3}\,d\rho_d$$

$$= 1.769R^2\left(\frac{2R}{r_0}\right)^{5/3}\kappa^{-11/3} \tag{4.54}$$

Using Eqs. (4.51) and (4.54) in Eq. (4.52) and performing the required integration finally yields[6]

$$\langle a_j a_k\rangle = 1.126\left(\frac{2R}{r_0}\right)^{5/3}(-1)^{(n+n'-2m)/2}\delta_{mm'}\sqrt{(n+1)(n'+1)} \times \left\{\frac{\Gamma(14/3)\Gamma[(n+n'-5/3)/2]}{2^{14/3}\{\Gamma[(n-n'+17/3)/2]\}^2\,\Gamma[(n+n'+23/3)/2]}\right\} \tag{4.55}$$

It is important to note that the expression given in Eq. (4.55) and subsequent related expressions differ by a factor of 3.44 from that given;[6] this difference stems from the arbitrary use of this numerical factor in the expressions of the MCF first adopted by Fried[3] solely for graphical convenience.

For the particular values needed by Eq. (4.49), one has from this general relation

$$\langle a_2^2\rangle = 0.1304\left(\frac{2R}{r_0}\right)^{5/3}$$

$$\langle a_2 a_8\rangle = -4.13\times 10^{-3}\left(\frac{2R}{r_0}\right)^{5/3}$$

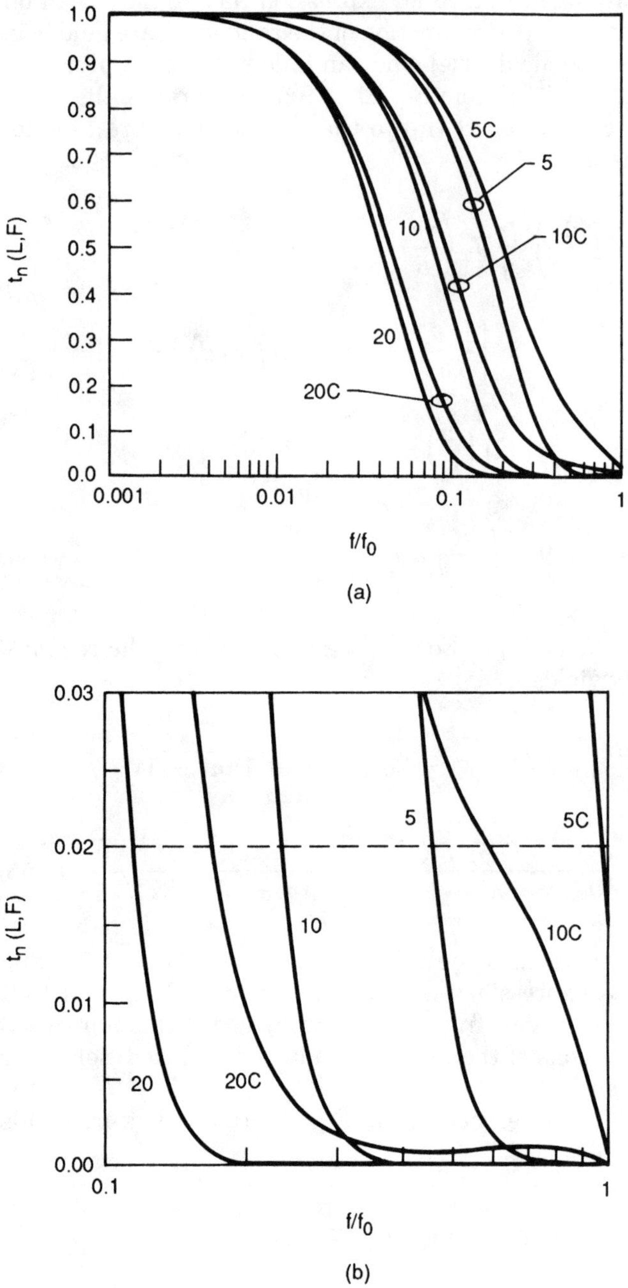

Figure 4.5 Plots of Eqs. (4.48) and (4.56).

Thus, using these expressions in Eq. (4.49) finally gives for the normalized short-term image transfer function

$$
\begin{aligned}
t_N(L, f) = {} & \exp\left[-\left(\frac{f}{f_c}\right)^2\right] \exp\left[-\left(\frac{f}{f_c}\right)^{5/3}\left(\frac{2R}{r_0}\right)^{5/3}\right] \\
& \times \left[1 + 0.8407\left(\frac{2R}{r_0}\right)^{5/3}\left(\frac{f}{f_c}\right)^2\right]^{-1} \\
& \times \exp\left[1.23\left(\frac{2R}{r_0}\right)^{5/3}\left(\frac{f}{f_c}\right)^2 - 0.279\left(\frac{2R}{r_0}\right)^{5/3}\left(\frac{f}{f_c}\right)^4\right]
\end{aligned}
\tag{4.56}
$$

Figure 4.5*a* shows a comparison of the long-exposure transfer function in the gaussian aperture approximation, i.e., Eq. (4.48), with the short-term transfer function of Eq. (4.56) for the values of the ratio $2R/r_0$ considered earlier. It is easily seen how the difference between the corrected transfer function and the long-term, uncorrected case diminishes as $2R/r_0$ increases in value. As discussed in Sec. 3.4.2, this is an affectation of the higher-order phase aberrations, neglected in the derivations leading to Eq. (4.56), that become prominent at higher turbulence levels.

The attendant improvement of image resolution, again at the $t_N(L, f) = 0.02$ level, is shown in Fig. 4.5*b*. Thus, at $2R/r_0 = 10$, the resolution improves by a factor of 2.57, and at $2R/r_0 = 20$, it improves by the factor 1.55. Continuing with the example used in Sec. 4.2.2, the maximum spatial frequency resolvable in the object plane in the short-exposure case is 10.20 line pairs/m for $2R/r_0 = 10$ and 3.38 line pairs/m for $2R/r_0 = 20$. Hence, in the first instance, short-exposure imaging allows the resolution of an object at the identification level that was previously at the recognition level; and in the second instance, it improves the resolution toward the recognition level while still remaining at the orientation level.

4.2.4 Wave structure function and MCF for an aerosol medium

One can now consider the analysis, within the context of the Rytov method, of propagation through an aerosol medium described by the corresponding spectral density given by Eq. (4.9). However, the instructive detailed analysis in terms of log-amplitude and phase structure functions, as appears in Sec. 4.2.1, will not be followed here. Instead, since the final result is the MCF of the propagation medium, only the wave structure function will be calculated here. Therefore, using Eq. (4.6), one has

$$D_\phi(L, \rho_d) \equiv D_\chi(L, \rho_d) + D_S(L, \rho_d)$$

$$= 2\pi^2 k^2 \int_0^L \int_0^\infty \left[1 - J_0\left(\frac{\kappa\rho_d x}{L}\right)\right] \Phi_\varepsilon(x, \kappa)\kappa \, d\kappa \, dx \quad (4.57)$$

which is, of course, the same result as that given by Eq. (1.135) with the assignments of Eqs. (4.4) and (4.5) taken into consideration. Substituting Eq. (4.9) for the spectral density into this relation, taking the deposition of the aerosol concentration to be homogeneous along the propagation path [i.e., $\vartheta(x) = \vartheta$], and simplifying yields

$$D_\phi(L, \rho_d) = \pi\vartheta\langle a\rangle^4 \frac{(\nu + 2)(\nu + 1)}{\nu^3} \int_0^L \int_0^\infty \left[1 - J_0\left(\frac{\kappa\rho_d x}{L}\right)\right] \times \exp\left(-\frac{\kappa^2}{K_a^2}\right) \kappa \, d\kappa \, dx \quad (4.58)$$

where

$$K_a^2 \equiv \frac{(2\nu)^2}{\langle a\rangle^2(\nu + 4)(\nu + 3)} \quad (4.59)$$

Following the procedure used to find the solution of Eq. (4.12) yields

$$D_\phi(L, \rho_d) = \frac{\pi}{2} \vartheta\langle a\rangle^4 \frac{(\nu + 2)(\nu + 1)}{\nu^3} \times K_a^2 \int_0^L \left[1 - \exp\left(-\frac{\rho_d^2 x^2 K_a^2}{L^2}\right)\right] dx$$

$$= \frac{\pi}{2} \vartheta L\langle a\rangle^4 \frac{(\nu + 2)(\nu + 1)}{\nu^3} K_a^2 \left[1 - \frac{\sqrt{\pi}}{2\rho_d K_a} \operatorname{erf}(\rho_d K_a)\right] \quad (4.60)$$

where erf (...) is the error function representing the probability integral. Substituting Eq. (4.59) into this result, rearranging factors, and forming the normalized image transfer function as was done in the deviation leading to Eq. (4.43), one obtains for the aerosol case

$$t_N(L, f) = \frac{2}{\pi}\left\{\cos^{-1}\left(\frac{f}{f_c}\right) - \left(\frac{f}{f_c}\right)\left[1 - \left(\frac{f}{f_c}\right)^2\right]^{1/2}\right\} \times \exp\left\{-\tau\left[1 - \frac{\sqrt{\pi}}{2\alpha}\left(\frac{\langle a\rangle}{2R}\right)\left(\frac{f_c}{f}\right) \operatorname{erf}\left(\alpha\frac{2R}{\langle a\rangle}\frac{f}{f_c}\right)\right]\right\} \quad (4.61)$$

where

$$\tau \equiv \pi \vartheta L \langle a \rangle^2 \left[\frac{(\nu + 2)(\nu + 1)}{\nu(\nu + 4)(\nu + 3)} \right] \qquad \alpha \equiv \frac{2\nu}{\sqrt{(\nu + 4)(\nu + 3)}} \tag{4.62}$$

The quantity τ is essentially the optical thickness of the aerosol medium. Also, it is easily seen that the average aerosol radius $\langle a \rangle$ is analogous with the coherence radius r_0 in the case of turbulence.

Figure 4.6 shows the behavior of the normalized transfer function for a Joss thunderstorm (J-T) raindrop size distribution as given in Table 2.3 for rain rates Rt in the range from 25 to 150 mm/h. Also displayed in the figure are the corresponding optical thicknesses as well as the diffraction-limited case ($Rt = 0$) for comparison. These results were derived using the fixed parameters $L = 1.6$ km and $2R/\langle a \rangle = 300$. It should be noted that this example represents an extreme case for propagation through rain, as it assumes homogeneity for the values characterizing the random medium; in particular, the rain rate is taken to be constant over distances $\sim$ 1.6 km in a thunderstorm. This is not true in most cases, especially at the larger rain rates, and could be better represented by taking the rain rate itself to be, e.g, a gaussian function of position over the propagation path.

As the figure shows, one can, even in this extreme homogeneous rain rate case, obtain approximately the same maximum spatial frequency

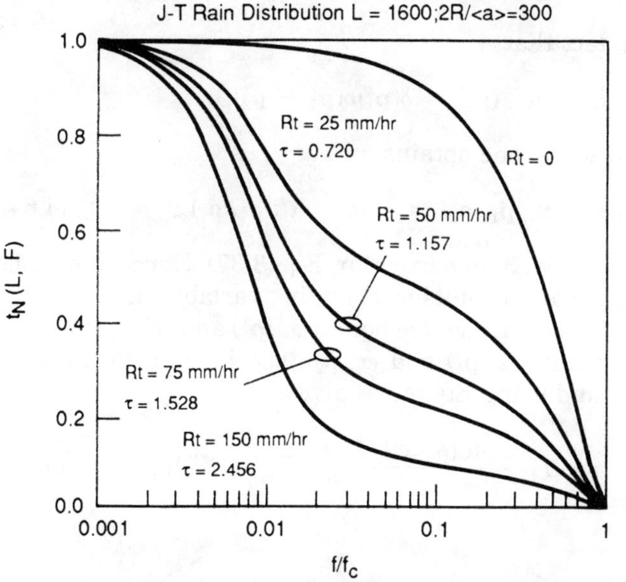

Figure 4.6 Plot of Eq. (4.61) for Joss thunderstorm rain case.

resolution capabilities at the contrast level $t_N(L, f) = 0.02$ for all the rain rate values considered. Most of the differentiation occurs in the range $0.01 \le f/f_c \le 0.1$. This circumstance is due to the fact that, as compared to the turbulence case considered earlier, there is an average value $\langle a \rangle$ characterizing the raindrop sizes after which the number density of such scatterers quickly diminishes as prescribed by the probability distribution of Eq. (2.13).

4.3 Application of the Parabolic Equation Method

As in Sec. 4.2, it is desired here to obtain the spherical wave MCF $\Gamma_{11}(L; \rho_d)$, but in this instance within the context of the parabolic wave equation, Eq. (1.91) or Eq. (1.94*b*), and the related expression for the general statistical moments of the field, viz., Eq. (1.83) with proper consideration given to the integral characteristic of Eq. (1.94*a*). However, at the outset, when calculating the MCF via the parabolic wave equation, one must first consider the relationship of Eq. (1.65) connecting the MCF Γ_{11} with that given by Eq. (1.83) for $n = m = 1$. Writing Eq. (1.65) in the more general form

$$\Gamma_{11}(x - x'; \boldsymbol{\rho}_1, \boldsymbol{\rho}_2) = \int_{-\infty}^{\infty} \int_{-\infty}^{\infty} \gamma_{11}(x - x'; \boldsymbol{\rho}_1 - \boldsymbol{\rho}_1', \boldsymbol{\rho}_2 - \boldsymbol{\rho}_2') \times U_0(\boldsymbol{\rho}_1')U_0^*(\boldsymbol{\rho}_2')\, d^2\rho_1'\, d^2\rho_2' \qquad (4.63)$$

and employing the fact that

$$U_0(\boldsymbol{\rho}_1')U_0^*(\boldsymbol{\rho}_2') = \delta(\boldsymbol{\rho}_1')\delta(\boldsymbol{\rho}_1' - \boldsymbol{\rho}_2')$$

for a unit spherical wave, one obtains

$$\Gamma_{11}(x - x'; \boldsymbol{\rho}_1, \boldsymbol{\rho}_2) = \gamma_{11}(x - x'; \boldsymbol{\rho}_1, \boldsymbol{\rho}_2) \qquad (4.64)$$

as was first noted in Chap. 3, in particular, Eq. (3.37). Hence, applying Eq. (1.83) at the observation points by changing variables in Eq. (1.83) from those of the intermediate source points $(x', \boldsymbol{\rho}_1')$ and $(x', \boldsymbol{\rho}_2')$ to those of the observation points $(x, \boldsymbol{\rho}_1)$ and $(x, \boldsymbol{\rho}_2)$, specializing to the case where $n = m = 1$, and using Eq. (4.64) gives

$$-\frac{\partial \Gamma_{11}}{\partial x} = -\Gamma_{11}\frac{\partial \ln [M(x)]}{\partial x} - \frac{i}{2k}(\nabla_{\rho_1}^2 - \nabla_{\rho_2}^2)\Gamma_{11} \qquad (4.65)$$

where

$$\Gamma_{11} \equiv \Gamma_{11}(x; \boldsymbol{\rho}_1, \boldsymbol{\rho}_2)$$

is the second-order moment of the atmospheric wave field and

$$M(x) = M(x; \boldsymbol{\rho}_1, \boldsymbol{\rho}_2) \equiv \left\langle \exp\left[-\frac{ik}{2} \int_0^x Q(x')\, dx' \right] \right\rangle \qquad (4.66)$$

is the characteristic functional of the random variable $Q(x')$, which is defined by

$$Q(x) = Q(x; \boldsymbol{\rho}_1, \boldsymbol{\rho}_2) \equiv \tilde{\varepsilon}(x, \boldsymbol{\rho}_1) - \tilde{\varepsilon}^*(x, \boldsymbol{\rho}_2) \qquad (4.67)$$

Hence, one can now be concerned solely with the solution of Eq. (4.65), remembering that, from the development given in Sec. 1.5 of Chap. 1, the solution involves the transverse coordinates that are defined by the characteristic curves defined by Eq. (1.94*a*), viz.,

$$\boldsymbol{\rho}_1'(x') = \boldsymbol{\rho}_1 \frac{x'}{x} \qquad \boldsymbol{\rho}_2'(x') = \boldsymbol{\rho}_2 \frac{x'}{x} \qquad (4.68)$$

as well as the boundary condition of Eq. (1.94*c*).

Just as was done for the Rytov method above, the parabolic equation will be applied to both turbulent and aerosol propagation scenarios. First, however, a general solution of Eq. (4.65) will be secured.

4.3.1 The general solution to the parabolic wave equation for the mutual coherence function

Following the development originally given by Tatarskii,[7] one defines the difference and "centroid" coordinates related to $\boldsymbol{\rho}_1$ and $\boldsymbol{\rho}_2$:

$$\boldsymbol{\rho}_c = \boldsymbol{\rho}_c(x) \equiv \frac{\boldsymbol{\rho}_1(x) + \boldsymbol{\rho}_2(x)}{2} \qquad \boldsymbol{\rho}_d = \boldsymbol{\rho}_d(x) \equiv \boldsymbol{\rho}_1(x) - \boldsymbol{\rho}_2(x)$$

At this point, the characteristic nature of the transverse coordinates will be suppressed for notational considerations as well as generality of the solution. Rearranging terms within Eq. (4.65) and employing these new variables yields

$$2ik \frac{\partial \Gamma_{11}(x, \boldsymbol{\rho}_c, \boldsymbol{\rho}_d)}{\partial x} - 2ik\Gamma_{11}(x, \boldsymbol{\rho}_c, \boldsymbol{\rho}_d) H(x, \boldsymbol{\rho}_d) + 2\nabla_{\rho_c} \cdot \nabla_{\rho_d} \Gamma_{11}(x, \boldsymbol{\rho}_c, \boldsymbol{\rho}_d) = 0 \qquad (4.69)$$

where

$$H(x, \boldsymbol{\rho}_d) \equiv \frac{\partial \ln [M(x, \boldsymbol{\rho}_1, \boldsymbol{\rho}_2)]}{\partial x} = \frac{\partial \ln [M(x, \boldsymbol{\rho}_d)]}{\partial x} \qquad (4.70)$$

is taken to be only a function of the difference coordinate due to statistical homogeneity (as well as isotropy) which the propagation medium is modeled (in Chap. 2) to possess. This circumstance then allows one to apply the Fourier transform pair

$$\Gamma_{11}(x, \boldsymbol{\rho}_c, \boldsymbol{\rho}_d) = \int_{-\infty}^{\infty} \Lambda(x, \boldsymbol{\kappa}_c, \boldsymbol{\rho}_d) \exp(i\boldsymbol{\kappa}_c \cdot \boldsymbol{\rho}_c)\, d^2\kappa_c$$

$$\Lambda(x, \boldsymbol{\kappa}_c, \boldsymbol{\rho}_d) = \left(\frac{1}{2\pi}\right)^2 \int_{-\infty}^{\infty} \Gamma_{11}(x, \boldsymbol{\rho}_c, \boldsymbol{\rho}_d) \exp(-i\boldsymbol{\kappa}_c \cdot \boldsymbol{\rho}_c)\, d^2\rho_c \quad (4.71)$$

Substituting the first relation of Eq. (4.71) into Eq. (4.69) and simplifying and rearranging terms once again gives

$$\left(\frac{\partial}{\partial x} + \frac{\boldsymbol{\kappa}_c}{k} \cdot \nabla_{\rho_d}\right) \Lambda(x, \boldsymbol{\kappa}_c, \boldsymbol{\rho}_d) - \Lambda(x, \boldsymbol{\kappa}_c, \boldsymbol{\rho}_d) H(x, \boldsymbol{\rho}_d) = 0 \quad (4.72)$$

Once again, one can consider the method of integral characteristics, used earlier in Sec. 1.5, in the solution of this first-order linear differential equation. To this end, one demands that

$$\frac{d\boldsymbol{\rho}_d(x')}{dx'} = \frac{\boldsymbol{\kappa}_c}{k}$$

with the boundary condition $\boldsymbol{\rho}_d(L) = \boldsymbol{\rho}_d$ requiring the characteristic to pass through the observation point. This then yields the relation

$$\boldsymbol{\rho}_d(x') = \boldsymbol{\rho}_d - \frac{\boldsymbol{\kappa}_c}{k}(x - x') \quad (4.73)$$

This then allows Eq. (4.72) to be written

$$\frac{d\Lambda(x', \boldsymbol{\kappa}_c, \boldsymbol{\rho}_d(x'))}{dx'} = \Lambda(x', \boldsymbol{\kappa}_c, \boldsymbol{\rho}_d(x')) H(x', \boldsymbol{\rho}_d(x'))$$

which straightforwardly integrates to give

$$\Lambda(x, \boldsymbol{\kappa}_c, \boldsymbol{\rho}_d(x)) = \Lambda(0, \boldsymbol{\kappa}_c, \boldsymbol{\rho}_d(0)) \exp\left[\int_0^x H(x', \boldsymbol{\rho}_d(x'))\, dx'\right] \quad (4.74)$$

Using the second relationship of Eq. (4.71) to represent the integration constant $\Lambda(0, \boldsymbol{\kappa}_c, \boldsymbol{\rho}_d(0))$, employing the expression for the characteristic,

Eq. (4.73), and substituting this intermediate result into the first relation of Eq. (4.71) gives

$$\Gamma_{11}(x, \boldsymbol{\rho}_c, \boldsymbol{\rho}_d) = \left(\frac{1}{2\pi}\right)^2 \int_{-\infty}^{\infty}\int_{-\infty}^{\infty} \Gamma_{11}\left(0, \boldsymbol{\rho}_c', \boldsymbol{\rho}_d - \frac{\boldsymbol{\kappa}_c}{k}x\right)$$
$$\times \exp\left[\int_0^x H\left(x', \boldsymbol{\rho}_d - \frac{\boldsymbol{\kappa}_c}{k}(x - x')\right) dx'\right]$$
$$\times \exp\left[i\boldsymbol{\kappa}_c \cdot (\boldsymbol{\rho}_c - \boldsymbol{\rho}_c')\right] d^2\rho_c'\, d^2\kappa_c$$

Finally, defining the auxiliary coordinate $\boldsymbol{\rho}_d' \equiv \boldsymbol{\rho}_d - \boldsymbol{\kappa}_c/k$, this expression yields for the general solution of Eq. (4.69) at observation points in the plane $x = L$

$$\Gamma_{11}(L, \boldsymbol{\rho}_c, \boldsymbol{\rho}_d) = \left(\frac{k}{2\pi L}\right)^2 \int_{-\infty}^{\infty}\int_{-\infty}^{\infty} \Gamma_{11}(0, \boldsymbol{\rho}_c', \boldsymbol{\rho}_d')$$
$$\times \exp\left[\int_0^L H\left(x', \boldsymbol{\rho}_d' - \frac{x'}{L}(\boldsymbol{\rho}_d - \boldsymbol{\rho}_d')\right) dx'\right]$$
$$\times \exp\left[i\frac{k}{L}(\boldsymbol{\rho}_d - \boldsymbol{\rho}_d') \cdot (\boldsymbol{\rho}_c - \boldsymbol{\rho}_c')\right] d^2\rho_c'\, d^2\rho_d' \quad (4.75)$$

This is the general solution for the second-order moment of a stochastic electric field of a propagating electromagnetic wave within the context of the parabolic equation approximation. Equation (4.75) can be used with any arbitrary boundary condition that may prevail.

However, the form of the parabolic wave equation with which one is concerned in this particular development is given by Eq. (1.91) and only takes the form of Eq. (4.75) contingent on the fact that one limits solutions along the characteristic curves given by Eq. (1.94*a*) and with the boundary condition of Eq. (1.94*c*). Thus, employing the related boundary condition for the MCF,

$$\Gamma_{11}(0, \boldsymbol{\rho}_c'(0), \boldsymbol{\rho}_d'(0)) = 1 \quad (4.76)$$

using Eq. (1.94*a*) in the definitions for the difference and centered coordinates, and introducing these into Eq. (4.75) yields the simple solution

$$\Gamma_{11}(L, \boldsymbol{\rho}_c, \boldsymbol{\rho}_d) = \Gamma_{11}(L, \boldsymbol{\rho}_d) = \exp\left[\int_0^L H\left(x', \boldsymbol{\rho}_d \frac{x'}{L}\right) dx'\right] \quad (4.77)$$

for the spherical wave MCF to be used in conjunction with the second-order moments generated by Eq. (1.93), which formed the basis for the development of the image transfer relationships of Chap. 3.

With this solution for the MCF obtained, it now remains to obtain expressions for the quantity $H(x, \boldsymbol{\rho}_d)$ defined by Eq. (4.70) for the cases of a turbulent and aerosol-laden atmosphere. This entails the evaluation of the characteristic functional $M(x)$ for these cases given by Eqs. (2.86) and (2.92) of Chap. 2.

4.3.2 Evaluation of the characteristic functional of atmospheric permittivity turbulence and the associated MCF

One must now return to Eq. (4.66) and evaluate it in the case of the turbulence model of Chap. 2 for use in Eq. (4.70). As shown by Eqs. (1.85) and (1.86), one has

$$M(x) = \Phi_\varepsilon(\chi_1, \chi_2) \tag{4.78}$$

Using this fact and Eq. (2.86), one has, for $n = m = 1$ and remembering that $\chi_3 \equiv \chi_1 + \chi_2$,

$$\begin{aligned} M(x) &= M(x, \boldsymbol{\rho}_1, \boldsymbol{\rho}_2) \\ &= \exp\left\{ -\frac{k^2}{8} \int_0^x \int_{-\infty}^{\infty} \int_0^x \int_{-\infty}^{\infty} \langle \tilde{\varepsilon}(x', \boldsymbol{\rho}')\tilde{\varepsilon}(x'', \boldsymbol{\rho}'')\rangle \right. \\ &\qquad \times [\delta(\boldsymbol{\rho}' - \boldsymbol{\rho}_1) - \delta(\boldsymbol{\rho}' - \boldsymbol{\rho}_2)][\delta(\boldsymbol{\rho}'' - \boldsymbol{\rho}_1) - \delta(\boldsymbol{\rho}'' - \boldsymbol{\rho}_2)] \\ &\qquad \left. \times\, d^2\rho''\, dx''\, d^2\rho'\, dx' \right\} \end{aligned}$$

Performing the multiplications of the product of the differences of the δ functions, integrating, and using Eq. (1.76), one obtains

$$M(x, \boldsymbol{\rho}_d) = \exp\left\{ -\frac{k^2}{4} \int_0^x [A(x', 0) - A(x', \boldsymbol{\rho}_d)]\, dx' \right\} \tag{4.79}$$

Substituting this expression into Eq. (4.70) yields the simple result

$$H(x, \boldsymbol{\rho}_d) = -\frac{k^2}{4} [A(x, 0) - A(x, \boldsymbol{\rho}_d)] \tag{4.80}$$

One can now make contact with the statistical model describing the turbulent fluctuations of the atmospheric permittivity field, i.e., the

Kolmogorov-Obukhov spectrum given by Eq. (2.61), through the relation given by Eq. (D1.3.19), viz.,

$$A(x, \boldsymbol{\rho}) = 4\pi^2 \int_0^\infty J_0(\kappa\rho)\Phi_\varepsilon(x, \kappa)\kappa\, d\kappa \tag{4.81}$$

Hence, using Eq. (4.81) in Eq. (4.80), one has for the function required in Eq. (4.77)

$$H\left(x, \rho_d \frac{x}{L}\right) = -(\pi k)^2 \int_0^\infty \left[1 - J_0\left(\kappa\rho_d \frac{x}{L}\right)\right] \Phi_\varepsilon(x, \kappa)\kappa\, d\kappa \tag{4.82}$$

Finally, using Eq. (4.8) and converting the permittivity structure parameter to the refractive index description via Eq. (2.105), Eq. (4.82) becomes

$$H\left(x, \rho_d \frac{x}{L}\right) = -4(\pi k)^2(0.033)C_n^2(x) \int_0^\infty \left[1 - J_0\left(\kappa\rho_d \frac{x}{L}\right)\right] \times (\kappa^2 + K_0^2)^{-11/6} \exp\left(\frac{-\kappa^2}{K_m^2}\right) \kappa\, d\kappa \tag{4.83}$$

As was the case with Eq. (4.10), one cannot explicitly integrate this expression and must resort to an approximate representation. However, unlike the case of Eq. (4.10), where the effect of the outer scale of turbulence was neglected, another approximation will be used here, first suggested by Ishimaru.[8] One first dissects the product of the second and third factors appearing in the integrand of Eq. (4.83) into a difference of three terms,

$$\phi(\kappa, K_0, K_m) \equiv (\kappa^2 + K_0^2)^{-11/6} \exp\left(\frac{-\kappa^2}{K_m^2}\right) = \phi_0(\kappa) - \phi_1(\kappa, K_0) - \phi_2(\kappa, K_0, K_m) \tag{4.84}$$

where

$$\phi_0(\kappa) \equiv \kappa^{-11/3} \tag{4.85a}$$

$$\phi_1(\kappa, K_0) \equiv \kappa^{-11/3} - (\kappa^2 + K_0^2)^{-11.6} \tag{4.85b}$$

and

$$\phi_2(\kappa, K_0, K_m) \equiv (\kappa^2 + K_0^2)^{-11/6} \left[1 - \exp\left(\frac{-\kappa^2}{K_m^2}\right)\right] \tag{4.85c}$$

It is now noted that the second factor of Eq. (4.85*c*) renders this expression negligible for spatial frequencies $\kappa < K_m$. For values of κ for which this factor is not negligible, one has $K_0 \ll \kappa$ and Eq. (4.85*c*) can be approximately written

$$\phi_2(\kappa, K_0, K_m) = \phi_2(\kappa, K_m) \approx \kappa^{-11/3}\left[1 - \exp\left(\frac{-\kappa^2}{K_m^2}\right)\right] \tag{4.85d}$$

One now has a description of the turbulent spectrum in the form of a linear combination of individual contributions of the effects of the inner scale and those of the outer scale. Substituting Eqs. (4.85*a*), (4.85*b*), and (4.85*d*) into Eq. (4.84) and this result into Eq. (4.83) allows one to perform the required integrations and obtain the left side of Eq. (4.83) in the form of three corresponding terms

$$H\left(x, \rho_d \frac{x}{L}\right) = -H_0\left(x, \rho_d \frac{x}{L}\right) + H_1\left(x, \rho_d \frac{x}{L}\right) + H_2\left(x, \rho_d \frac{x}{L}\right) \tag{4.86}$$

where

$$H_0\left(x, \rho_d \frac{x}{L}\right) \equiv 1.457 k^2 C_n^2(x) \left(\rho_d \frac{x}{L}\right)^{5/3} \tag{4.87a}$$

$$H_1\left(x, \rho_d \frac{x}{L}\right) \equiv k^2 C_n^2(x) \left[1.457\left(\rho_d \frac{x}{L}\right)^{5/3} - 0.782 L_0^{5/3} - 0.777\left(L_0 \rho_d \frac{x}{L}\right) K_{-5/6}\left(\frac{\rho_d}{L_0}\frac{x}{L}\right)\right] \tag{4.87b}$$

and

$$H_2\left(x, \rho_d \frac{x}{L}\right) \equiv k^2 C_n^2(x) \left\{1.457\left(\rho_d \frac{x}{L}\right)^{5/3} - 4.351 K_m^{-5/3} \times \left[{}_1F_1\left(-\frac{5}{6}, 1; -\frac{\rho_d^2 x^2 K_m^2}{4L^2}\right) - 1\right]\right\} \tag{4.87c}$$

Equations (4.86) and (4.87*a*) to (4.87*c*) yield a general expression that, when substituted into Eq. (4.77), yields an expression for the spherical wave MCF over the entire range for the transverse separation, $0 < \rho_d < \infty$. However, a useful piecewise model can be obtained by examining the characteristics of the general expression by considering the three important regions defined by the relative magnitude of ρ_d with respect to the scales l_0 and L_0. In particular, one obtains the following:

$$H\left(x, \rho_d \frac{x}{L}\right) = \begin{cases} -1.64k^2C_n^2(x)l_0^{-1/3}\left(\rho_d \dfrac{x}{L}\right)^2 & \rho < l_0 \\ -1.457k^2C_n^2(x)\left(\rho_d \dfrac{x}{L}\right)^{5/3} & l_0 < \rho < L_0 \\ -0.7818k^2C_n^2(x)L_0^{5/3} & \rho > L_0 \end{cases} \tag{4.88}$$

Taking the case of a homogeneous atmosphere where $C_n^2(x) = C_n^2$ and using this piecewise model in Eq. (4.77), one has that for $\rho < l_0$,

$$\Gamma_{11}(L, \rho_d) = \exp\left(-0.546k^2C_n^2l_0^{-1/3}\rho_d^2L\right) \tag{4.89}$$

for $l_0 < \rho < L_0$,

$$\Gamma_{11}(L, \rho_d) = \exp\left(-0.546k^2C_n^2\rho_d^{5/3}L\right) \tag{4.90}$$

and for $\rho > L_0$,

$$\Gamma_{11}(L, \rho_d) = \Gamma_{11}(L) = \exp\left(-0.782k^2C_n^2L_0^{5/3}L\right) \tag{4.91}$$

Equation (4.90) agrees with the results obtained in Sec. 4.2.2 for the case of a homogeneous atmosphere, viz., Eqs. (4.41) and (4.42). Equation (4.91) demonstrates the effect of the outer scale of turbulence,

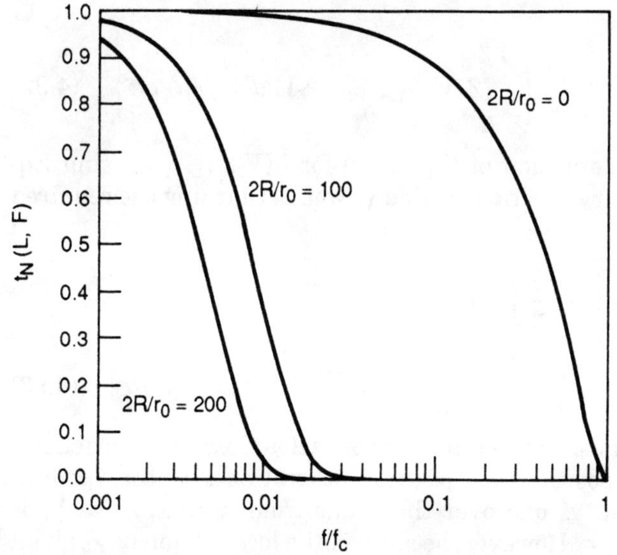

Figure 4.7 Plot of MCF for $2R/r_0 = 100$ and 200.

which was neglected in the development of Sec. 4.2.1. Here, it is seen how the values of the MCF saturate beyond the range $\rho > L_0$. It is interesting to note that even though Eq. (4.41) is a result of the application of the Rytov method and is thus limited in its region of applicability, it is formally identical to the result of Eq. (4.90) and thus can formally be extended to the cases not able to be addressed in Table 4.1. This is due to the fact that in the formation of the MCF within the context of the Rytov theory, terms within the log-amplitude and phase expressions mutually cancel, and it is those terms that reflect the limitation imposed by a first-order solution to the problem. Hence, the two cases listed in Table 4.1 as N/A can be calculated using Eq. (4.90), which is identical to the formally limited result given by Eq. (4.41). The result for the case $C_n^2 = 1.0 \times 10^{-14}\ \text{m}^{-2/3}$ and $L = 5$ km is $r_0 = 8.704 \times 10^{-3}$ m and for $L = 10$ km is $r_0 = 5.742 \times 10^{-3}$ m. As these values are ~10 times those in Table 4.1 and used in the examples, one can now consider cases where $2R/r_0 = 100$ to 200.

Figure 4.7 displays the results for these severe turbulence cases. For purposes of comparison, the diffraction-limited case is also shown.

4.3.3 Evaluation of the characteristic functional of atmospheric aerosols and the associated MCF

As in the case of turbulence statistics above, one must again return to the key results of Chap. 2 in the evaluation of Eq. (4.66) for the aerosol propagation case. In particular, using Eq. (2.92) in the relationship of Eq. (4.78) yields

$$M(x) = \exp\left\{\vartheta \int_0^\infty \int_{-\infty}^\infty [Z(V_0, \chi_1, \chi_2) - 1]\, d\mathbf{R}_0\, p(a)\, da\right\} \tag{4.92}$$

Remembering the definition of Eq. (2.90) for $Z(V_0, \chi_1, \chi_2)$, using Eq. (1.86) for the auxiliary functions χ_1 and χ_2, and performing the required integrations yields

$$Z(V_0, \chi_1, \chi_2) = \exp\left\{-i\frac{k}{2}\int_0^x [V_0(\boldsymbol{\rho}_1 - \boldsymbol{\rho}_0, \xi - x_0; a) - V_0^*(\boldsymbol{\rho}_2 - \boldsymbol{\rho}_0, \xi - x_0; a)]\, d\xi\right\} \tag{4.93}$$

One notes that because of the finite size of each aerosol inhomogeneity about the possible positions $(x_0, \boldsymbol{\rho}_0)$, the integration over the ξ coordinate can be written as one over the range from x_0 to x, since V_0 is nonzero for $\xi - x_0 < a$. However, because of the localization of V_0 about each possible location, one can let $x \to \infty$. Using this development and assuming the case where the condition $kV_0x \ll 1$, one can expand the exponential function in Eq. (4.93) retaining second-order terms and obtain

$$Z(V_0, \chi_1, \chi_2) \approx 1 - i\frac{k}{2}\int_0^\infty D_V(\boldsymbol{\rho}_1, \boldsymbol{\rho}_2, \boldsymbol{\rho}_0, \xi)\, d\xi$$

$$-i\frac{k}{8}\int_0^\infty\int_0^\infty D_V(\boldsymbol{\rho}_1, \boldsymbol{\rho}_2, \boldsymbol{\rho}_0, \xi_1)D_V(\boldsymbol{\rho}_1, \boldsymbol{\rho}_2, \boldsymbol{\rho}_0, \xi_2)\, d\xi_1\, d\xi_2 \quad (4.94)$$

where

$$D_V(\boldsymbol{\rho}_1, \boldsymbol{\rho}_2, \boldsymbol{\rho}_0, \xi) \equiv V_0(\boldsymbol{\rho}_1 - \boldsymbol{\rho}_0, \xi; a) - V_0^*(\boldsymbol{\rho}_2 - \boldsymbol{\rho}_0, \xi; a) \quad (4.95)$$

Substituting Eq. (4.94) into Eq. (4.92) and this intermediate result into Eq. (4.70) yields

$$H(x, \boldsymbol{\rho}_1, \boldsymbol{\rho}_2) = \frac{\partial}{\partial x}\left\{\vartheta(x_0)\int_0^\infty\int_0^x\int_{-\infty}^\infty\left[-\frac{ik}{2}\int_0^\infty D_v(\boldsymbol{\rho}_1, \boldsymbol{\rho}_2, \boldsymbol{\rho}_0, \xi)\, d\xi\right.\right.$$

$$\left.-\frac{k^2}{8}\int_0^\infty\int_0^\infty D_v(\boldsymbol{\rho}_1, \boldsymbol{\rho}_2, \boldsymbol{\rho}_0, \xi_1)D_v(\boldsymbol{\rho}_1, \boldsymbol{\rho}_2, \boldsymbol{\rho}_0, \xi_2)\, d\xi_1\, d\xi_2\right]$$

$$\left.\times\, p(a)\, d\boldsymbol{\rho}_0\, dx_0\, da\right\} \quad (4.96)$$

where the position coordinate $\mathbf{R}_0$ has been dissociated into its longitudinal and transverse components and the general case of the inhomogeneous distribution of aerosol concentration $\vartheta(x_0)$ has been admitted. Specializing the situation to a homogeneous atmosphere, $\vartheta(x_0) = \vartheta$, performing the required differentiation, using Eq. (4.95) and multiplying the V_0 terms, employing the δ function approximation of Eq. (1.75), and performing the required integrations finally gives

$$H(x, \boldsymbol{\rho}_1, \boldsymbol{\rho}_2) = -\frac{ik}{2}\vartheta\int_0^\infty\int_0^\infty\int_{-\infty}^\infty [V_0(\boldsymbol{\rho}_1 - \boldsymbol{\rho}_0, \xi; a)$$

$$- V_0^*(\boldsymbol{\rho}_2 - \boldsymbol{\rho}_0, \xi; a)]p(a)\, d\boldsymbol{\rho}_0\, da\, d\xi$$

$$-\frac{k^2}{8}\vartheta\int_0^\infty\int_0^\infty\int_{-\infty}^\infty [V_0^2(\boldsymbol{\rho}_1 - \boldsymbol{\rho}_0, \xi; a)$$

$$- 2V_0(\boldsymbol{\rho}_1 - \boldsymbol{\rho}_0, \xi; a)V_0^*(\boldsymbol{\rho}_2 - \boldsymbol{\rho}_0, \xi; a)$$

$$+ V_0^{*2}(\boldsymbol{\rho}_2 - \boldsymbol{\rho}_0, \xi; a)]p(a)\, d\boldsymbol{\rho}_0\, da\, d\xi \quad (4.97)$$

Using Eqs. (2.93) and (2.94), one finds that each term within the first member of Eq. (4.97) is equal to the constant

$$i\left(\frac{4\pi}{k}\right)\int_0^\infty a^2 p(a)\, da$$

thus allowing these terms within the first member to mutually cancel (at least under the assumptions employed here). Similarly, taking V_0 to be real-valued quantities and using Eqs. (2.98) and (2.99), one has that

$$\int_0^\infty\int_0^\infty\int_{-\infty}^\infty [V_0(\boldsymbol{\rho}_1 - \boldsymbol{\rho}_0, \xi; a)V_0(\boldsymbol{\rho}_2 - \boldsymbol{\rho}_0, \xi; a)]p(a)\, d\boldsymbol{\rho}_0\, da\, d\xi$$

$$= 8\pi\left(\frac{1}{k}\right)^2\int_0^\infty\int_0^\infty\left[\frac{J_1(\kappa a)}{\kappa a}\right]^2 a^4 p(a) J_0(\kappa\rho_d)\, da\, \kappa\, d\kappa \quad (4.98)$$

where $\rho_d \equiv |\boldsymbol{\rho}_1 - \boldsymbol{\rho}_2|$. Using these results, Eq. (4.97) then becomes

$$H(x, \rho_d) = -(\pi k)^2\int_0^\infty [1 - J_0(\kappa\rho_d)]\Phi_{\varepsilon_p}(\kappa)\kappa\, d\kappa \quad (4.99)$$

analogous to Eq. (4.82) for the turbulence case where $\Phi_{\varepsilon_p}(\kappa)$ is given by Eq. (2.100) and approximated by Eq. (2.110), thus demonstrating the same results as obtained in Sec. 4.2.4, with, of course, the value of ρ_d replaced with the characteristic defined by Eq. (4.68).

4.3.4 Derivation of the small scattering angle approximation of the radiative transfer equation from the parabolic equation for the second-order moment

It will now be demonstrated how the equation for the second-order electric field moment, Eq. (4.69), can be made to yield the well-known form of the radiative transfer equation within the context of the small scattering angle approximation.[9] To this end, it is convenient to define the Fourier transform pair in the difference coordinate $\boldsymbol{\rho}_d$,

$$\Gamma_{11}(x, \boldsymbol{\rho}_c, \boldsymbol{\rho}_d) = \int_{-\infty}^\infty J(x, \boldsymbol{\rho}_c, \boldsymbol{\kappa}_d)\exp(i\boldsymbol{\kappa}_d\cdot\boldsymbol{\rho}_d)\, d^2\kappa_d$$

$$J(x, \boldsymbol{\rho}_c, \boldsymbol{\kappa}_d) = \left(\frac{1}{2\pi}\right)^2\int_{-\infty}^\infty \Gamma_{11}(x, \boldsymbol{\rho}_c, \boldsymbol{\rho}_d)\exp(-i\boldsymbol{\kappa}_d\cdot\boldsymbol{\rho}_d)\, d^2\rho_d \quad (4.100)$$

Substituting the first relation into Eq. (4.69) and simplifying gives

$$\frac{\partial J(x, \boldsymbol{\rho}_c, \boldsymbol{\kappa}_d)}{\partial x} + \frac{\boldsymbol{\kappa}_d}{k} \cdot \nabla_{\rho_c} J(x, \boldsymbol{\rho}_c, \boldsymbol{\kappa}_d) = \left(\frac{1}{2\pi}\right)^2 \int_{-\infty}^{\infty} \int_{-\infty}^{\infty} J(x, \boldsymbol{\rho}_c, \boldsymbol{\kappa}_d') H(x, \boldsymbol{\rho}_d) \times \exp\,[i(\boldsymbol{\kappa}_d' - \boldsymbol{\kappa}_d) \cdot \boldsymbol{\rho}_d]\, d^2\kappa_d'\, d^2\rho_d \quad (4.101)$$

As shown by the results of Eqs. (4.82) and (4.99), one has, for both turbulence and aerosols, the result given by Eq. (4.80), viz.,

$$H(x, \boldsymbol{\rho}_d) = \frac{k^2}{4}\,[A(x, \boldsymbol{\rho}_d) - A(x, 0)] \quad (4.102)$$

where, in general,

$$A(x, \boldsymbol{\rho}_d) = 2\pi \int_{-\infty}^{\infty} \Phi(x, \boldsymbol{\kappa}) \exp\,(i\boldsymbol{\kappa} \cdot \boldsymbol{\rho}_d)\, d^2\kappa \quad (4.103)$$

Substituting Eq. (4.102) into Eq. (4.101) and simplifying yields

$$\frac{\partial J(x, \boldsymbol{\rho}_c, \boldsymbol{\kappa}_d)}{\partial x} + \frac{\boldsymbol{\kappa}_d}{k} \cdot \nabla_{\rho_c} J(x, \boldsymbol{\rho}_c, \boldsymbol{\kappa}_d) = 2\pi \left(\frac{k^2}{4}\right) \int_{-\infty}^{\infty} J(x, \boldsymbol{\rho}_c, \boldsymbol{\kappa}_d') \Phi(x, \boldsymbol{\kappa}_d' - \boldsymbol{\kappa}_d)\, d^2\kappa_d' - \left(\frac{k^2}{4}\right) A(x, 0) J(x, \boldsymbol{\rho}_c, \boldsymbol{\kappa}_d) \quad (4.104)$$

Finally, rearranging terms and defining $\mathbf{n}_\perp \equiv \boldsymbol{\kappa}_d/k$ as the transverse component of the unit vector defining the direction of propagation, Eq. (4.104) can be written as

$$\left[\frac{\partial}{\partial x} + \mathbf{n}_\perp \cdot \nabla_{\rho_c} + \varepsilon(x)\right] J(x, \boldsymbol{\rho}_c, \mathbf{n}_\perp) = \int_{-\infty}^{\infty} J(x, \boldsymbol{\rho}_c, \mathbf{n}_\perp') X(x, \mathbf{n}_\perp' - \mathbf{n}_\perp)\, d^2\mathbf{n}_\perp' \quad (4.105)$$

where

$$X(x, \mathbf{n}_\perp' - \mathbf{n}_\perp) \equiv \frac{\pi k^4}{2} \Phi(x, k\mathbf{n}_\perp' - k\mathbf{n}_\perp)$$

has the interpretation of the effective scattering function, just as

$$\varepsilon(x) \equiv \left(\frac{k^2}{4}\right) A(x, 0)$$

plays the role of the effective extinction coefficient. Equation (4.105) is of the form of the small scattering angle approximation of the radiative transfer equation. Here, $J(x, \boldsymbol{\rho}_c, \mathbf{n}_\perp)$ is the field intensity at a point $(x, \boldsymbol{\rho}_c)$ in a plane due to radiative propagating in a direction whose transverse unit normal vector is $\mathbf{n}_\perp$.

Within the confines of photometry, upon which the radiative transfer equation is based, the quality of an image is relegated to a quantity called the spatial spectral (or simply, frequency) contrast ratio. The essence of this quantity can be conveyed by considering an approximate treatment and solution of the transfer equation, Eq. (4.105). As has been shown elsewhere,[10] one can approximately geometrically transform the field intensity in the case to an "on-axis" imaging system in such a way that Eq. (4.105) can be written in the form

$$\left[\frac{\partial}{\partial x} + \varepsilon(x)\right] J(x, \boldsymbol{\rho}_c) = \int_{-\infty}^{\infty} J(x, \boldsymbol{\rho}_c - x\mathbf{n}_\perp) X(x, \mathbf{n}_\perp)\, d^2 n_\perp \quad (4.106)$$

Defining for this field intensity the Fourier transform

$$J(x, \boldsymbol{\rho}_c, 0) \equiv J(x, \boldsymbol{\rho}_c) = \int_{-\infty}^{\infty} F(x, \boldsymbol{\kappa}) \exp(i\boldsymbol{\kappa} \cdot \boldsymbol{\rho}_c)\, d^2\kappa$$

and applying it to Eq. (4.106) gives

$$F(L, \boldsymbol{\kappa}) = F(0, \boldsymbol{\kappa}) \exp\left\{-\int_0^L [\varepsilon(x) - P(x, \boldsymbol{\kappa}x)]\, dx\right\} \quad (4.107)$$

where

$$P(x, \mathbf{q}) \equiv \int_{-\infty}^{\infty} X(x, \mathbf{n}_\perp) \exp(-i\mathbf{q} \cdot \mathbf{n}_\perp)\, d^2 n_\perp$$

is the Fourier transform of the scattering function. The spatial spectral contrast ratio $M(L, \boldsymbol{\kappa})$ is defined as

$$M(L, \boldsymbol{\kappa}) \equiv \frac{F(L, \boldsymbol{\kappa})}{F(0, \boldsymbol{\kappa})} \quad (4.108)$$

which, via Eq. (4.107), becomes for this case

$$M(L, \boldsymbol{\kappa}) = \exp\left\{-\int_0^L [\varepsilon(x) - P(x, \boldsymbol{\kappa}x)]\, dx\right\} \quad (4.109)$$

Comparing the derivation of the classical photometric result of Eq. (4.105), its use in the well-known definition of Eq. (4.108), and its connection to the rigorous diffraction theory (sometimes called "wave" theory) from which Eq. (4.69) obtains, one has a motivation for the contrast assessment of image resolution as discussed in Sec. 3.6 of Chap 3. As Eq. (4.109) demonstrates within the context of radiative transfer theory, spatial frequencies become attenuated through the connection between the function $P(x, \boldsymbol{\kappa}x)$ and $\Phi(x, k\mathbf{n}'_{\perp} - k\mathbf{n}_{\perp})$ through the scattering function $X(x, \mathbf{n}'_{\perp} - \mathbf{n}_{\perp})$. This definition of image resolution is to be distinguished from that of Chap. 3 in that the former is a measure of spatial resolution, whereas the latter is a measure of angular resolution.[11]

4.4 Analysis of the Temporal Spectrum of Wavefront Tilt and Astigmatism

Using the results of the foregoing developments, a rigorous analysis can be applied to the temporal spectrum of the stochastic wavefront entering the lens of an optical system, thus augmenting the qualitative discussions given in Sec. 3.4 concerning the point of demarcation between long-term and short-term image exposures. In addition to this, it is also now possible to quantitatively examine the differences in the behavior of the various orders of wavefront aberrations as presented in Sec. 3.4.2.

Returning to Eq. (3.52) and admitting a temporal dependence to the phase field which is represented by the Zernike polynomial expansion, one has

$$S_1(\boldsymbol{\rho}, t) = \sum_{j=1}^{\infty} \Phi_j(\boldsymbol{\rho}, t) \qquad \Phi_j(\boldsymbol{\rho}, t) = a_j(t) Z_j\left(\frac{\boldsymbol{\rho}}{R}\right) \tag{4.110}$$

where the expansion parameter $a_j(t)$ is now a function of time t. Hence, Eq. (3.58) now becomes

$$a_j(t) = \left(\frac{1}{R^2}\right) \int_{-\infty}^{\infty} W_N\left(\frac{\boldsymbol{\rho}}{R}\right) S(\boldsymbol{\rho}, t) Z_j\left(\frac{\boldsymbol{\rho}}{R}\right) d^2\rho \tag{4.111}$$

Similar to the formation of the second-order correlation function of the expansion parameters as given by Eq. (3.65), one can define a temporal correlation function of these parameters via Eq. (4.111) over an arbitrary time interval τ, i.e.,

$$\langle a_j(t) a_k(t + \tau)\rangle \equiv B_{a_j a_k}(\tau) = \left(\frac{1}{R}\right)^4 \int_{-\infty}^{\infty} \int_{-\infty}^{\infty} W_N\left(\frac{\boldsymbol{\rho}}{R}\right) W_N^*\left(\frac{\boldsymbol{\rho}'}{R}\right) \langle S(\boldsymbol{\rho}, t) \times S^*(\boldsymbol{\rho}', t + \tau)\rangle Z_j\left(\frac{\boldsymbol{\rho}}{R}\right) Z_k\left(\frac{\boldsymbol{\rho}'}{R}\right) d^2\rho\, d^2\rho' \tag{4.112}$$

This expression employs the space-time correlation function of the phase field defined by

$$B_S(\boldsymbol{\rho} - \boldsymbol{\rho}', \tau) \equiv \langle S(\boldsymbol{\rho}, t)S(\boldsymbol{\rho}', t + \tau)\rangle \tag{4.113}$$

where homogeneity and stationarity of the field are assumed. Hence, making use of the property exhibited by Eq. (D1.1.19), one can connect the spatial portion of this correlation with the temporal portion through the velocity $\mathbf{v}$ of the atmosphere, thus allowing Eq. (4.112) to be written

$$B_{a_ja_k}(\tau) = \left(\frac{1}{R}\right)^4 \int_{-\infty}^{\infty}\int_{-\infty}^{\infty} W_N\left(\frac{\boldsymbol{\rho}}{R}\right) W_N^*\left(\frac{\boldsymbol{\rho}'}{R}\right) \times B_S(\boldsymbol{\rho} - \boldsymbol{\rho}' - \mathbf{v}\tau)Z_j\left(\frac{\boldsymbol{\rho}}{R}\right) Z_k\left(\frac{\boldsymbol{\rho}'}{R}\right) d^2\rho\, d^2\rho' \tag{4.114}$$

Finally, converting the integrand of this relation into its Fourier representation, as was done in Sec. 4.2.3 with the analogous result of Eq. (3.65), one obtains

$$B_{a_ja_k}(\tau) = \left(\frac{2\pi}{R}\right)^2 \int_{-\infty}^{\infty} Q_j(\boldsymbol{\kappa})Q_k^*(\boldsymbol{\kappa})\beta_S\left(\frac{\boldsymbol{\kappa}}{R}, \tau\right) d^2\kappa \tag{4.115}$$

where

$$B_S\left(\frac{\boldsymbol{\kappa}}{R}, \tau\right) \equiv \int_{-\infty}^{\infty} B_S(\boldsymbol{\rho}_d - \mathbf{v}\tau) \exp\left(\frac{i\boldsymbol{\kappa}\cdot\boldsymbol{\rho}_d}{R}\right) d^2\rho_d \tag{4.116}$$

from which the temporal frequency content can be obtained by applying Eq. (D1.3.11), viz.,

$$W_{a_ja_k}(\omega) \equiv \left(\frac{1}{2\pi}\right)\int_{-\infty}^{\infty} \exp(-i\omega\tau)B_{a_ja_k}(\tau)\, d\tau = \left(\frac{1}{\pi}\right)\int_0^{\infty} \cos(\omega\tau)\, B_{a_ja_k}(\tau)\, d\tau \tag{4.117}$$

where the last result is obtained by considering only positive values of time τ.

In order to keep the results as general as possible as well as to highlight the approximations employed, the phase correlation function will be obtained directly from the results of Rytov theory. In particular, using Eq. (1.127) with the results given by Eqs. (4.3) to (4.5), one has for a spherical wave propagating in a homogeneous atmosphere

$$B_S(L, \boldsymbol{\rho}_d) = \frac{k^2\pi}{4} \int_0^L \int_{-\infty}^{\infty} \mathrm{Re}\left(\left\{1 + \exp\left[-\frac{i\kappa^2}{k}\frac{x}{L}(L - x)\right]\right\}\right.$$
$$\left. \times \exp\left(i\,\frac{\boldsymbol{\kappa}\cdot\boldsymbol{\rho}_d x}{L}\right)\right) \Phi_\varepsilon(\kappa)\, d^2\kappa\, dx \quad (4.118)$$

This expression can be simplified by assuming that the constraint $L < l^2/\lambda$ prevails such that only the first term of the expansion of the exponential function within the braces can be retained. This defines the geometric optical region approximation of Eq. (4.118). Applying this approximation to Eq. (4.118) and converting the spectrum of permittivity to that of refractive index, i.e., $\Phi_\varepsilon(\kappa) = 4\Phi_n(\kappa)$, one finally has for the desired phase correlation function

$$B_S(\boldsymbol{\rho}_d - \mathbf{v}\tau) = 2\pi k^2 L \int_{-\infty}^{\infty} \exp\,(i\boldsymbol{\kappa}\cdot\boldsymbol{\rho}_d - i\boldsymbol{\kappa}\cdot\mathbf{v}\tau)\Phi_n(\boldsymbol{\kappa})\, d^2\boldsymbol{\kappa} \quad (4.119)$$

Substituting this result into Eq. (4.116) and performing the required integrations yields for the Fourier transform of the phase correlation

$$\beta_s\left(\frac{\boldsymbol{\kappa}}{R}, \tau\right) = (2\pi)^3 k^2 L \exp\left(-i\,\frac{\boldsymbol{\kappa}}{R}\cdot\mathbf{v}\tau\right)\Phi_n\left(\frac{\boldsymbol{\kappa}}{R}\right) \quad (4.120)$$

Considering now only those cases where $j = k$, substituting Eqs. (4.51) and (4.120) into Eq. (4.115) and converting the integration to one in plane polar coordinates gives for the Zernike coefficient correlation

$$B_{a_j a_j}(\tau) = \frac{16\pi(n + 1)}{R^2}\, k^2 L \int_0^{2\pi}\int_0^{\infty} \frac{J_{n+1}^2(\kappa R)}{\kappa}$$
$$\times \begin{pmatrix}\cos^2\theta \\ \sin^2\theta\end{pmatrix} \exp\,(-i\kappa v\tau\cos\theta)\Phi_n(\kappa)\, d\kappa\, d\theta \quad (4.121)$$

where the $\cos^2\theta$ factor is used if j is even; otherwise the $\sin^2\theta$ factor is selected. However, since the turbulence spectrum is taken to be isotropic, one finds that by expanding the remaining exponential function into a series of Bessel functions and integrating in the angular variable, these trigonometric functions give the same result, and one finally obtains

$$B_{a_j a_j}(\tau) = \frac{16\pi^2(n+1)}{R^2} k^2 L \int_0^\infty \frac{J_{n+1}^2(\kappa R)}{\kappa} J_0(\kappa v \tau) \Phi_n(\kappa)\, d\kappa \quad (4.122)$$

To facilitate an analytical result from Eq. (4.122) when used with even the simplest form of the turbulence spectrum, i.e., the Kolmogorov-Obukhov form of Eq. (2.50), it is found to be expeditious to take advantage of the fact that $\kappa R \gg 1$ within most of the integration region and for typical values of lens aperture radii. Ignoring at the outset the oscillatory behavior of the Bessel function within which this product appears and dwelling only on the behavior of its accompanying "envelope," one surmises that

$$J_\nu^2(z) \approx N_\nu(z) \exp(-a_\nu^2 z^2) \quad (4.123)$$

with appropriate selections for the function N_ν and the constant a_ν, both of which can be dependent on the order ν. Expanding both functions of Eq. (4.123) and demanding that the first two terms of each ascending series coincide, one finds that

$$a_\nu = \frac{1}{2(\nu+1)} \qquad N_\nu(z) = N_\nu' \left(\frac{z}{2}\right)^{2\nu} \left(\frac{1}{\nu!}\right)^2 \quad (4.124)$$

where N_ν' is another constant added to account for the overall effect of the oscillatory behavior of the Bessel function. In what is to follow, ratios of quantities formed from Eq. (4.122) will be considered, thus eliminating the need for the calculation of such a constant here.

Using Eq. (4.124) with Eq. (4.123) in Eq. (4.122) as well as Eq. (2.50) (with, of course, $C_{uv} \equiv C_n^2$), one can perform the resulting integration. Forming the ratio $B_{a_2 a_2}(\tau)/B_{a_2 a_2}(0)$, one has for the general case

$$\frac{B_{a_j a_j}(\tau)}{B_{a_j a_j}(0)} = {}_1F_1\left[\frac{2n - 5/3}{2}; 1; -\left(\frac{n+2}{2}\right)\frac{v^2\tau^2}{R^2}\right] \quad (4.125)$$

Hence, for phase tilts, one has the mode number $j = 2$ or 3 with a radial degree $n = 1$, thus giving

$$\frac{B_{a_2 a_2}(\tau)}{B_{a_2 a_2}(0)} = {}_1F_1\left(\frac{1}{6}, 1; -\frac{3v^2\tau^2}{2R^2}\right) \quad (4.126)$$

the limiting forms of which show that for $\tau < R/v$,

$$\frac{B_{a_2 a_2}(\tau)}{B_{a\,a}(0)} \sim 1 - \frac{1}{4}\frac{v^2\tau^2}{R^2}$$

and for the opposite case where $\tau > R/v$,

$$\frac{B_{a_2a_2}(\tau)}{B_{a_2a_2}(0)} \sim 0.828 \left(\frac{v\tau}{R}\right)^{-1/3}$$

Similarly for the case of astigmatism, the mode numbers are $j = 5$ or 6 with radial degree $n = 2$. One therefore has

$$\frac{B_{a_5a_5}(\tau)}{B_{a_5a_5}(0)} = {}_1F_1\left(\frac{7}{6}; 1; -\frac{2v^2\tau^2}{R^2}\right) \tag{4.127}$$

with the limiting forms

$$\frac{B_{a_5a_5}(\tau)}{B_{a_5a_5}(0)} \sim 1 - \frac{7}{3}\frac{v^2\tau^2}{R^2} \qquad \tau < \frac{R}{v}$$

and

$$\frac{B_{a_5a_5}(\tau)}{B_{a_5a_5}(0)} \sim -0.066 \left(\frac{v\tau}{R}\right)^{-7/3} \qquad \tau > \frac{R}{v}$$

Comparison of the temporal correlation of the Zernike coefficients for the tilt and astigmatism components shows that the former tends to increase more slowly than the latter at times smaller than the characteristic time v/R. The same behavior is exhibited for times larger than v/R; the correlation of the tilt coefficients decreases more slowly than that of the astigmatism coefficients. In addition, the astigmatism coefficients tend to zero at large correlation times from the negative region. This comparative behavior demonstrates the relative contributions of the tilt and astigmatism components to the total phase structure in the temporal ranges below and above the characteristic time v/R, which, as discussed in Sec. 3.4, is taken to be the point of demarcation between short-term and long-term imaging.

To be able to explicitly derive the characteristic times that are associated with phase tilt and astigmatism, it is instructive to consider the temporal spectra associated with the above temporal correlations. Substituting Eq. (4.125) into Eq. (4.117) also results in an integral with an analytical result,[12] viz.,

$$\begin{aligned} \frac{W_{a_ja_k}(\omega)}{B_{a_ja_k}(0)} \equiv \left(\frac{1}{\pi}\right)^{1/2} \left(\frac{2}{n+2}\right)^{n-5/6} \left(\frac{1}{2}\right)^{2n-8/3} \left[\frac{1}{2\Gamma(n-5/6)}\right] \\ \times \left(\frac{R}{v}\right)\left(\frac{R\omega}{v}\right)^{2n-8/3} \exp\left[-\frac{R^2\omega^2}{2(n+2)v^2}\right] \\ \times\, U\left(1/2, n-1/3, \frac{R^2\omega^2}{2(n+2)v^2}\right) \end{aligned} \tag{4.128}$$

where $U(a, b; z)$ is a Kummer function which is related to a corresponding confluent hypergeometric function ${}_1F_1(a, b; z)$. An analysis of this equation in the appropriate limits for the case of phase tilt yields for $f < v/2\pi R$

$$\frac{fW_{a_2a_2}(f)}{B_{a_2a_2}(0)} \sim 0.028 \left(\frac{2\pi fR}{v}\right)^{1/3}$$

and similarly for $f > v/2\pi R$

$$\frac{fW_{a_2a_2}(f)}{B_{a_2a_2}(0)} \sim 0.029 \left(\frac{2\pi fR}{v}\right)^{-2/3} \exp\left[-\frac{(2\pi fR/v)^2}{6}\right]$$

A numerical analysis of the general expression of Eq. (4.128) for phase tilt (i.e., with $n = 1$) shows that the normalized spectra $fW(f)/B(0)$ has a single maximum at $f_{\max} = 0.078v/R$, the inverse of which defines the transition between short- ($\tau_{ex} < f_{\max}^{-1}$) and long- ($\tau_{ex} > f_{\max}^{-1}$) exposure images. Similarly, for phase astigmatism, when $f < v/2\pi R$,

$$\frac{fW_{a_5a_5}(f)}{B_{a_5a_5}(0)} \sim 0.166 \left(\frac{2\pi fR}{v}\right)$$

and when $f > v/2\pi R$,

$$\frac{fW_{a_5a_5}(f)}{B_{a_5a_5}(0)} \sim 0.154 \left(\frac{2\pi fR}{v}\right)^{4/3} \exp\left[-\frac{(2\pi fR/v)^2}{8}\right]$$

where the maximum frequency that separates these two regimes is found from Eq. (4.128) (with $n = 2$) to be $f_{\max} = 0.349v/R$, which is approximately 4.5 times larger than that for the tilt component. These relations clearly demonstrate the semi-qualitative model predictions made in Sec. 3.4, i.e., the fact that the temporal frequency behavior as well as the spectral content of the astigmatic component is greater than for the tilt component.

Equations (4.125) and (4.128) also provide the necessary basis upon which an adaptive optics wavefront correcting scheme must be designed so as to respond to the relevant component aberrations.

References

1. V. I. Tatarskii, *The Effects of the Turbulent Atmosphere on Wave Propagation.* U.S. Dept. of Commerce, TT-68-50464, Springfield, Va., 1971. Chap. 3.
2. J. W. Strohbehn, "Modern Theories in the Propagation of Optical Waves in a Turbulent Medium," *Laser Beam Propagation in the Atmosphere,* J. W. Strohbehn, ed. Springer-Verlag, Berlin, 1978. Chap. 3.

3. D. L. Fried, "Optical Resolution through a Randomly Inhomogeneous Medium for Very Long and Very Short Exposures," *J. Opt. Soc. Am.* **56** (10), pp. 1372–1379 (1966).
4. R. J. Noll, "Zernike Polynomials and Atmospheric Turbulence," *J. Opt. Soc. Am.* **66** (3), pp. 207–211 (1976).
5. M. Born and E. Wolf, *Principles of Optics*. Pergamon, New York, 1965. Sec. 9.2.
6. J. Y. Wang and J. K. Markey, "Modal Compensation of Atmospheric Turbulence Phase Distortion," *J. Opt. Soc. Am.* **68** (1), pp. 78–87 (1978).
7. Ref. 1, Sec. 70.
8. A. Ishimaru, "The Beam Wave Case and Remote Sensing," *Laser Beam Propagation in the Atmosphere,* J. W. Strohbehn, ed. Springer-Verlag, Berlin, 1978. Chap. 5.
9. Ref. 1, Sec. 65.
10. R. M. Manning, "New Method for Image Propagation through a Turbid Atmosphere: Radiative Image Transfer Theory," Proc. SPIE 926, *Optical, Infrared, and Millimeter Wave Propagation Engineering,* pp. 248–257 (1988).
11. L. A. Apresyan and Yu. A. Kravtsov, "Photometry and Coherence: Wave Aspects of the Theory of Radiation Transport," *Sov. Phys. Usp.* **27** (4), pp. 301–313 (1984).
12. I. S. Gradshteyn and I. M. Ryzhik, *Table of Integrals, Series, and Products*. Academic Press, New York, 1980. Eq. (7.642).

Chapter 5

Methods of Compensation for Image Degradation due to the Atmosphere

5.1 Introduction

Now that the mechanisms by which the formation of the image of an extended object is degraded by the intervening atmosphere have been examined, this chapter will present derivations and/or results of some methods by which one can compensate for such deleterious effects. Many *ad hoc* methods have been applied to the problem and can be broadly categorized into the two groups of *passive* and *active* compensation techniques. In the case of active image compensation, an independent measurement must be made so as to secure an estimate of the statistical state of the atmospheric propagation path (essentially, a measurement of the instantaneous realization of the atmosphere). For example, in addition to the image, a deterministic reference source may be required, with known phase and coherence properties, so as to provide an indication of relevant phase or coherence statistics parameters. Such schemes, by their very design, do not need to draw on the results of the descriptive modeling of the atmospheric propagation path as afforded by the results of Chap. 4. Active image compensation schemes find a majority of their usefulness in those applications that can be specifically suited to their requirements. However, for more general imaging situations that can be encountered where one cannot, *a priori,* establish propagation conditions, one must employ passive compensation techniques. These methods make use of whatever information is available and conveyed to the imaging optics in the reception of the image of a target of arbitrary size, shape, and reflective properties. In this case, the *a priori* information concerning the atmospheric propagation path and/or target, which is only available to an active compensation technique, is lost and must essentially be replaced by whatever information is afforded by the descriptive modeling process. It is the subject of such passive image compensation techniques that will be dealt with in this chapter.

The material presented in this chapter is an attempt to provide a representative basis for all of the various passive methods that exist in the literature. Although no attempt is made to give an exhaustive account of each method, the general developments given in this chapter will be related to the more specific applications in the literature where possible. The hardware implementation of specific techniques will not be covered here, as excellent overviews of this subject can be found elsewhere.[1] Not surprisingly, many of these implementations as well as the development of the compensation methods that they perform have been motivated by astronomical applications.[2]

In what is to follow, three major categories of image reconstruction will be addressed. After a presentation in Sec. 5.2 of the two forms that can be assumed by the unaveraged (instantaneous) transfer function that necessarily forms the basis for compensation algorithms, image compensation by phase compensation within the imaging optics is treated in Sec. 5.3. Section 5.4 considers the use of speckle phenomena[2] and related image reconstruction. Finally, Sec. 5.5 treats the general and rigorous problem formulation of image intensity estimation by maximum likelihood considerations.

5.2 Two Forms of the Instantaneous Image Transfer Function

In what is to follow, the results involving the use of the instantaneous point spread function as derived in Chap. 3 will be needed and will prove to be most useful in one of two formulations. Rather than employ the form given by Eqs. (3.69) and (3.70), which, by the way, are specific to only a Zernike polynomial expansion of the wavefront with aberrations beyond those of the tilt component neglected, one has for the more general case, from Eqs. (3.31) to (3.33) for nonisoplanatic imaging,

$$\langle I(\boldsymbol{\rho}_i)\rangle = C\int_{-\infty}^{\infty}\int_{-\infty}^{\infty}\exp\left[\frac{ik}{l}\left(\boldsymbol{\rho}_i+\frac{l\boldsymbol{\rho}_0}{L}\right)\cdot\boldsymbol{\rho}_d\right]K_\gamma(\boldsymbol{\rho}_d,\boldsymbol{\rho}_0)$$

$$\times\ \langle|R(\boldsymbol{\rho}_0)|^2\rangle_o\, d^2\rho_0\, d^2\rho_d \quad (5.1)$$

with

$$K_\gamma(\boldsymbol{\rho}_d, \boldsymbol{\rho}_0) \equiv \int_{-\infty}^{\infty} \gamma_{11}(L, \boldsymbol{\rho}' - \boldsymbol{\rho}_0, \boldsymbol{\rho}' - \boldsymbol{\rho}_0 - \boldsymbol{\rho}_d) \times W(\boldsymbol{\rho}')W^*(\boldsymbol{\rho}' - \boldsymbol{\rho}_d)\, d\boldsymbol{\rho}' \quad (5.2)$$

and where C is a constant. One now evokes the same approximation that was made in Sec. 3.4.3 which allows one to neglect the dependence of Eq. (5.2) on the object coordinate ρ_0 yet keep the function γ_{11} nonisoplanatic, viz.,

$$\gamma_{11}(L, \boldsymbol{\rho}' - \boldsymbol{\rho}_0, \boldsymbol{\rho}' - \boldsymbol{\rho}_0 - \boldsymbol{\rho}_d) \approx \gamma_{11}(L, \boldsymbol{\rho}', \boldsymbol{\rho}' - \boldsymbol{\rho}_d)$$

which leads to the development

$$K_\gamma(\boldsymbol{\rho}_d, \boldsymbol{\rho}_0) = K_\gamma(\boldsymbol{\rho}_d) \equiv \int_{-\infty}^{\infty} \gamma_{11}(L, \boldsymbol{\rho}', \boldsymbol{\rho}' - \boldsymbol{\rho}_d) \times W(\boldsymbol{\rho}')W^*(\boldsymbol{\rho}' - \boldsymbol{\rho}_d)\, d'\boldsymbol{\rho}' \quad (5.3)$$

Finally, using Eq. (1.63) and admitting only phase variations, thus deeming negligible the attendant log-amplitude variations (i.e., admitting the phase approximation of the MCF), one has that

$$\gamma_{11}(L, \boldsymbol{\rho}', \boldsymbol{\rho}' - \boldsymbol{\rho}_d) \approx \langle \exp\{i[S(L, \boldsymbol{\rho}') - S(L, \boldsymbol{\rho}' - \boldsymbol{\rho}_d)]\}\rangle \quad (5.4)$$

Substituting Eq. (5.4) into (5.3) and this intermediate result into Eq. (5.1) and using the equivalent object coordinate $\boldsymbol{\rho}_0' \equiv -\boldsymbol{\rho}_0 M$, which is related to the magnification factor M as discussed in Sec. 3.2, allows one to write the convolutional relationship

$$\langle I(\boldsymbol{\rho}_i)\rangle = C' \int_{-\infty}^{\infty} \langle T(\boldsymbol{\rho}_i - \boldsymbol{\rho}_0')\rangle \langle |R(\boldsymbol{\rho}_0')|^2\rangle_o \, d\boldsymbol{\rho}_0' \quad (5.5)$$

where $\langle\ldots\rangle_o$ indicates the formation of the ensemble average over the reflectivity distribution of the object (as discussed in Chap. 3) and where $\langle T(\boldsymbol{\rho}_i - \boldsymbol{\rho}_0')\rangle$ is still the instantaneous nonisoplanatic point spread function but, with the simplifications noted above, is modified from that of Chap. 3 and is now given by

$$\langle T(\boldsymbol{\rho}_i - \boldsymbol{\rho}_0')\rangle \equiv \int_{-\infty}^{\infty} \exp\left[\frac{ik}{l}(\boldsymbol{\rho}_i - \boldsymbol{\rho}_0')\cdot \boldsymbol{\rho}_d\right] \times \int_{-\infty}^{\infty} \langle \exp\{i[S(L, \boldsymbol{\rho}') - S(L, \boldsymbol{\rho}' - \boldsymbol{\rho}_d)]\}\rangle \times W(\boldsymbol{\rho}')W(\boldsymbol{\rho}' - \boldsymbol{\rho}_d)\, d^2\rho'\, d^2\rho_d \quad (5.6)$$

As discussed in Chap. 3, the ensemble average of the intensity in the image plane indicated on the left side of Eq. (5.5) is connected with two such averages on the right side, in particular, an average over the realizations of the propagation medium given by that shown over the phase factor in Eq. (5.6), and an average over the realizations of the incoherent intensity distributions in the object plane, $|R(\boldsymbol{\rho}_0')|^2$. In what is to follow, it is desired to model the actual formation of the ensemble average of the image plane intensity distribution, even taking into account measurement noise, such that Eq. (5.5) must be generalized even further and be written to yield a *particular realization* of the image as it propagates through the atmosphere. This implies that one should retain the average over the intensity distribution in the object plane (a process the time scale of which is much smaller than that of the atmosphere), but not the average over the realizations of the propagation medium. Thus, for the purposes of analysis of adaptive optics situations, Eqs. (5.5) and (5.6) can be written

$$I(\boldsymbol{\rho}_i) = C' \int_{-\infty}^{\infty} T(\boldsymbol{\rho}_i - \boldsymbol{\rho}_0') \langle |R(\boldsymbol{\rho}_0')|^2 \rangle_o \, d^2\rho_0' + N(\boldsymbol{\rho}_i) \qquad (5.7)$$

where $T(\boldsymbol{\rho}_i - \boldsymbol{\rho}_0')$ is a *specific* realization of that given by Eq. (5.6) without the ensemble average over the exponentials containing the random phase terms and where $N(\boldsymbol{\rho}_i)$ is a "measurement noise" term that represents the uncertainty in the particular measurement of the intensity distribution at the point ρ_i in the image plane.

Another form that is useful in the spatial frequency formulation. Using the Fourier transform relationship

$$\langle i(\boldsymbol{\kappa}_i) \rangle = \left(\frac{1}{2\pi} \right)^2 \int_{-\infty}^{\infty} \langle I(\boldsymbol{\rho}_i) \rangle \exp\,(-i\boldsymbol{\kappa}_i \cdot \boldsymbol{\rho}_i) \, d^2\rho_i$$

applying this to Eq. (5.5), using Eq. (5.6), and simplifying gives

$$\langle i(\boldsymbol{\kappa}_i) \rangle = C'' K_\gamma \left(\frac{l}{k} \boldsymbol{\kappa}_i \right) \left\langle r \left(\frac{l}{L} \boldsymbol{\kappa}_i \right) \right\rangle \qquad (5.8)$$

Proceeding with the same reasoning that led from Eq. (5.5) to Eq. (5.7) above, Eq. (5.8) can be written

$$i(\mathbf{f}_i) = C'' \langle r(\mathbf{f}_i) \rangle \int_{-\infty}^{\infty} \exp\,\{i[S(L, \boldsymbol{\rho}') - S(L, \boldsymbol{\rho}' - l\lambda \mathbf{f}_i)]\} \times W(\boldsymbol{\rho}') W(\boldsymbol{\rho}' - l\lambda \mathbf{f}_i) \, d^2\rho' + n(\mathbf{f}_i) \qquad (5.9)$$

where $n(\mathbf{f}_i)$ is the random noise associated with the measurement of the image spectrum or otherwise the random error attendant on its calculation. In what is to follow, the "realization-specific" expressions of Eqs. (5.7) and (5.9), both of which transcend the related short-term expressions of Chap. 3, will be applied to the adaptive cancellation of the deleterious turbulence effects the source of which is the random phase differences appearing in the exponential functions of these relations.

5.3 Object Intensity Distribution Estimation by Phase Compensation within the Imaging Optics

As noted at the end of Sec. 3.2, the pupil function $W(\boldsymbol{\rho})$ is the only deterministic element of the atmosphere/optics imaging system that can be modified in such a way as to counter the random phase perturbations induced by the atmosphere on the electromagnetic wave conveying the image of an object. In particular, Eq. (5.9) shows how the random phase perturbations combine with the pupil functions to further degrade the reproduction $i(\mathbf{f}_i)$ in the image plane of the spatial spectrum of the object intensity distribution $\langle r(\mathbf{f}_i)\rangle$ and how, by the same relation, the pupil function is a prime candidate to aid in the cancellation of the phase effects.

One way in which the pupil function of an optical system can be modified is to use, in place of a lens, a flexible parabolic reflector or mirror. At optical wavelengths, the transformation of the complex amplitude at the input aperture of a lens into its focal plane is mathematically isomorphic to that of a parabolic reflector.[3,4] However, the flexible surface of the parabolic reflector can be deformed in prescribed ways by a system of piston actuators acting from the back of the reflector.[1]

Deformations of the parabolic reflector surface give rise to phase perturbations of the reflected field. More succinctly, if a square array of i rows $1 \le i \le N$, and j columns $1 \le j \le N$ of such actuators is placed over the surface behind the parabolic reflector, each actuator imparts a phase change θ_{ij} to the transformed image fields. The effects of this process can be appended to the pupil function defining the optical system; in particular, one has

$$W(\boldsymbol{\rho}) \longrightarrow W(\boldsymbol{\rho}) \exp\,[i\theta_{ij}\,\tilde{W}(\boldsymbol{\rho} - \boldsymbol{\rho}_{ij}^{0})] \tag{5.10}$$

where $\tilde{W}(\boldsymbol{\rho}')$ is a "subaperture function" defined over the subaperture areas, centered at the coordinates $\boldsymbol{\rho}_{ij}^{0}$ of the actuator positions, that are induced on the reflector surface by the actuators, viz.,

$$\tilde{W}(\boldsymbol{\rho} - \boldsymbol{\rho}_{ij}^{0}) \equiv \begin{cases} 1 & \{\boldsymbol{\rho} - \boldsymbol{\rho}_{ij}^{0}\} \in \sigma_{ij} \\ 0 & \{\boldsymbol{\rho} - \boldsymbol{\rho}_{ij}^{0}\} \notin \sigma_{ij} \end{cases} \tag{5.11}$$

where σ_{ij} is the effective area that defines the subaperture.

Consider now the following adaptive problem defined by Eqs. (5.9) to (5.11): given that the only information that one has concerning the propagation path is the resulting degraded real-time spectrum of an image $i(\mathbf{f}_i)$, it is of interest to be able to adaptively adjust the phase perturbations induced on the reflector in such a way that the effect of the random phase factor appearing in Eq. (5.9) is mitigated by the phase perturbations θ_{ij}. To this end, it is of interest to obtain an expression relating the adaptive phase perturbations to the resulting change in the image spectrum. At the outset, taking the idealistic case where $n(\mathbf{f}_i) = 0$ in Eq. (5.9) and using Eq. (5.10), one has for the change $\Delta_i(\mathbf{f}_i, \theta_{ij})$ in the spatial spectrum due to an arbitrary phase perturbation θ_{ij}, after some straightforward simplifications,

$$\begin{aligned}\Delta_i(\mathbf{f}_i, \theta_{ij}) &= 2C''|\langle r(\mathbf{f}_i)\rangle| \\ &\quad\times \exp\left\{\frac{i}{2}\left[S(L, \boldsymbol{\rho}_{ij}^0 + l\lambda\mathbf{f}_i) - S(L, \boldsymbol{\rho}_{ij}^0 - l\lambda\mathbf{f}_i)\right] + i \arg\left[\langle r(\mathbf{f}_i)\rangle\right]\right\} \\ &\quad\times \{\cos\left[\Sigma_{ij}(\mathbf{f}_i) + \theta_{ij}\right] - \cos\left[\Sigma_{ij}(\mathbf{f}_i)\right]\}\end{aligned} \tag{5.12}$$

where

$$\begin{aligned}\Sigma_{ij}(\mathbf{f}_i) \equiv -\frac{1}{2}\{&[S(L, \boldsymbol{\rho}_{ij}^0 + l\lambda\mathbf{f}_i) - S(L, \boldsymbol{\rho}_{ij}^0)] \\ &- [S(L, \boldsymbol{\rho}_{ij}^0) - S(L, \boldsymbol{\rho}_{ij}^0 - l\lambda\mathbf{f}_i)]\}\end{aligned} \tag{5.13}$$

It is important to note that the derivation of the above relations assumed that the condition

$$\rho_{ij}^0 > l\lambda f_i \tag{5.14}$$

prevails over the magnitudes of the participating vector quantities, thus placing a limitation on the effective subaperture area and/or the maximum spectral frequency content of the image.

The meaning of the quantity given by Eq. (5.13) is very important and gives this method its effectiveness; it is the second difference (i.e., the difference of the differences) between the random instantaneous phase perturbations of the propagating image induced by the atmosphere. Since this quantity is related to the variation of the image spectrum $\Delta i(\mathbf{f}_i, \theta_{ij})$ by Eq. (5.12), it is possible to be able to passively determine such propagation parameters via predetermined adaptive phase settings, then, after such a determination has been made, to reset the adaptive phase settings to counter the estimated atmospheric distortions. Thus, in this particular case of noise-free observations, one has from Eq. (5.12) for the second difference for the atmospheric phase distortions determined from the changes of the image spectrum

$$\hat{\Sigma}_{ij}(\mathbf{f}_i) = \tan^{-1}\left[\frac{\Delta_i(\mathbf{f}_i, \pi/2) - \Delta i(\mathbf{f}_i, -\pi/2)}{\Delta i(\mathbf{f}_i, \pi/2) + \Delta i(\mathbf{f}_i, -\pi/2)}\right] \tag{5.15}$$

Hence, by intentionally inducing phase perturbations of $\pm\pi/2$ over a subaperture, one can, without needing to know any *a priori* information about the object spectrum $\langle r(\mathbf{f}_i)\rangle$, obtain the prevailing second difference of the atmospheric phase statistics over that subaperture.

Having obtained the estimated values $\hat{\Sigma}_{ij}(\mathbf{f}_i)$, it now remains to calculate estimates $\hat{S}(L, \boldsymbol{\rho}_{ij}^0)$ for the separate component atmospheric phase distortions. Following the results of Hudgin,[5] one can implement the following algorithm:

$$\begin{aligned}\hat{S}(L, \boldsymbol{\rho}_{ij}^0) = {} & \tfrac{1}{3}(S_{i,j-1} + S_{i,j+1} + S_{i-1,j} + S_{i+1,j}) \\ & - \tfrac{1}{12}(S_{i+2,j} + S_{i-2,j} + S_{i,j-2} + S_{i,j+2}) \\ & + \tfrac{1}{6}(2\Sigma_{i,j}^{(1)} - \Sigma_{i-1,j}^{(1)} - \Sigma_{i-1,j}^{(1)} + 2\Sigma_{i,j}^{(2)} - \Sigma_{i,j+1}^{(2)} - \Sigma_{i,j-1}^{(2)})\end{aligned} \tag{5.16}$$

where

$$\Sigma_{i,j}^{(1)} \equiv S_{ij} - \frac{(S_{i+1,j} + S_{i-1,j})}{2} \qquad \Sigma_{i,j}^{(2)} \equiv S_{ij} - \frac{(S_{i,j+1} + S_{i,j-1})}{2} \tag{5.17}$$

and the initial conditions for commencement of the calculational procedure are $S_{ij} = \Sigma_{i,j}^{(1)} = \Sigma_{i,j}^{(2)} = 0$ for $i, j > N$ or $i, j < 1$.

The phase distortion estimates obtained from Eq. (5.16) can then be employed to adaptively cancel those distortions that aid in image degradation simply by setting $\theta_{ij} = -\hat{S}(L, \boldsymbol{\rho}_{ij}^0)$ and recapturing the image spectrum via Eqs. (5.9) and (5.10).

In the more realistic and general case where uncertainties are attendant on the observation of $i(\mathbf{f}_i)$, i.e., $n(\mathbf{f}_i) \neq 0$ in Eq. (5.9), the problem becomes one of the optimal estimation of Σ_{ij} given the statistics that govern the noise. In particular, it is convenient (yet realistic) to assume that, at the particular instants of time t_k at which the image spectrum is obtained, the measurement uncertainty is described by gaussian white noise, i.e.,

$$\langle n(\mathbf{f}_i, t_k)\rangle = 0 \qquad \langle n(\mathbf{f}_i, t_1)n(\mathbf{f}_i, t_2)\rangle = N_0(\mathbf{f}_i)\delta(t_1 - t_2) \tag{5.18}$$

where the noise spectral density $N_0(\mathbf{f}_i)$ is, in general, a function of the frequency vector $\mathbf{f}_i$. This, however, will not be discussed here, as it is not the purpose of this chapter to develop such estimation procedures; the final result of any such estimation will ultimately be employed with the algorithmic procedure of Eq. (5.16) and the object spectrum obtained via Eqs. (5.9) and (5.10). For development of statistical optimal estimation procedure as applied to this problem, the reader is referred to the appropriate literature.[6]

5.4 Object Intensity Distribution Reconstruction via Speckle Imaging and Interferometry

This section will be devoted to the adaptive optics method of target intensity reconstruction using the phenomena of image speckles that result from spectrally filtered short-exposure imaging of objects that cannot be resolved by the atmosphere-optics imaging channel. The first part of this section will present a very brief description of speckle imaging and the associated speckle interferometry. The second part will present a derivation of a passive speckle interferometry technique that can be employed in the reconstruction of extended objects observed through the atmosphere.

5.4.1 A brief description of speckle phenomena and their use in object intensity reconstruction

The desire to obtain high-resolution optical images of astronomical objects observed through the atmosphere has given rise to a class of techniques employing *speckle* images.[2] These are short-term images formed by an imaging system, such as a telescope, which is not able to totally resolve the object in question because of atmospheric turbulence. Hence, the technique is applicable in situations involving strong turbulence conditions. Each short-term image is made up of an assembly of bright patches called speckles, each of which has a size that is on the order of an Airy disk that corresponds to the telescope aperture. Such speckle phenomena have their origin in the very short-term constructive and destructive interference induced in the electromagnetic radiation making up the object image as it propagates through the atmosphere. In addition, the image is usually spectrally filtered before it enters the optical system. Hence, the recorded image is limited to a wavelength range $\Delta\lambda$ about the nominal wavelength λ_{nom} such that a value of the prevailing coherence length $l_{coh} = \lambda_{nom}^2/\Delta\lambda$ can be established that corresponds to the largest difference in the random optical path distance through which the image must propagate through the random medium to reach the input aperture. (Of course, this is also a random quantity that varies from one short-term exposure to another. This suggests that it would be expedient to use only one such exposure. This will be discussed further below.)

If the object being imaged has a maximum resolvable size that is smaller than the capability of the imaging optics, as can be the case in astronomical applications, the object can be considered a point object. In this instance, the speckles are bright areas with well-defined centers that are essentially the Airy pattern of the optical instrument. In the other case, where the object cannot be considered as a point, which of course encompasses situations involving extended objects in terrestrial imaging, the speckles are diffuse distributions of light that correspond to the sharper speckles, as described above, in the limit as the extended object approaches a point. The statistical analysis of such short-term spectrally filtered intensity distributions, i.e., *speckle interferograms*, falls into two categories. The first type[7] is *speckle interferometry*, which employs the autocorrelation of the intensity distribution of the object and from which, unless some other *a priori* information about the object is available, the actual intensity distribution of the object cannot be obtained. The other type[8] is *speckle imaging*, which can result in the direct recovery of the entire intensity distribution of the object.

When imaging extended objects of any type, whether they are targets on the earth and viewed entirely through the atmosphere or astronomical objects existing outside the atmosphere, each intensity distribution obtained in the short-exposure case is an ensemble of an arbitrary number of N speckles, each essentially an image of the object degraded by both the atmosphere and the diffraction limit of the optics. Each such member of this speckle distribution is a single representation of the statistical propagation process of the target image and is therefore statistically independent from other speckles within the total short-term image.[8] It therefore becomes expedient to consider the possibility of using a large number of short-term images, each consisting of N speckles, and superposing all the individual speckles obtained so as to remove the distortions present in these individual speckles by the use of an averaging technique[9] or by Fourier transform methods.[10] However, among other drawbacks of these methods, e.g., needing to have *a priori* information concerning the object, such as its geometrical shape and orientation, the major obstacle in their implementation is the need for a reference source near the extended object. In what is to follow, a speckle imaging technique will be developed that allows for the reconstruction of the incoherent intensity distribution of an extended object viewed through the turbulent atmosphere without the need for a reference source or any *a priori* object information. In addition, this technique employs only one such speckle interferogram so as to assure the satisfaction of the coherence length versus propagation path difference length as discussed above.

5.4.2 Application of speckle interferometry in the reconstruction of an extended object

As is demonstrated by the coordinate representation of Eq. (5.7) for the short-term imaging process, neglecting the contribution of the measurement noise, the image in the focal plane is simply the convolution of a realization of the nonisoplanatic point spread function (incorporating the action of the imaging optics as well as the instantaneous action of the phase perturbations induced by the atmosphere) with the intensity distribution of the object. From what was said in the above description of the speckle image phenomena, the positions (but not the actual intensity distribution) of the arbitrary number of N speckles in a short-term image of an extended object correspond to those for a point object. Hence, at the outset in obtaining a model of the speckle imaging process, the instantaneous point spread function of Eq. (5.7) can be represented by the summation of the N speckles, either from one short-term image or from several such images, i.e.,

$$T(\boldsymbol{\rho}_i - \boldsymbol{\rho}_0') = \sum_{j=1}^{N} \iota_p^{(j)}(\boldsymbol{\rho}_i - \boldsymbol{\rho}_{i,j} - \boldsymbol{\rho}_0') \tag{5.19}$$

where

$$\iota_p^{(j)}(\boldsymbol{\rho}_i - \boldsymbol{\rho}_{i,j} - \boldsymbol{\rho}_0') \equiv \int_{-\infty}^{\infty} \exp\left[\frac{ik}{l}(\boldsymbol{\rho}_i - \boldsymbol{\rho}_{i,j} - \boldsymbol{\rho}_0') \cdot \boldsymbol{\rho}_d\right] \times \int_{-\infty}^{\infty} \exp\{i[S^{(j)}(L, \boldsymbol{\rho}') - S^{(j)}(L, \boldsymbol{\rho}' - \boldsymbol{\rho}_d)]\} \times W(\boldsymbol{\rho}')W(\boldsymbol{\rho}' - \boldsymbol{\rho}_d)\, d^2\rho'\, d^2\rho_d \tag{5.20}$$

is the point spread profile of the jth speckle in the image plane *due to a point object* and $\rho_{i,j}$ is its position. Substituting Eq. (5.19) into Eq. (5.7) and employing the conditions of the problem, i.e., neglecting the noise term, one can write

$$I(\boldsymbol{\rho}_i) = \sum_{j=1}^{N} \iota_{\text{ex}}^{(j)}(\boldsymbol{\rho}_i - \boldsymbol{\rho}_{i,j}) \tag{5.21}$$

where

$$\iota_{\text{ex}}^{(j)}(\boldsymbol{\rho}_i - \boldsymbol{\rho}_{i,j}) \equiv C' \int_{-\infty}^{\infty} \iota_p^{(j)}(\boldsymbol{\rho}_i - \boldsymbol{\rho}_{i,j} - \boldsymbol{\rho}_0') \times \langle |R(\boldsymbol{\rho}_0')|^2 \rangle_o\, d^2\rho_0' \tag{5.22}$$

is the point spread profile of the jth speckle in the image *due to an extended object.*

Equation (5.21) is the basis of the proposed methods[9,10] for reconstruction of an extended object image from a single speckle image. What has been proposed in Refs. 9 and 10 involves calculating from the measured speckle image the average intensity profile $\iota_p^{(j)}(\boldsymbol{\rho}_i -$

$\boldsymbol{\rho}_{i,j} - \boldsymbol{\rho}_0')$ of the N brightest speckles. This presupposes that the individual positions of these speckles $\boldsymbol{\rho}_{i,j}$ can be accurately measured. However, the ever-present uncertainty in the determination of these positions contributes to the measurement noise, which has deleterious effects when the processing associated with the inverse transformation of Eq. (5.22), in conjunction with Eq. (5.21), is performed so as to yield a measure of $\langle|R(\boldsymbol{\rho}_0')|^2\rangle_o$. This, along with the attendant need for *a priori* knowledge of the symmetry of the object, is one of the major problems encountered in the use of just one speckle image for a reconstruction of an extended object.

A method that alleviates these drawbacks was advanced by Bakut et al.[11] The structure of Eq. (5.22) suggests that for each of the N brightest speckles chosen, their respective intensity distributions are centered about a common origin, thus essentially giving N speckle intensity distributions, each of the form

$$\iota_{\text{ex}}^{(j)}(\boldsymbol{\rho}_i) \equiv C' \int_{-\infty}^{\infty} \iota_p^{(j)}(\boldsymbol{\rho}_i - \boldsymbol{\rho}_0')\langle|R(\boldsymbol{\rho}_0')|^2\rangle_o \, d^2\rho_0' \tag{5.23}$$

Taking the Fourier transform of these speckle intensity distributions and employing the linear spatial frequency $\mathbf{f}$ yields

$$\hat{\iota}_{\text{ex}}^{(j)}(\mathbf{f}) \equiv C'\hat{\iota}_p^{(j)}(\mathbf{f})\langle r(\mathbf{f})\rangle_o \tag{5.24}$$

where the transform $\hat{\iota}_p^{(j)}(\mathbf{f})$ of the jth speckle profile due to a point source is, by the definition given by Eq. (5.19) and the use of Eq. (5.6),

$$\hat{\iota}_p^{(j)}(\mathbf{f}) = \left(\frac{l}{k}\right)^2 \int_{-\infty}^{\infty} \exp\{i[S(L, \boldsymbol{\rho}') - S(L, \boldsymbol{\rho}' - \lambda l\mathbf{f})]\} \times W(\boldsymbol{\rho}')W(\boldsymbol{\rho}' - \lambda l\mathbf{f}) \, d^2\rho' \tag{5.25}$$

For reasons that will be apparent shortly, it is convenient at this point to decompose the transform of the object image into its amplitude $|\langle r(\mathbf{f})\rangle|$ and phase $\phi(\mathbf{f})$ components, viz.,

$$\langle r(\mathbf{f})\rangle = |\langle r(\mathbf{f})\rangle| \exp[\phi(\mathbf{f})] \tag{5.26}$$

In order to obtain the absolute phases from a single measurement, one needs to proceed as in Sec. 5.3 above and consider application of Eqs. (5.24) and (5.26) at two different spatial frequencies, i.e., form the autocorrelation of the known quantity $\hat{\iota}_{\text{ex}}^{(j)}(\mathbf{f})$ over an arbitrary (usually the minimum) increment $\Delta\mathbf{f}$ in the spatial frequency averaged over the N speckles,

$$\langle\hat{\iota}_{\text{ex}}^{(j)}(\mathbf{f})\hat{\iota}_{\text{ex}}^{(j)}(\mathbf{f} + \Delta\mathbf{f})\rangle_N = C'^2\langle\hat{\iota}_p^{(j)}(\mathbf{f})\hat{\iota}_p^{(j)}(\mathbf{f} + \Delta\mathbf{f})\rangle \times |\langle r(\mathbf{f})\rangle||\langle r(\mathbf{f} + \Delta\mathbf{f})\rangle| \exp\{i[\phi(\mathbf{f}) - \phi(\mathbf{f} + \Delta\mathbf{f})]\} \tag{5.27}$$

This formulation allows one to replace the *a priori* information that is needed in the earlier methods with the modeling results of this book. In particular, using Eq. (5.25), one has

$$\langle \hat{\imath}_p^{(j)}(\mathbf{f})\hat{\imath}_p^{(j)}(\mathbf{f}+\Delta\mathbf{f})\rangle = \left(\frac{l}{k}\right)^4 \int_{-\infty}^{\infty}\int_{-\infty}^{\infty} \langle \exp\,(i\{S(L,\boldsymbol{\rho}') - S(L,\boldsymbol{\rho}' - \lambda l\mathbf{f})$$

$$- S(L,\boldsymbol{\rho}'') + S[L,\boldsymbol{\rho}'' - \lambda l(\mathbf{f}+\Delta\mathbf{f})]\})\rangle W(\boldsymbol{\rho}')$$

$$\times\, W(\boldsymbol{\rho}'' - \lambda l\mathbf{f})W(\boldsymbol{\rho}'')W[\boldsymbol{\rho}'' - \lambda l(\mathbf{f}+\Delta\mathbf{f})]\; d^2\rho'\, d^2\rho'' \qquad (5.28)$$

As per the discussion leading up to the application of Eq. (1.131) in Sec. 1.6, linear combination of the phase terms within the exponential of Eq. (5.28) can, in this case,[12,13] be treated as a composite gaussian random variable with a zero mean. Using Eq. (1.131), multiplying out the phase terms, placing the results in the form of phase structure functions, and recentering the coordinates and defining $\rho_s \equiv \rho' + \rho''$ and $\rho_d \equiv \rho' - \rho''$, one has

$$\langle \hat{\imath}_p^{(j)}(\mathbf{f})\hat{\imath}_p^{(j)}(\mathbf{f}+\Delta\mathbf{f})\rangle$$

$$= \left(\frac{l}{k}\right)^4 \int_{-\infty}^{\infty}\int_{-\infty}^{\infty} W\left(\frac{\boldsymbol{\rho}_s + \boldsymbol{\rho}_d + \lambda l\mathbf{f}}{2}\right)$$

$$\times\, W^*\left(\frac{\boldsymbol{\rho}_s + \boldsymbol{\rho}_d - \lambda l\mathbf{f}}{2}\right) W\left[\frac{\boldsymbol{\rho}_s - \boldsymbol{\rho}_d + \lambda l(\mathbf{f}+\Delta\mathbf{f})}{2}\right]$$

$$\times\, W^*\left[\frac{\boldsymbol{\rho}_s - \boldsymbol{\rho}_d - \lambda l(\mathbf{f}+\Delta\mathbf{f})}{2}\right] d^2\rho_s \times \exp\left(-\frac{1}{2}\left\{D_s(\lambda l\mathbf{f})\right.\right.$$

$$+ D_s[\lambda l(\mathbf{f}+\Delta\mathbf{f})] + D_s\left(\boldsymbol{\rho}_d + \frac{\lambda l\mathbf{f}}{2}\right)$$

$$+ D_s\left(\boldsymbol{\rho}_d - \frac{\lambda l\mathbf{f}}{2}\right) - D_s\left[\lambda l\left(\mathbf{f}+\frac{\Delta\mathbf{f}}{2}\right) + \boldsymbol{\rho}_d\right]$$

$$\left.\left. - D_s\left[\lambda l\left(\mathbf{f}+\frac{\Delta\mathbf{f}}{2}\right) - \boldsymbol{\rho}_d\right]\right\}\right) d^2\rho_d \qquad (5.29)$$

where the phase structure functions $D_s(\rho) = D_s(|\rho|)$ are as given in Chap. 4. In particular, for the case considered here, large spatial frequencies near the cutoff frequency f_c of the aperture will be considered. Hence, one has $\lambda l|\mathbf{f}| \sim \lambda l f_c \sim 2R$ upon using the definition of Eq. (4.44). Taking the condition $4R^2 \gg \lambda L$ to prevail (i.e., the "near field" condition as defined by Fried[14]), it then becomes appropriate to use for $D_s(\ldots)$ the expression given by Eq. (4.24). In terms of the coherence length defined by Eq. (4.42) for a homogeneous atmosphere, one has

$$D_s(\lambda l\mathbf{f}) = 2\left(\frac{\lambda l f}{r_0}\right)^{5/3} \qquad \text{near field case} \tag{5.30}$$

and similarly for the other structure functions that enter into Eq. (5.29). Specializing these results further by assuming the case $\lambda lf \gg r_0$, one can implement a series expansion of the structure functions and an associated approximation procedure as detailed in Section 3.B.3 of Ref. 13 and place Eq. (5.29) into the form

$$\begin{aligned}
&\langle \hat{\iota}_p^{(j)}(\mathbf{f})\hat{\iota}_p^{(j)}(\mathbf{f} + \Delta\mathbf{f})\rangle \\
&= \left(\frac{l}{k}\right)^4 \int_{-\infty}^{\infty} W\left(\frac{\boldsymbol{\rho}_s + \lambda l\mathbf{f}}{2}\right) W^*\left(\frac{\boldsymbol{\rho}_s - \lambda l\mathbf{f}}{2}\right) \\
&\quad \times W\left[\frac{\boldsymbol{\rho}_s + \lambda l(\mathbf{f} + \Delta\mathbf{f})}{2}\right] W^*\left[\frac{\boldsymbol{\rho}_s - \lambda l(\mathbf{f} + \Delta\mathbf{f})}{2}\right] d^2\rho_s \\
&\quad \times \int_{-\infty}^{\infty} \exp\left\{-\left[\left(\frac{|\boldsymbol{\rho}_d + \lambda l\,\Delta\mathbf{f}/2|}{r_0}\right)^{5/3}\right.\right. \\
&\qquad\qquad \left.\left. + \left(\frac{|\boldsymbol{\rho}_d - \lambda l\,\Delta\mathbf{f}/2|}{r_0}\right)^{5/3}\right]\right\} d^2\rho_d
\end{aligned} \tag{5.31}$$

where the integration over the coordinate $\boldsymbol{\rho}_s$ only involves the product of the aperture functions. Finally, admitting values for the frequency increment $\Delta\mathbf{f}$ such that the condition $\lambda l\,|\Delta\mathbf{f}| \ll r_0$ obtains, one has the use of the approximation $|\boldsymbol{\rho}_d \pm \lambda l\,\Delta\mathbf{f}/2| \approx |\boldsymbol{\rho}_d| + |\lambda l\,\Delta\mathbf{f}/2|$ for the arguments of the exponentials. Incorporating these considerations into Eq. (5.31), simplifying, and remembering the definition of Eq. (3.16) finally gives the relation

$$\begin{aligned}
\langle \hat{\iota}_p^{(j)}(\mathbf{f})\hat{\iota}_p^{(j)}(\mathbf{f} + \Delta\mathbf{f})\rangle &= \alpha(r_0)K^2\left[\lambda l\left(\mathbf{f} + \frac{\Delta\mathbf{f}}{2}\right)\right] \\
&\quad \times \exp\left[-2\left(\frac{\lambda l\,\Delta\mathbf{f}}{2r_0}\right)^{5/3}\right]
\end{aligned} \tag{5.32}$$

where $$\alpha(r_0) \equiv 2\pi\left(\frac{l}{k}\right)^4 r_0^2 \int_0^{\infty} \exp\left(-2x^{5/3}\right)x\,dx \tag{5.33}$$

is a parameter of the problem that is a function of the coherence length; the numerical value of the integral in Eq. (5.33) is 0.239.

Returning to the problem at hand, viz., the analysis of Eq. (5.27), one now has a prescription provided by Eq. (5.32) for separating the modulus and phase of the object distribution from the Fourier transform of the observed data, i.e.,

$$\langle \hat{\imath}_{\mathrm{ex}}^{(j)}(\mathbf{f})\hat{\imath}_{\mathrm{ex}}^{(j)}(\mathbf{f} + \Delta\mathbf{f})\rangle_N \equiv \frac{1}{N}\sum_{j=1}^{N} [\hat{\imath}_{\mathrm{ex}}^{(j)}(\mathbf{f})\hat{\imath}_{\mathrm{ex}}^{(j)}(\mathbf{f} + \Delta\mathbf{f})] \tag{5.34}$$

In particular, considering the case where $\Delta\mathbf{f} = 0$, one has, upon substituting Eq. (5.32) into Eq. (5.27) and using Eq. (5.34),

$$\frac{1}{N}\sum_{j=1}^{N} |\hat{\imath}_{\mathrm{ex}}^{(j)}(\mathbf{f})|^2 = C_2(r_0)K^2(\lambda l\mathbf{f})|\langle\tilde{r}(\mathbf{f})\rangle|^2 \tag{5.35}$$

which involves only the modulus of the estimated spatial spectrum $\langle\tilde{r}(\mathbf{f})\rangle$ of the quantity desired, i.e., the intensity distribution in the object plane, as well as the aperture function of the imaging optics. The coefficient $C_2(r_0)$ convolves all the intermediate parameters of the problem as well as the coherence length through the incorporation of Eq. (5.33).

The calculation of the phase $\phi(\mathbf{f})$ required by Eq. (5.26) is straightforward and is commenced by establishing a square grid in frequency space across the two-dimensional spectral distribution of the image; the length in frequency of such a grid is determined by the aperture cutoff frequency $\mathbf{f}_c$ as defined by Eq. (4.44), and the grid spacing is set by the minimum frequency increment $\Delta\mathbf{f}$ employed. Hence, one must consider M separate frequency increments, where

$$M = \left(\frac{f_c}{\Delta f}\right)^2 = \left(\frac{2R}{\lambda l\,\Delta f}\right)^2 \tag{5.36}$$

over which the difference of the estimated phases $\tilde{\phi}(\mathbf{f})$ of the spatial spectrum of the object is calculated for each pair of frequencies separated by the distance $|\Delta\mathbf{f}|$. Thus, for each such spatial frequency $\mathbf{f}_k$ in the range $1 \le k \le M$, one has

$$\arg\left\{\frac{1}{N}\sum_{j=1}^{N} [\hat{\imath}_{\mathrm{ex}}^{(j)}(\mathbf{f}_k)\hat{\imath}_{\mathrm{ex}}^{(j)}(\mathbf{f}_k + \Delta\mathbf{f})]\right\} = [\tilde{\phi}(\mathbf{f}_k) - \tilde{\phi}(\mathbf{f}_k + \Delta\mathbf{f})] \tag{5.37}$$

By summing and thus matching the resulting phase increments, taking $\phi(0) = 0$, one finally has for the phase of the reconstructed object

$$\tilde{\phi}(\mathbf{f}_M) = \sum_{k=1}^{M} [\tilde{\phi}(\mathbf{f}_k) - \tilde{\phi}(\mathbf{f}_k + \Delta\mathbf{f})] \tag{5.38}$$

Finally, using Eqs. (5.35) and (5.38) with Eq. (5.26), one can obtain an estimate of the diffraction-limited image of the reflectivity of the object, viz.,

$$\langle|\tilde{R}(\boldsymbol{\rho}_0)|^2\rangle = \int_{-\infty}^{\infty} \langle\tilde{r}(\mathbf{f})\rangle \exp\,(i2\pi\mathbf{f}\cdot\boldsymbol{\rho}_0)\, d^2f \qquad (5.39)$$

The accuracy of this image compensation is limited by the ability to reconstruct the phase component of the image frequency spectrum. Implicit in the formation of Eq. (5.38) is the fact that the phase is a continuous function and that only its potential principal values within the interval $(-\pi, \pi)$ need be used, so long as the spectral interval $\Delta\mathbf{f}$ is chosen to be sufficiently small. Although this condition is satisfied for the large class of target geometries that do not have step-function-like components in their images,[15] one must employ a modified phase matching process for those that do.[16]

The introduction of specific *ad hoc* techniques to minimize the errors that are inherent in the spectral phase matching process is one of several such indications of the fact that this image reconstruction technique was developed on a heuristic basis. In particular, the formation of Eq. (5.22) as well as Eq. (5.23) is based on results that are extrapolated from limiting forms. The rigorous and unified treatment of the image reconstruction problem, which includes the measurement noise component and transcends the use of tacit assumptions noted above, will now be reviewed in the following section.

5.5 Application of the Maximum Likelihood Method to the Analysis and Subsequent Reconstruction of Speckle Images

The statistical nature of the image compensation problem is succinctly formulated within the context of the maximum likelihood method (MLM) for the mathematically rigorous estimation of the intensity distribution in the object plane. Since it is not within the purview of this chapter to cover such well-known statistical optimization methods, only a brief review and the conditions and results of the pertinent investigations[16,17] will be given.

At the outset, it is important to first secure the statistical process that governs the behavior of the image intensity in the focal plane. In general, a gamma probability distribution can be used to describe such a process. However, in the limit of the incoherent imaging process of an extended object considered here, the probability distribution for the instantaneous intensity transforms into the well-known gaussian distribution. Second, although the general case could be considered, the radius of the input aperture of the optical system and the prevailing coherence length will be taken to be such that the realistic condition $2R \gg r_0$ holds.

Given the circumstance that each spectrally filtered, short-term intensity distribution $\iota_{\text{ex}}^{(j)}(\boldsymbol{\rho}_i)$ is a random gaussian function, one can write for a conditional probability density governing the gaussian image function in the image plane $I(\rho_i)$, given an intensity function $I_0(\boldsymbol{\rho}_0) \equiv \langle |R(\boldsymbol{\rho}_0)|^2 \rangle_o$ describing the target being imaged,

$$\Pr\,[I(\boldsymbol{\rho}_i)|I_0(\boldsymbol{\rho}_0)] = A[I_0(\boldsymbol{\rho}_0)]\exp\left\{-\frac{1}{2}\sum_{j=1}^{N}\int_{-\infty}^{\infty}\int_{-\infty}^{\infty} R_I(\boldsymbol{\rho}_{i1}, \boldsymbol{\rho}_{i2}) \right.$$

$$\left. \times\,[\iota_{\text{ex}}^{(j)}(\boldsymbol{\rho}_{i1}) - \langle I(\boldsymbol{\rho}_{i1})\rangle][\iota_{\text{ex}}^{(j)}(\boldsymbol{\rho}_{i2}) - \langle I(\boldsymbol{\rho}_{i2})\rangle]\,d^2\rho_{i1}\,d^2\rho_{i2}\right\} \tag{5.40}$$

where $R_I(\boldsymbol{\rho}_{i1}, \boldsymbol{\rho}_{i2})$ is the function that is inverse to the spatial correlation function $B_I(\boldsymbol{\rho}_{i1}, \boldsymbol{\rho}_{i2})$ of the intensity between two points in the image plane; this function is formally found from the solution of the integral equation

$$\int_{-\infty}^{\infty} B_I(\boldsymbol{\rho}_{i1}, \boldsymbol{\rho})R_I(\boldsymbol{\rho}, \boldsymbol{\rho}_{i2})\,d^2\rho = \delta(\boldsymbol{\rho}_{i1} - \boldsymbol{\rho}_{i2}) \tag{5.41}$$

The indicated integrations in both Eqs. (5.40) and (5.41) are over the entire area of the image plane. The average intensities $\langle I(\boldsymbol{\rho}_i)\rangle$ appearing in Eq. (5.40) are defined by employing Eqs. (5.21) and (5.23),

$$\langle I(\boldsymbol{\rho}_i)\rangle = \frac{1}{N}\sum_{j=1}^{N}\iota_{\text{ex}}^{(j)}(\boldsymbol{\rho}_i) \tag{5.42}$$

where, as before, the intensity distribution of the individual speckles $\iota_{\text{ex}}^{(j)}(\boldsymbol{\rho}_i)$ is related to the atmospheric and optical system parameters by Eqs. (5.20) and (5.22). Finally, $A[I_0(\boldsymbol{\rho}_0)]$ is a functional, the form of which is not important in the development to follow.

The following problem can now be succinctly defined: given the "passive" information inputs $\iota_{\text{ex}}^{(j)}(\boldsymbol{\rho}_i)$ regarding the image of the target and using the *a priori* descriptions that can be obtained from the modeling results of Chaps. 3 and 4 for $\langle I(\boldsymbol{\rho}_i)\rangle$ and $B_I(\boldsymbol{\rho}_{i1}, \boldsymbol{\rho}_{i2})$, obtain an optimal estimate $\tilde{I}_0(\boldsymbol{\rho}_0)$ of the original (i.e., unperturbed) intensity distribution $I_0(\boldsymbol{\rho}_0)$ of the target. The probabilistic formulation of such a problem given by Eq. (5.40) allows one to interpret an optimal estimate as the most probable one. In this instance, one can form an expression for such a maximum likelihood by requiring the derivative of the logarithm of Eq. (5.40) to be an extremum for the function $\tilde{I}_0(\boldsymbol{\rho}_0)$ taken to be the optimal, most likely one for representation of the target image.

One thus has the condition given by the functional derivative equation, i.e., the maximum likelihood equation,

$$\left.\frac{\delta \ln \{\Pr [I(\boldsymbol{\rho}_i)|I_0(\boldsymbol{\rho}_0)]\}}{\delta I_0(\boldsymbol{\rho}_0)}\right|_{I_0(\mathbf{r}_0)=\hat{I}_0(\mathbf{r}_0)} = 0 \qquad (5.43)$$

in addition to Eqs. (5.40) and (5.41), to complete the system of equations needed to obtain a complete and rigorous problem definition.

It now only remains to obtain the *a priori* expressions for $\langle I(\boldsymbol{\rho}_i)\rangle$ and $B_I(\boldsymbol{\rho}_{i1}, \boldsymbol{\rho}_{i2})$, using Eq. (5.7). Admitting the measurement noise term into the process, Eq. (5.7) gives for the average intensity

$$\langle I(\boldsymbol{\rho}_i)\rangle = C' \int_{-\infty}^{\infty} \langle T(\boldsymbol{\rho}_i - \boldsymbol{\rho}_0')\rangle \times I_0(\boldsymbol{\rho}_0)\, d^2\rho_0 + \langle N(\boldsymbol{\rho}_i)\rangle \qquad (5.44)$$

where $I_0(\boldsymbol{\rho}_0) \equiv \langle |R(\boldsymbol{\rho}_0)|^2\rangle_o$. Further, from the results of Eq. (5.6),

$$\langle T(\boldsymbol{\rho}_i - \boldsymbol{\rho}_0')\rangle = \int_{-\infty}^{\infty} \exp\left[-\left(\frac{\rho_i}{r_0}\right)^{5/3}\right] \times \exp\left[\frac{ik}{l}(\boldsymbol{\rho}_i - \boldsymbol{\rho}_0')\cdot\boldsymbol{\rho}\right] K(\boldsymbol{\rho})\, d^2\rho \qquad (5.45)$$

Similarly, for the intensity correlation, one has, upon using the same approximation process as was applied to the result of Eq. (5.29),

$$B_I(\boldsymbol{\rho}_{i1}, \boldsymbol{\rho}_{i2}) = \int_{-\infty}^{\infty}\int_{-\infty}^{\infty} \alpha(\boldsymbol{\rho}_{0d})\beta(\boldsymbol{\rho}_{ic} - \boldsymbol{\rho}_{0c})\gamma(\boldsymbol{\rho}_{id} - \boldsymbol{\rho}_{0d}) \times I_0(\boldsymbol{\rho}_{01})I_0(\boldsymbol{\rho}_{02})\, d^2\rho_{01}\, d^2\rho_{02} + N_0\delta(\boldsymbol{\rho}_{id}) \qquad (5.46)$$

where $\quad \alpha(\boldsymbol{\rho}_{0d}) \equiv \exp\left[-2\left(\frac{\rho_i}{r_0}\right)^{5/3}\right]$

$$\beta(\boldsymbol{\rho}_{ic} - \boldsymbol{\rho}_{0c}') \equiv \langle T(\boldsymbol{\rho}_{ic} - \boldsymbol{\rho}_{0c}')\rangle^2$$

$$\gamma(\boldsymbol{\rho}_{id} - \boldsymbol{\rho}_{0d}) \equiv \left|\int_{-\infty}^{\infty} W(\boldsymbol{\rho}) \exp\left[\frac{ik}{l}(\boldsymbol{\rho}_{id} - \boldsymbol{\rho}_{0d}')\cdot\boldsymbol{\rho}\right] d^2\rho\right|^2$$

with $\quad \boldsymbol{\rho}_{ic} \equiv \dfrac{\boldsymbol{\rho}_{i1} + \boldsymbol{\rho}_{i2}}{2} \qquad \rho_{id} \equiv \boldsymbol{\rho}_{i1} - \boldsymbol{\rho}_{i2}$

$$\boldsymbol{\rho}_{0c} \equiv \frac{\boldsymbol{\rho}_{01} + \boldsymbol{\rho}_{02}}{2} \qquad \boldsymbol{\rho}_{0d} \equiv \boldsymbol{\rho}_{01} - \boldsymbol{\rho}_{02}$$

Given these circumstances, one can now obtain a solution to the integral equation Eq. (5.41) for the inverse correlation function $R_I(\boldsymbol{\rho}_{i1}, \boldsymbol{\rho}_{i2})$. In the realistic approximation where the radius R of the lens of the optical system and the coherence length r_0 are such that $2R \gg r_0$, one has

$$R_I(\boldsymbol{\rho}_{i1}, \boldsymbol{\rho}_{i2}) = \frac{1}{N_0^2} \int_{-\infty}^{\infty} \int_{-\infty}^{\infty} \alpha(\boldsymbol{\rho}_{0d}) \beta_N(\boldsymbol{\rho}_{ic} - \boldsymbol{\rho}_{0c}) \times \gamma(\boldsymbol{\rho}_{id} - \boldsymbol{\rho}_{0d}) I_0(\boldsymbol{\rho}_{01}) I_0(\boldsymbol{\rho}_{02})\, d^2\rho_{01}\, d^2\rho_{02} + \frac{1}{N_0}\, \delta(\boldsymbol{\rho}_{id}) \tag{5.47}$$

the form of which is identical to that of Eq. (5.46) with the exception of the noise term and the integrand term

$$\beta_N(\boldsymbol{\rho}_{ic} - \boldsymbol{\rho}_{0c}) \equiv \frac{\beta(\boldsymbol{\rho}_{ic} - \boldsymbol{\rho}_{0c})}{1 + \sigma\beta(\boldsymbol{\rho}_{ic} - \boldsymbol{\rho}_{0c})} \tag{5.48}$$

which employs a signal-to-noise ratio σ of the image defined by

$$\sigma \equiv \frac{1}{N_0} \int_{-\infty}^{\infty} \int_{-\infty}^{\infty} I_0\left(\boldsymbol{\rho}_{01} + \frac{\boldsymbol{\rho}_{02}}{2}\right) I_0\left(\boldsymbol{\rho}_{01} - \frac{\boldsymbol{\rho}_{02}}{2}\right) \times \alpha(\boldsymbol{\rho}_{0d})\, d^2\rho_{01}\, d^2\rho_{02} \tag{5.49}$$

The approximation of the general case used above, viz.,

$$r_0 \ll 2R$$

is analogous to the spatial frequency condition

$$\frac{r_0}{\lambda l} < |\mathbf{f}| < \frac{2R}{\lambda l}$$

and one has, upon applying the foregoing to the maximum likelihood equation, Eq. (5.43),

$$\int_{-\infty}^{\infty} \int_{-\infty}^{\infty} \left[\frac{\delta B_I(\boldsymbol{\rho}_{i1}, \boldsymbol{\rho}_{i2})}{\delta I_0(\boldsymbol{\rho}_0)} R(\boldsymbol{\rho}_{i1}, \boldsymbol{\rho}_{i2}) + \frac{\delta R_I(\boldsymbol{\rho}_{i1}, \boldsymbol{\rho}_{i2})}{\delta I_0(\boldsymbol{\rho}_0)} L_I(\boldsymbol{\rho}_{i1}, \boldsymbol{\rho}_{i2}) \right] \Bigg|_{I_0(\mathbf{r}_0) = \tilde{I}_0(\mathbf{r}_0)} d^2\rho_{i1}\, d^2\rho_{i2} = 0 \tag{5.50}$$

where

$$L_I(\boldsymbol{\rho}_{i1}, \boldsymbol{\rho}_{i2}) \equiv \frac{1}{N} \sum_{j=1}^{N} [\iota_{\text{ex}}^{(j)}(\boldsymbol{\rho}_{i1}) - \langle I(\boldsymbol{\rho}_{i1})\rangle] \times [\iota_{\text{ex}}^{(j)}(\boldsymbol{\rho}_{i2}) - \langle I(\boldsymbol{\rho}_{i2})\rangle] \tag{5.51}$$

Functionally differentiating Eq. (5.41) and using the result in Eq. (5.50) finally yields the result

$$\int_{-\infty}^{\infty}\int_{-\infty}^{\infty} \frac{\delta R_I(\boldsymbol{\rho}_{i1}, \boldsymbol{\rho}_{i2})}{\delta I_0(\boldsymbol{\rho}_0)} [B_I(\boldsymbol{\rho}_{i1}, \boldsymbol{\rho}_{i2}) - L_I(\boldsymbol{\rho}_{i1}, \boldsymbol{\rho}_{i2})]|_{I_0(\mathbf{r}_0)=\tilde{I}_0(\mathbf{r}_0)} \times d^2\rho_{i1}\, d^2\rho_{i2} = 0 \quad (5.52)$$

This fundamental result forms the basis for the general optimal (in the maximum likelihood sense) estimation $\tilde{I}_0(\boldsymbol{\rho}_0)$ of a nonisoplanatic image, the reception of which has attendant measurement noise. In the most general situation, this equation must be numerically solved for $\tilde{I}_0(\boldsymbol{\rho}_0)$ with the use of *a priori* information concerning the target in question. What is more, however, a quasi-optimal solution of this equation can be considered in the case where the entire image is decomposed into isoplanatic fragments. The resulting analytic solution is identical with those well-known *ad hoc* results proposed in Refs. 7 and 8, thus placing these latter developments on a rigorous basis.

References

1. J. W. Hardy, "Active Optics: A New Technology for the Control of Light," *Proc. IEEE* **66** (6), pp. 651–697 (1978).
2. F. Roddier, "The Effect of Atmospheric Turbulence in Optical Astronomy," in *Progress in Optics,* E. Wolf, ed., North Holland, Amsterdam, 1981. Vol. 19.
3. R. M. Manning, "Theoretical Investigation of Millimeter Wave Propagation through a Clear Atmosphere," Proc. SPIE **410**, *Laser Beam Propagation in the Atmosphere,* J. C. Leader, ed., pp. 119–136 (1983).
4. J. Goodman, *Introduction to Fourier Optics*. McGraw-Hill, New York, 1968.
5. R. H. Hudgin, "Wave-front Reconstruction for Compensated Imaging," *J. Opt. Soc. Am.* **67** (3), pp. 375–378 (1977).
6. A. V. Anufriev, Yu. A. Zimin, and A. I. Tolmachev, "Adaptive Compensation of Atmospheric Phase Distortions Using the Spatial Spectrum of Images," *Sov. J. Quantum Electron.* **17** (10), pp. 1352–1357 (1987).
7. A. Labeyrie, "Attainment of Diffraction-limited Resolution in Large Telescopes by Fourier-analyzing Speckle Patterns in Star Images," *Astron. Astrophys.* **6**, p. 85–87 (1970).
8. K. T. Knox and B. J. Thompson, "Recovery of Images from Atmospherically Degraded Short Exposure Photographs," *Astrophys. J.* **193**, pp. L45–L48 (1974).
9. C. R. Lynds, S. P. Worden, and J. W. Harvey, "Digital Image Reconstruction Applied to Alpha Orionis," *Astrophys. J.* **207**, pp. 174–180 (1976).
10. M. J. McDonnell and R. H. T. Bates, "Digital Restoration of an Image of Betelgeuse," *Astrophys. J.* **208**, pp. 443–452 (1976).
11. P. A. Bakut, I. N. Matveev, K. N. Sviridov, N. D. Ustinov, and N. Yu. Khomich, "Possibility of Reconstructing the Image of an Object Undistorted by the Atmosphere from a Single Speckle Interferogram," *Opt. Spectrosc. (USSR)* **57** (1), pp. 82–84 (1984).
12. D. Korff, "Analysis of a Method for Obtaining Near-Diffraction-Limited Information in the Presence of Atmospheric Turbulence," *J. Opt. Soc. Am.* **63** (8), pp. 971–980 (1973).
13. O. von der Lühe, "Signal Transfer Function of the Knox-Thompson Speckle Imaging Technique," *J. Opt. Soc. Am.* **A5** (5), pp. 721–729 (1988).

14. D. L. Fried, "Optical Resolution through a Randomly Inhomogeneous Medium for Very Long and Very Short Exposures," *J. Opt. Soc. Am.* **56** (10), pp. 1372–1379 (1966).
15. P. A. Bakut, E. N. Kuklin, A. D. Ryakhin, K. N. Sviridov, and N. D. Ustinov, "Comparative Analysis of Phase Reconstruction Methods for the Spatial Spectrum of an Astronomical Object Using a Series of Short-Exposure Images Distorted by the Atmosphere," *Opt. Spectrosc. (USSR)* **58** (6), pp. 806–808 (1985).
16. P. A. Bakut, S. D. Pol'skikh, A. D. Ryakhin, K. N. Sviridov, and N. D. Ustinov, "Statistical Synthesis of Algorithms for the Optimum Processing of the Image of an Astronomical Object," *Radioeng. and Electron. Physics* **29** (9), pp. 104–110 (1984).
17. P. A. Bakut, S. D. Pol'skikh, K. N. Sviridov, and N. Yu. Komich, "Statistical Synthesis of Optimum Image Processing Algorithms Spatially Noninvariant under Atmospheric Distortions," *Radioeng. and Electron. Physics* **33** (3), pp. 154–159 (1989).

Appendix A

Comparison of the Depolarized and Scattered Fields due to Random Atmospheric Permittivity Fluctuations

In the transition from Eq. (1.15) to Eq. (1.16) in Chap. 1, it was shown, via an "order of magnitude" argument, that the depolarization term of Eq. (1.15) could be neglected for the case $\lambda \ll l_0$. It is the purpose of this appendix to show in a rigorous fashion just what the contributions are of the depolarized field to the total power contained within the propagating wave field. The development given below is an expansion of that given by Tatarskii.[1]

In particular, considering the vector wave equation in question, viz.,

$$\nabla^2 \mathbf{E} + k^2 \mathbf{E} = -k^2 \tilde{\varepsilon} \mathbf{E} - \nabla(\mathbf{E} \cdot \nabla \tilde{\varepsilon}) \tag{A.1}$$

and decomposing the total random electric field vector $\mathbf{E}$ into its incident $\mathbf{E}_0$, scattered (or copolarized) $\mathbf{E}_\parallel$, and depolarized $\mathbf{E}_\perp$ field components, i.e.,

$$\mathbf{E} = \mathbf{E}_0 + \mathbf{E}_\parallel + \mathbf{E}_\perp \tag{A.2}$$

one has, after substituting Eq. (A.2) into Eq. (A.1), neglecting the second-order terms in $\tilde{\varepsilon}$ (such as $\mathbf{E}_\parallel \cdot \nabla \tilde{\varepsilon}$), and separating out the various vector components,

$$\nabla^2 \mathbf{E}_0 + k^2 \mathbf{E}_0 = 0 \tag{A.3a}$$

$$\nabla^2 \mathbf{E}_\parallel + k^2 \mathbf{E}_\parallel = -k^2 \tilde{\varepsilon} \mathbf{E}_\parallel - \nabla_\parallel(\mathbf{E}_0 \cdot \nabla \tilde{\varepsilon}) \tag{A.3b}$$

$$\nabla^2\mathbf{E}_\perp + k^2\mathbf{E}_\perp = -\nabla_\perp(\mathbf{E}_0 \cdot \nabla\tilde{\varepsilon}) \tag{A.3c}$$

where $\nabla_\parallel$ is the component parallel to $\mathbf{E}_0$ and $\nabla_\perp$ is that perpendicular to $\mathbf{E}_0$. Equation (A.3*a*) describes the propagation of the unperturbed incident field and is uninteresting for the purposes of this treatment. Equations (A.3*b*) and (A.3*c*) describe, respectively, the propagation of the scattered and depolarized field components. Let the direction of wave propagation be along the x axis of a coordinate system; thus, $\mathbf{E}_0 = \hat{\mathbf{z}}E_0$, $\mathbf{E}_\parallel = \hat{\mathbf{z}}E_\parallel$, and $\mathbf{E}_\perp = \hat{\mathbf{y}}E_\perp$. Equations (A.3*b*) and (A.3*c*) become

$$\nabla^2E_\parallel + k^2E_\parallel = -\left[k^2\tilde{\varepsilon}E_0 + \frac{\partial}{\partial z}\left(E_0\frac{\partial\tilde{\varepsilon}}{\partial z}\right)\right] \tag{A.4a}$$

$$\nabla^2E_\perp + k^2E_\perp = -\frac{\partial}{\partial y}\left(E_0\frac{\partial}{\partial z}\tilde{\varepsilon}\right) \tag{A.4b}$$

In what is to follow, the contribution of the energy flux represented by the depolarized field component $E_\perp$ will be compared to that of the scattered field component.

Noting that one can neglect the second term relative to the first within the brackets of Eq. (A.4*a*) and that the solution to the resulting equation as well as Eq. (A.4*b*) takes the form as given by Eqs. (1.17) to (1.20), applying the paraxial approximation for Eq. (1.20) as given in Sec. 1.4.1, and taking the incident wave field to be that of a plane wave field, i.e.,

$$\mathbf{E}_0 \equiv \mathbf{E}_0(\mathbf{r}) = A_0 \exp(ikx)$$

one has

$$E_\parallel(L, \boldsymbol{\rho}) = -A_0 \exp(ikL)\int_0^L\int_{-\infty}^{\infty} G_p(L, \boldsymbol{\rho}; x', \boldsymbol{\rho}') \times [k^2\tilde{\varepsilon}(\boldsymbol{\rho}')]\, d^2\rho'\, dx' \tag{A.5a}$$

and

$$E_\perp(L, \boldsymbol{\rho}) = -A_0 \exp(ikL)\int_0^L\int_{-\infty}^{\infty} G_p(L, \boldsymbol{\rho}; x', \boldsymbol{\rho}') \times \left[\frac{\partial^2\tilde{\varepsilon}(\boldsymbol{\rho}')}{\partial y'\partial z'}\right] d^2\rho'\, dx' \tag{A.5b}$$

where, as per the results of Sec. 1.4.1, the integrations are taken over the scattering volume bounded by the length of the propagation path $0 \le x' \le L$ and the semi-infinite transverse surface $-\infty \le |\boldsymbol{\rho}'| \le \infty$, where $\boldsymbol{\rho}' \equiv y\hat{\mathbf{y}} + z\hat{\mathbf{z}}$ is an intermediate point in the plane transverse to the direction of propagation at x'; the vector $\boldsymbol{\rho}$ is that at the field reception point $x' = L$.

The associated magnetic fields of $E_\parallel$ and $E_\perp$ are readily found via one of the Maxwell equations, which yields, for the plane wave case,

$$\mathbf{H} = \frac{1}{ik}\nabla \times \mathbf{E} \tag{A.6}$$

For the transverse components of the electric field considered above, assuming that one is in the far (or radiation) zone of the propagation wave, one has magnetic field components

$$H_\perp = -\frac{1}{ik}\frac{\partial E_\parallel}{\partial x'} \approx -E_\parallel \tag{A.7a}$$

and

$$H_\parallel = \frac{1}{ik}\frac{\partial E_\perp}{\partial x'} \approx E_\perp \tag{A.7b}$$

One can now calculate the ensemble average of the random Poynting vector due to the fluctuating fields:

$$\begin{aligned}\langle S\rangle &= \frac{c}{8\pi}\langle \mathrm{Re}\,(\mathbf{E}\times\mathbf{H}^*)\rangle \\ &= \hat{\mathbf{x}}\frac{c}{8\pi}(\langle E_\perp H_\parallel^*\rangle - \langle E_\parallel H_\perp^*\rangle) \\ &= \hat{\mathbf{x}}(\langle S_{x\perp}\rangle + \langle S_{x\parallel}\rangle)\end{aligned} \tag{A.8}$$

where

$$\langle S_{x\parallel}\rangle \equiv \frac{c}{8\pi}|E_\parallel|^2 \qquad \langle S_{x\perp}\rangle \equiv \frac{c}{8\pi}|E_\perp|^2 \tag{A.9}$$

are, respectively, the average values of the copolarized and depolarized contributions to the Poynting vector.

Using Eqs. (A.5a) and (A.5b) in the relations of Eq. (A.9), one obtains

$$\begin{aligned}S_{x\parallel}(L) = \frac{cA_0^2}{8\pi}\int_0^L\int_{-\infty}^{\infty}\int_0^L\int_{-\infty}^{\infty} & G_p(L, \boldsymbol{\rho}; x', \boldsymbol{\rho}') \\ & \times G_p(L, \boldsymbol{\rho}; x'', \boldsymbol{\rho}'')k^4 \\ & \times \langle\tilde{\varepsilon}(x', \boldsymbol{\rho}')\tilde{\varepsilon}(x'', \boldsymbol{\rho}'')\rangle\, d^2\rho'\, dx\, d^2\rho''\, dx''\end{aligned} \tag{A.10a}$$

and

$$S_{x\perp}(L) = \frac{cA_0^2}{8\pi} \int_0^L \int_{-\infty}^{\infty} \int_0^L \int_{-\infty}^{\infty} G_p(L, \boldsymbol{\rho}; x', \boldsymbol{\rho}')$$

$$\times\, G_p(L, \boldsymbol{\rho}; x'', \boldsymbol{\rho}'') \left(\frac{\partial^4}{\partial y'^2\, \partial z'^2}\right)$$

$$\times\, \langle \tilde{\varepsilon}(x', \boldsymbol{\rho}')\tilde{\varepsilon}(x'', \boldsymbol{\rho}'')\rangle\, d^2\rho'\, dx'\, d^2\rho''\, dx'' \qquad \text{(A.10}b\text{)}$$

where the derivatives within the brackets of Eq. (A.5b) were convolved to those over the coordinates y' and z' on the basis of assumed statistical homogeneity. Substituting Eqs. (1.43), and (1.120) to (1.122) into Eqs. (A.10a) and (A.10b), performing the integrations in rectangular coordinates over x'' as well as $\boldsymbol{\rho}'$ and $\boldsymbol{\rho}''$ [i.e., (y', z') and (y'', z'')], and converting the remaining integrations to those over plane polar coordinates, making use of the isotropy of the spectral density $\Phi_\varepsilon(\kappa)$ governing the permittivity fluctuations, yields for the respective energy flux densities

$$S_{x\parallel}(L) = \frac{\pi c A_0^2 L k^2}{2} \int_0^{\infty} \Phi_\varepsilon(\kappa)\kappa\, d\kappa \qquad \text{(A.11}a\text{)}$$

and

$$S_{x\perp}(L) = \frac{\pi c A_0^2 L}{16k^2} \int_0^{\infty} \kappa^4 \Phi_\varepsilon(\kappa)\kappa\, d\kappa \qquad \text{(A.11}b\text{)}$$

One can now form the ratio of energy density scattered in the orthogonal direction to that in the copolarized direction, viz.,

$$\frac{\langle S_{x\perp}(L)\rangle}{\langle S_{x\parallel}(L)\rangle} = \left(\frac{1}{k^4}\right) \frac{\displaystyle\int_0^{\infty} \kappa^5 \Phi_\varepsilon(\kappa)\, d\kappa}{\displaystyle\int_0^{\infty} \kappa \Phi_\varepsilon(\kappa)\, d\kappa} \qquad \text{(A.12)}$$

In the case of atmospheric turbulence governed by the spectral density given by Eq. (4.8), one has the general relation

$$\frac{\langle S_{x\perp}(L)\rangle}{\langle S_{x\parallel}(L)\rangle} = \frac{K_0^4 U(3, 13/6\,; K_0^2/K_m^2)}{4k^4 U(1, 1/6\,; K_0^2/K_m^2)} \qquad \text{(A.13)}$$

where $U(\alpha, \beta; \zeta)$ is a Kummer function. In most atmospheric cases, however, one has the condition $K_m^2 \gg K_0^2$, allowing the Kummer functions to be represented by the first terms of their series expansions,[2] and thus finally giving

$$\frac{\langle S_{x\perp}(L)\rangle}{\langle S_{x\parallel}(L)\rangle} = \frac{0.235 K_m^{1/3} K_0^{11/3}}{k^4} \tag{A.14}$$

Using the representative values $K_0 = 0.628\ \text{m}^{-1}$ and $K_m = 5.92 \times 10^3\ \text{m}^{-1}$, and considering propagation at a nominal wavelength of $\lambda = 0.63\ \mu\text{m}$, one has for the random depolarized wave contribution relative to that of the copolarized

$$\frac{\langle S_{x\perp}(L)\rangle}{\langle S_{x\parallel}(L)\rangle} = 7.815 \times 10^{-29} \qquad \text{turbulence case}$$

indicating the insignificant role of depolarization in this case.

In the case of a turbid atmosphere, one employs in Eq. (A.12) the spectrum given by Eq. (4.9), with the result that

$$\frac{\langle S_{x\perp}(L)\rangle}{\langle S_{x\parallel}(L)\rangle} = \frac{32\nu^4}{k^4(\nu + 4)^2(\nu + 3)^2\langle a\rangle^4} \tag{A.15}$$

Considering propagation, at the same nominal wavelength as above, through an advective fog and using the values given in Table 2.2, one finds that

$$\frac{\langle S_{x\perp}(L)\rangle}{\langle S_{x\parallel}(L)\rangle} = 2.379 \times 10^{-7} \qquad \text{advective fog case}$$

Although this result is 22 orders of magnitude larger than that obtained with atmospheric turbulence, it also indicates a negligible depolarized wave component contribution relative to that of the copolarized wave.

Similarly, considering propagation through rain described by the Joss thunderstorm (J-T) distribution with the parameters as given in Table 2.3 and taking the prevailing rain rate to be $R = 100$ mm/h, one has from Eq. (A.15)

$$\frac{\langle S_{x\perp}(L)\rangle}{\langle S_{x\parallel}(L)\rangle} = 2.18 \times 10^{-16} \qquad \text{J-T rain, } R = 100\ \text{mm/h}$$

Thus, the neglect of the depolarization term, i.e., the second term on the right side of Eq. (A.1), in the transition from Eq. (1.15) to Eq. (1.16) is justified in the propagation scenarios considered in this book.

References

1. V. I. Tatarskii, "Depolarization of Light by Turbulent Atmospheric Inhomogeneities," *Radiophys. Quant. Electron.* **10** (12), pp. 987–988 (1967).
2. M. Abramowitz and I. A. Stegun, Eds., *Handbook of Mathematical Functions*. Dover Publications, New York, 1972. Eq. (13.5.10).

Appendix

B

Analysis of the Applicability of the Stochastic Parabolic Wave Equation

The transition from the full wave equation of Eq. (1.33), viz.,

$$\nabla_\rho^2 U(\mathbf{r}) + 2ik\,\frac{\partial U(\mathbf{r})}{\partial x} + k^2\tilde{\varepsilon}(\mathbf{r})U(\mathbf{r}) = -\frac{\partial^2 U(\mathbf{r})}{\partial x^2} \tag{B.1}$$

to the parabolic wave equation of Eq. (1.34), viz.,

$$\nabla_\rho^2 U(\mathbf{r}) + 2ik\,\frac{\partial U(\mathbf{r})}{\partial x} + k^2\tilde{\varepsilon}(\mathbf{r})U(\mathbf{r}) = 0 \tag{B.2}$$

necessarily introduces an error into the field statistics obtained from the application of Eq. (B.2), which thus determines the validity of its use. Here, the validity of such an approximation will be demonstrated in the case of the simplest field moment, i.e., the mean field of a propagating plane wave in a turbulent atmosphere.

Once again, following a method detailed by Tatarskii,[1] one treats the term on the right side of the full wave equation Eq. (B.1) as a small perturbation addition to the parabolic equation of Eq. (B.2). Thus, one commences the analysis by representing the total stochastic wave field $U(\mathbf{r})$ as a perturbation series

$$U(\mathbf{r}) = U_1(\mathbf{r}) + U_2(\mathbf{r}) \tag{B.3}$$

where the first-order field $U_1(\mathbf{r})$ is the solution to Eq. (B.2) and the second-order field contribution $U_2(\mathbf{r})$ is of the same order of smallness as the quantity $\partial U_1(\mathbf{r})/\partial x$. Substituting Eq. (B.3) into Eq. (B.1) and equating terms of the same order, one obtains the system of equations

$$\nabla_\rho^2 U_1(\mathbf{r}) + 2ik \frac{\partial U_1(\mathbf{r})}{\partial x} + k^2 \tilde{\varepsilon}(\mathbf{r}) U_1(\mathbf{r}) = 0 \tag{B.4}$$

$$\nabla_\rho^2 U_2(\mathbf{r}) + 2ik \frac{\partial U_2(\mathbf{r})}{\partial x} + k^2 \tilde{\varepsilon}(\mathbf{r}) U_2(\mathbf{r}) = -\frac{\partial^2 U_1(\mathbf{r})}{\partial x^2} \tag{B.5}$$

with the boundary conditions given in terms of the mean values $\langle U_1(0, \boldsymbol{\rho})\rangle \equiv \langle U_1(\boldsymbol{\rho})\rangle$ and $\langle U_2(0, \boldsymbol{\rho})\rangle = 0$ where $\mathbf{r} = (x, \boldsymbol{\rho})$.

These stochastic differential equations are of the same form as Eq. (1.68) for the general statistical moment of the wave field. Hence, following the same procedure used in the transition from the form of Eq. (1.68) to that of Eq. (1.78), which entails the application of the Markov approximation, one obtains from Eqs. (B.4) and (B.5) the relationships governing the statistical first-order moments of the fields,

$$2ik\langle U_1(x, \boldsymbol{\rho})\rangle - 2ik\langle U_1(\boldsymbol{\rho})\rangle \left\langle \exp\left[\frac{ik}{2}\int_0^x \tilde{\varepsilon}(x, \boldsymbol{\rho})\, d^2\rho\right]\right\rangle + \int_0^x \left\langle \exp\left[\frac{ik}{2}\int_{x_1}^x \tilde{\varepsilon}(x', \boldsymbol{\rho})\, dx'\right]\right\rangle \nabla_\rho^2 \langle U_1(x_1, \boldsymbol{\rho})\rangle\, dx_1 = 0 \tag{B.6}$$

and

$$2ik\langle U_2(x, \boldsymbol{\rho})\rangle + \int_0^x \left\langle \exp\left[\frac{ik}{2}\int_{x_1}^x \tilde{\varepsilon}(x', \boldsymbol{\rho})\, dx'\right]\right\rangle \times \left[\nabla_\rho^2 \langle U_2(x_1, \boldsymbol{\rho})\rangle + \frac{\partial^2 \langle U_1(x_1, \boldsymbol{\rho})\rangle}{\partial x_1^2}\right] dx_1 = 0 \tag{B.7}$$

In the case of a turbulent medium, one has the fact that the stochastic permittivity function $\tilde{\varepsilon}(x', \boldsymbol{\rho})$ is governed by gaussian statistics. Hence, one can use for the expectation values of the exponential functions in Eqs. (B.6) and (B.7) the relation given by Eq. (1.131) and, in conjunction with Eqs. (1.75) and (1.76), obtain

$$\left\langle \exp\left[\frac{ik}{2}\int_{x_1}^x \tilde{\varepsilon}(x', \boldsymbol{\rho})\, dx'\right]\right\rangle = \exp\left[-\frac{k^2 A(0)}{8}(x - x_1)\right] \tag{B.8}$$

thus allowing Eqs. (B.6) and (B.7) to be written as

$$2ik\langle U_1(x, \boldsymbol{\rho})\rangle \exp\left[\frac{k^2A(0)}{8}\,x\right] = 2ik\langle U_1(\boldsymbol{\rho})\rangle$$

$$-\int_0^x \exp\left[\frac{k^2A(0)}{8}\,x_1\right] \nabla_\rho^2\langle U_1(x_1, \boldsymbol{\rho})\rangle\, dx_1 = 0 \quad \text{(B.9)}$$

and

$$2ik\langle U_2(x, \boldsymbol{\rho})\rangle \exp\left[\frac{k^2A(0)}{8}\,x\right] = -\int_0^x \exp\left[\frac{k^2A(0)}{8}\,x_1\right]$$

$$\times \left[\nabla_\rho^2\langle U_2(x_1, \boldsymbol{\rho})\rangle + \frac{\partial^2\langle U_1(x_1, \boldsymbol{\rho})\rangle}{\partial x_1^2}\right] dx_1 = 0 \quad \text{(B.10)}$$

Finally, differentiating these expressions with respect to x and simplifying yields

$$2ik\,\frac{\partial\langle U_1(x, \boldsymbol{\rho})\rangle}{\partial x} + \nabla_\rho^2\langle U_1(x, \boldsymbol{\rho})\rangle + \frac{ik^3A(0)}{8}\,\langle U_1(x, \boldsymbol{\rho})\rangle = 0 \quad \text{(B.11)}$$

for the mean value of the first-order perturbation field and

$$2ik\,\frac{\partial\langle U_2(x, \boldsymbol{\rho})\rangle}{\partial x} + \nabla_\rho^2\langle U_2(x, \boldsymbol{\rho})\rangle + \frac{ik^3A(0)}{4}\,\langle U_2(x, \boldsymbol{\rho})\rangle = -\frac{\partial^2\langle U_1(x, \boldsymbol{\rho})\rangle}{\partial x^2} \quad \text{(B.12)}$$

for the differential equation governing the mean of the second-order field. It should be noted that Eq. (B.11) can also be directly derived from Eqs. (1.83) and (1.84) where $n = 1, m = 0$. Hence, a comparison of the solution of this equation, which is the basis of much of the theory presented here, Eq. (B.12) will provide an indication of the conditions which must implicitly prevail when the second-order "correction" field, given by Eq. (B.12), is assumed to be negligible.

The solution to Eq. (B.11) is straightforward and is given by

$$\langle U_1(x, \boldsymbol{\rho})\rangle = \langle U_1(\boldsymbol{\rho})\rangle \exp\left(-\frac{i\alpha}{2}\,x\right) \quad \text{(B.13)}$$

where the attenuation coefficient α is given by

$$\alpha \equiv \frac{k^2A(0)}{4}$$

Note that the form of Eq. (B.13) is typical for odd-order statistical moments of wave fields, as was discussed in the paragraph leading to

Eq. (1.61); in particular, the average or first-order moment of the wave field is oscillatory and is a function of the product αx. If an imaginary term, representing an absorptive component, were admitted into the expression for the stochastic permittivity (as in the case for an aerosol medium), one would find that this component would rapidly oscillate to zero. In what is to follow, only the range of values of these parameters will be taken to be such that $\alpha x < 1$.

Substituting Eq. (B.13) into Eq. (B.12) and solving the resulting differential equation, noting the boundary condition $\langle U_2(0, \boldsymbol{\rho})\rangle = 0$, yields, after simplification,

$$\langle U_2(x, \boldsymbol{\rho})\rangle = \langle U_1(x, \boldsymbol{\rho})\rangle \left(\frac{i\alpha^2}{8k}\right) x \qquad \text{(B.14)}$$

Finally, one can write the perturbation expansion for the total field by substituting Eqs. (B.13) and (B.14) into Eq. (B.3), viz.,

$$\langle U(x, \boldsymbol{\rho})\rangle = \langle U_1(x, \boldsymbol{\rho})\rangle \left(1 + \frac{i\alpha^2}{8k}\, x\right) \qquad \text{(B.15)}$$

This provides an obvious condition for the negligibility of the second-order correction term relative to the first-order perturbation term, which is the solution of the parabolic wave equation, i.e., the use of the parabolic wave equation approximation will provide negligible error as compared to the solution issuing from the full wave equation so long as

$$1 \gg \frac{\alpha^2}{8k}\, x$$

Using the condition $\alpha x < 1$ adopted above, one finally arrives at the condition given by Eq. (1.55),

$$\tfrac{1}{4}k\langle\tilde{\varepsilon}^2\rangle \ll 1 \qquad \text{(B.16)}$$

As it is only the mean of the wave field that is considered in the above, this result can only rigorously be taken to hold for such a case, thus necessitating a similar analysis for the second-order moment of the wave field, which is the quantity of most interest in this book. Such an analysis has been performed,[1] and although it is analogous to that above, it is much more lengthy and therefore will not be repeated here. Suffice it to say, however, that the condition of Eq. (B.16), along with others of Sec. 1.4.1, is sufficient for application of the parabolic wave equation to atmospheric propagation problems.

Finally, it is interesting to note that the conditions governing the applicability of the Markov approximation[2] are identical to those for the parabolic wave equation. Thus, one has simultaneous applicability of both the parabolic wave equation and the Markov approximation in stochastic propagation problems.

References

1. V. I. Tatarskii, *The Effects of the Turbulent Atmosphere on Wave Propagation*. U.S. Dept. of Commerce, TT-68-50464, Springfield, Va., 1971. Sec. 68.
2. Ref. 1, Sec. 67.

Index

Aberrations (*see* Phase aberrations)
Aerosol number, Poisson distribution model of, 74
Aerosol size distribution, Deirmendjian modified gamma model of, 75
Aperture cutoff frequency, 164
Aperture planes of lens (*see* Imaging system, lens aperture planes)
Apodization, 126
Atmospheric permittivity (refractive index) field, 70
  dry continuum, 71–72
  fluctuations:
    due to aerosols or hydrometeors (*see* Turbidity)
    due to temperature and humidity variations (*see* Turbulence)
    unified spectral model, 106–108
  structure parameter, typical values and models of, 92–94
  water vapor continuum, 72

Backscattering, 21, 41
Bethe-Salpeter equation, 21

Characteristic functionals:
  general discussion, 36–38
  of random permittivity field, 38–39, 99–103, 175
    evaluation for turbulence, 178
    evaluation for aerosols, 182–184
Characteristic functions, 36
Coherence function (*see* Mutual coherence function)
Coherence length, 163

Depolarization, 7–8
  due to turbulence and aerosols, analysis, 215–219
Diagrammatic method, 15–21
  strongly connected graphs, 18
  weakly connected graphs, 19
Diffusion approximation, 22, 34
Dynamic causality, 33
Dyson equation, 20

Ensemble averages, 10, 12
  propagation statistics vs. target reflection statistics, 120–121
Ergodic theorem and application to stochastic wave fields, 12
Extended Huygens-Fresnel principle, 41, 59
  phase approximation, 60
    validity of, 61–66

Far field solution of log-amplitude and phase structure functions, 155
Fokker-Planck-Kolmogorov diffusion equation, 34
Fraunhoffer diffraction zone, 27
Fresnel approximation, 24
Fresnel diffraction zone, 27
Frozen space-time random fields, 14, 127, 188

Geometrical optics zone, 27, 189

Homogeneous (spatial) statistics, 13
  in wide sense, 13

Image compensation techniques (adaptive optics methods):
  instantaneous image transfer functions for, 196–199
  passive vs. active, 195
  phase compensation, 199–202
  speckle imaging and maximum-likelihood methods, 209–213
  speckle interferometry, 202–209
Image transfer function, 124–125
  instantaneous forms, 196–199
  long term, 136
    for aerosols, 172–173
    for turbulence, 164
  short term, 136
    for turbulence, 171
Imaging system, 112
  equivalent image and object coordinates, 116

Imaging system (*Cont.*):
image formation, 115–118
target reflection statistics and propagation statistics, 120–121
image transfer function, 124–125
lens input aperture plane, 113
lens law, 116
lens output aperture plane, 113
lens pupil function, 113
lens transmission function, 113
linear system formulation, 125
magnification factor, 116
resolution, 118, 139
minimum requirements, 146–147
Inertial subrange, 84
Inner scale of turbulence, 84
Isotropic random field, 14, 50

Karhunen-Loeve integral and related phase expansion, 135
Kolmogorov-Obukhov ⅔ law, 87
spectrum, 89
Kolmogorov similarity hypotheses, 84

Log-amplitude, 46

Markov (δ-function) approximation, 33–34, 42, 51
validity of, 32–33, 225
Modulation transfer function (MTF), 125
Mutual coherence function (MCF), 125
coherent, 141
derived from first Rytov approximation:
for aerosol case, 171–172
for turbulence case, 162–163
derived from parabolic equation method:
for aerosol case, 182–184
for turbulence case, 178–182
incoherent, 141
long term (long exposure), 126, 135–136
short term (short exposure), 126, 135–136
wave structure function, relation to, 58

Optical transfer function (OTF), 125
Outer scale of turbulence, 84

Parabolic equation method:
application to propagation scenarios, 178–184
derivation, 27–42
general solution for MCF, 175–178
derivation of radiative transfer equation for small angle scattering, 184–187
Parabolic wave equation (*see* Stochastic wave equations, scalar parabolic)
Passive (atmospheric) additives, 79
Permittivity field (*see* Atmospheric permittivity field)
Phase aberrations, 126
short term vs. long term, 129
tilt corrected MCF, 132–134
Zernike polynomial expansions, 129–130
Phase compensation via subapertures, 199
Phase fluctuation, 46
Point spread function, 122–124

Radiative transfer equation, small scattering angle approximation, 185–186
Raindrop fall velocity, Gunn-Kinzer model, 76
Raindrop size distributions, Marshal-Palmer, Joss thunderstorm and drizzle, 75–76
Random electric field, general moment expression for, 14, 29
Random fields:
general discussion of, 10–14
homogeneous, 13
isotropic, 14, 50
locally homogeneous and isotropic, general discussion of, 53–56
moments and correlation functions, 10–11
spectral representation of, 48–50
Refractive index field (*see* Atmospheric permittivity field)
Resolution, 118, 139
from coherent and incoherent MCF, 141–145
improvement due to phase tilt removal, 171
minimum requirements for human vision as related to MCF, 146–147

Rytov method (method of weak fluctuations):
  first Rytov approximation:
    application to propagation scenarios, 150–174
    derivation, 47–58
    limitation of applicability, 58, 163
  general derivation, 42–47
  second Rytov approximation, 58
Rytov transformation, 45, 59

Speckle contrast, 137
Speckle imaging, 203
Speckle interferometry, 203
  reconstruction of an extended object, 203–209
Spectral amplitude, 48, 51
Spectral density, spatial and temporal, 49–51
  two-dimensional, 49–50, 150
Stationary (temporal) statistics, 12
  in wide sense, 13
Statistical moments, general discussion of, 10–14
Structure functions, 54
  log-amplitude and phase fluctuation, 56
    derived from first-order Rytov approximation, 150–157
  wave, 58
    derived from first-order Rytov approximation:
      aerosol case, 172
      turbulence case, 162
Stochastic wave equations:
  scalar, 8
    parabolic, 22
      Green function for linear systems approach, 39–42
      Helmholtz-Kirchoff integral, 27
      validity of, 22–27, 221–225

Stochastic wave equations, scalar, parabolic (*Cont.*):
      Helmholtz-Kirchoff integral, 27
      validity of, 22–27, 221–225
  vector, 7
    general form for atmosphere, 4

Temporal correlation time, 121
Turbidity (aerosols and hydrometeors), 72
  fog, 75
  number distribution, 74
  rain, 75–76
  two-parameter size distribution model, 74–75
  scattering model for effective refractive index, 94–99
Turbulence (temperature and humidity fluctuations), 77–79
  energy dissipation, 82–83
  fundamental statistical fluid mechanics of, 79–92
  inertial subrange (temperature), 84
  intertial-convective subrange (humidity), 86
  inner scale, 84
  input subrange, 91
  outer scale, 84
  spectral density of fluctuations, 89–91
  thermal and molecular diffusivity, 80

Wave structure function, 58
Weak fluctuation method (*see* Rytov method)

Zernike polynomials, 129–131
  formation of tilt-corrected (short-term) MCF, 132–135, 166–171
  tilt and astigmatism, temporal behavior, 187–192

ABOUT THE AUTHOR

Robert M. Manning is staff physicist at NASA's Lewis Research Center in Cleveland, Ohio, and adjunct professor of physics at both Cleveland State University and the Ohio Aerospace Institute. Dr. Manning holds a Ph.D. from Case Western Reserve University. His principal research interests include stochastic electromagnetic wave propagation in turbulent and turbid media, adaptive optics, microwave and optical communications link modeling and analysis, and remote sensing of atmospheric propagation parameters.